Drought

Past Problems and Future Scenarios

Justin Sheffield and Eric F. Wood

Routledge
Taylor & Francis Group
LONDON AND NEW YORK

earthscan
from Routledge

First published in paperback 2024

First published by Earthscan in the UK and USA in 2011

For a full list of publications please contact:
Earthscan
4 Park Square, Milton Park, Abingdon, Oxon OX14 4RN
605 Third Avenue, New York, NY 10158

Earthscan is an imprint of the Taylor & Francis Group, an informa business

Publisher's Note
The publisher has gone to great lengths to ensure the quality of this reprint but points out that some imperfections in the original copies may be apparent.

A catalogue record for this book is available from the British Library

Library of Congress Cataloging-in-Publication Data

Sheffield, Justin.
Drought : past problems and future scenarios / Justin Sheffield and Eric F. Wood.
p. cm.
Includes bibliographical references and index.
ISBN 978-1-84971-082-4 (hardback)
1. Droughts–United States. 2. Drought relief–United States. 3. Water quality management. 4. Water-supply–Management. I. Wood, Eric F. II. Title.
QC929.24.S54 2011
363.34'929–dc22

2010045894

ISBN: 978-1-84971-082-4 (hbk)
ISBN: 978-1-03-292526-4 (pbk)
ISBN: 978-1-84977-525-0 (ebk)

DOI: 10.4324/9781849775250

Typeset by Domex e-Data, India
Cover design by Adam Bohannon

Contents

Figures, Tables, Boxes and Plates

Figures

Tables

Boxes

Plates

Preface

It is August 2009, and I (Justin) am in the Central American country of Belize, standing on top of a Mayan temple that pokes just over the top of the lush forest. The air is thick with moisture, the bugs are fierce and the sun beats down. The view is spectacular. Behind me, dense jungle stretches out for miles to the Mayan highlands and Guatemala. To my right, another temple pierces the forest canopy. In front of me is a lagoon, with wetlands surrounding it. White ibises drift above the lagoon and jabiru storks perch in the grasses. Dropping below me are the 33m of vertiginous steps that I have just climbed. This temple is part of the ancient Mayan city of Lamanai, which translates as 'submerged crocodile', an apt name given the largest inhabitants of the lagoon. The city was built in the pre-Classic age of the Mayas and last modified during the Classic age of 250–900 AD. At its peak the city supported a population of 30,000–60,000, and it is a testament to the sophisticated cultural, scientific and engineering advancements made by this great civilization. By 900 AD the inhabitants of many of the great Mayan cities, such as Tikal, Copan, Palenque and Caracol, had all but disappeared. Why did this happen? Numerous theories have been put forward, including social revolution, foreign invasions and the collapse of trade routes. Increasingly, the occurrence of widespread and lengthy drought is implicated as a major contributing factor.

The debate on how the Mayan civilization demised is ongoing. There are considerable uncertainties about how climate varied hundreds or thousands of years ago. The social mechanisms that eventually lead to demise are also highly complex, whether they are climate-driven or not. Nevertheless, the fate of the Mayans is reflected in the fate of ancient civilizations elsewhere, such as the Anasazi culture in southwestern US and the Akkadian empire of Mesopotamia. Clues are emerging from tree rings, cave deposits, and lake and ocean sediments that hint at climatic factors coinciding with dramatic changes in these civilizations. In recent times, there is more tangible and direct evidence of the impact of drought on human activity and the environment. The Dustbowl drought that affected much of the central US in the 1930s, which was epitomized in the writings of John Steinbeck, led to mass migration and a continuation of the economic downturn of the Great Depression. The African Sahel droughts of the 1970s and 1980s impacted millions of people, with an estimated 600,000 deaths due to widespread famine – arguably one of the greatest natural disasters of the 20th century in humanitarian terms. More recent droughts in southeast Australia, the western US, Argentina, the Amazon, northeastern China and Russia are causing ongoing problems for agriculture and water supply for consumption, power generation and transport. As temperatures increase, the prospect of climate-change-induced increases in drought frequency and severity over the 21st century is increasingly worrisome for a planet already under considerable environmental stress.

This book is about drought. Of all natural disasters, drought is arguably the most problematic, causing immense social, economic and ecological damage. Yet it is one of the most under-rated and least understood of disasters. This book explores how and why drought has occurred over the past millennia, and the prospects for changes in drought occurrence under future climate change. In particular we focus on the 20th and early 21st centuries, for which we have the best observational data and the most confidence in our representations of past events. The goal of the book is to provide the reader with an understanding of the mechanisms of how drought occurs and impacts, to apply this to historic droughts and to understand why future droughts may change. In doing so we have deemed it necessary to provide background material that covers the science behind drought (the how and why), drawing from the latest scientific research and data at the intersection of hydrology and climatology. Explanations are given on how variations in climate and hydrology conspire to generate drought conditions, and its propagation through the hydrologic cycle to impact ecological and human systems. The intended audience are undergraduate and graduate students with an interest in climate, hydrology, natural hazards, and human and ecosystem vulnerability to drought, although the relatively low level of technical content makes it well suited to the general reader. It will provide documentation of past droughts, including their impacts and causes, as well as a primer on the potential changes in the future.

The idea for the book developed from Justin's PhD thesis, which explored how drought has occurred and changed over the past 50 years and how it might change in the future. The ever-increasing interest in climate change and its impacts provided incentive to put this research into the broader context of past and future global change. We have therefore expanded the original concept of the thesis to compare contemporary experience of drought with the longer term of the past millennia and the epoch of major human development (the Holocene: approximately the last 12,000 years). Furthermore, the abundance of data, research, and improvement in understanding of drought that have emerged in the past several years has allowed us to provide a more comprehensive discussion of the topic. Although there are many good books available on drought, and especially methods for drought monitoring and assessment, there is a gap in the literature on global drought and its variation in the past and potential changes in the future. There is also an ever-expanding and vast literature on climate change, much of which touches on the possible consequences, including drought. However, a comprehensive treatment of climate and drought is lacking, especially given the potential for global warming to exacerbate the occurrence and severity of drought. Recent drought events are often associated in the popular press with climate change, but the scientific evidence for this link is tentative and remains a subject of debate. This book presents the latest research into how droughts have changed in recent decades, and discusses whether global warming has had an influence in the past and how it might impact in the future.

Justin Sheffield and Eric F. Wood, Princeton, NJ

Acknowledgements

This book is an accumulation of scientific findings and knowledge derived from many years of research at the Hydroclimatology Group at Princeton University and by our collaborators. Without the incremental efforts made over the years by many graduate students and researchers, none of the content of this book would have been possible. We must also acknowledge the vast contribution of the hydrologic, climate and social science research communities who have made significant progress in recent years towards understanding the mechanisms and impacts of drought. In particular, we would like to thank the many individuals and research teams who have made their data available, upon which much of our research has been based. Our thanks to Tim Hardwick, editor at Earthscan, and the publication team for instigating and seeing this endeavour through to completion. Finally, I (Justin) am thankful to my wife for allowing me the indulgence of writing this book. I hope I allow her the same opportunities to express herself.

1

Introduction

Drought Today

2009 was a worrisome year for drought, nowhere more so than in eastern Africa. In September, the media were reporting a looming humanitarian disaster in Kenya. Aid agencies had been warning of the danger for several months, since rains had failed to arrive early in the growing season. By the middle of 2009, 20 million people across East Africa were estimated to be suffering from food insecurity. In October 2009, the Ethiopian government requested international emergency food aid to feed more than 6 million people. As reports were coming in of the illegal selling of grain by government officials, the staple crops of corn (maize) and wheat had become so expensive that people had resorted to eating cactus in some regions. Ominously, conflicts were brewing between tribes in the north. A year later, when the writing of this book was completed, the drought was relieved in some parts by much welcomed rains. Yet recovery has been hampered by the lingering effects of several years of crop failures and the continuation of conflict and insecurity in the region (FEWS-NET, 2010). Across much of eastern Africa, this story has been retold, in Tanzania, Burundi, Uganda, Malawi and Mozambique, and even further south, in Zambia, Zimbabwe and South Africa.

For Africa, such events have tragically become commonplace, fuelled by the vulnerability of the people and governments. Even for Kenya, which is one of the most developed countries in Africa, the drought took, and continues to take, a devastating toll on the people and environment. But Africa is not alone, as similar stories have emerged across the world during the past couple of years of drought, causing suffering, hardship and economic loss. In southeast Australia, the Murray-Darling River basin, which drains 14 per cent of the land mass but grows more than 40 per cent of the nation's food, has been suffering from a drought for more than ten years, which is likely to have been the worst in 117 years of records. The river stopped flowing at its outlet and its lower lakes became so critically low that there was a chance of acidification. Outbreaks of potentially toxic blue-green algae were also reported. The city of Melbourne, which consumes 1 billion litres of water a day, endured raging wildfires just outside the city in the summer of 2008–2009. Its reservoir levels reached record lows in early 2009 of 514 billion litres (BL), which is 30 per cent of capacity, exposing the rivers that had existed before the reservoirs were built. Water restrictions and conservation programmes have reduced demand (hotel bathroom signs now encourage the use of four-minute showers, or even *sailor showers*, where water is only used for rinsing), but with no end in sight to the drought, further measures are required. The ongoing drought conditions led the Australian government to form the Murray-Darling Basin Authority in 2008 to oversee the management of efforts to protect and conserve water supplies in the region. However, it was forecast that drier than average conditions and the drought were set to continue in the area.

In northern and central China, seven provinces that cover the major wheat-growing areas suffered severe drought during 2008–2009, perhaps the worst in 50 years and the worst in recorded history in some northern regions. The drought affected up to 100,000km^2 and more than 4 million people. With threats to the summer harvest (half the wheat crop in eight provinces), the government announced the drought was a level 1 emergency (the highest level), in February 2009. The government began pouring billions of dollars worth of relief money into the regions and local authorities resorted to cloud seeding with rockets and cannon in the hope of producing rain. Henan, which is China's largest crop-producing region, received 80 per cent less rain than normal, the worst in 50 years, and Jiangsu Province had already lost one fifth of its wheat. Diversions of water were taking place to combat the drought, with Hebei Province transferring 100 million cubic metres of water from outside the region.

In the US, drought was affecting California, Texas and parts of the southeast of the country. California declared a state of emergency for the first time in 20 years, with expected losses of US$3 billion in 2009. Many acres of agricultural land were left fallow, with no prospect of available irrigation water. This was partly a result of declining snowpack in the mountains because of less precipitation and more rain than snow, but also due to increasing demand from a burgeoning population. More than 80 per cent of Texas was experiencing dry conditions, with 20 per cent facing extreme conditions. Crops and cattle were suffering from a lack of water. In the southeast, the La Niña weather pattern was forcing dry conditions from Georgia through to the eastern seaboard and down to Florida, where crops were being affected by wintertime drought.

In South America, Paraguay and Uruguay declared agricultural emergencies. A similar story emerged in Chile, which was also suffering because of its heavy reliance on hydroelectric power. Brazil's corn harvest had been reduced drastically and Argentina was suffering the worst drought in 50 years, with wheat yields down by 50 per cent. Ranchers were forced to sell their cattle and some converted to growing wheat and soy. The Argentinian pampas grasslands, which cover 1 million km^2, were once one of the most agriculturally productive regions in the world. Since the drought, 1.5 million cattle have died. Coupled with government pricing policies to keep national prices low, the country has gone from being the second largest exporter of beef to the seventh largest. The drought has turned much of the Argentinian pampas into a dry and desolate wasteland.

In 2009, Europe was suffering from dry conditions in the east. Wildfires ravaged Greece in August, only two years after devastating wildfires in 2007 killed 70 people and destroyed 6 per cent of tree cover. In 2009, more than 12,000ha of forest, farms and olive groves were destroyed. Athens was threatened as high winds drove flames to the northern outskirts of the city. Several of Russia's grain-producing regions were suffering from a drought that affected the summer harvest of winter crops, particularly in southern Siberia, which lies to the east of the Ural Mountains. The year before, grain yields stood at record levels (108 million tonnes). Although this is not the main grain belt of Russia (which is in the west), drought has had a great impact, with 80 per cent of the harvest destroyed in Chelyabinsk District. The national shortage reached 11 million tonnes, out of an expected total of 90 million, with consequences for exports to the rest of Siberia.

In India, where half of the 1.1 billion population depends on agriculture, the failure of the monsoon rains (down by 25 per cent in 2009) has led to drought hardship. At the same time, groundwater supplies, which form the backbone of much of India's irrigation and agriculture, are now dwindling, with 1000 out of India's 5723 districts either overexploited or existing in a critical condition. This is partly a result of the recent abundance of cheap electric pumps for accessing groundwater.

In the Middle East, drought can also be a major problem because water generally does not come directly from rain but instead from underground or from an upstream nation. Agriculture is highly dependent on irrigation, which in turn depends on these indirect supplies. In 2008–2009, drought spread across much of the Middle East and Central Asia, with restrictions on water usage across Turkey, Syria, Iraq and Iran, due to low reservoir levels. Wheat production fell by about 45–60 per cent in Iraq and Afghanistan. In the rain-fed north, wheat production was down by 95 per cent. In addition to years of war and ongoing problems of internal conflict, Iraq has been suffering from drought in part because of low flows in the Euphrates and Tigris rivers, which supply 98 per cent of Iraq's water. Part of the problem is due to dam-building upstream in Turkey and Syria. These countries were making some releases, although not enough, and since there is no official agreement between the countries, there is no guarantee that this will increase or even continue. These factors are transforming a food-exporting nation into one of the world's largest importers of wheat and rice.

In 2009, this global picture of drought was worrisome. It seemed that drought was affecting every continent, with impacts ranging from wildfires to famine (Plate 1.1). But was 2009 an unusual year for drought? The words on many people's lips were that global warming was showing itself through drought.

But was it a manifestation of global warming or just a natural swing in climate variability? Would it be a harbinger of things to come? By late 2009, conditions had improved in most places and the media had lost interest. Yet drought has emerged in many other places. The key question is whether drought is becoming more severe and more frequent, and if this will accelerate as a result of climate change. Many scientists think so, and politicians are starting to come to the same realization, especially when coupled with increasing demands for agriculture and water supplies.

An oft-used saying is 'Water is becoming the new oil.' These words not only capture the likely increase in price of a currently cheap commodity that is becoming scarcer, but also the economic benefits of improving access to clean water through improved levels of health and sanitation. President Laurent Gbagbo of the Côte d'Ivoire recently summarized this at a national conference on water policy in August 2009: 'In 50 years, potable water will be harder to find than gasoline. Water is at risk of becoming more expensive.' In this book, we will explore the natural hazard of drought, to understand when and why drought has occurred in the past and to try to answer questions about what may happen in the future.

Drought and Global Change

Changes to the global environment and its link with human activity, whether as a cause or an impact, are major issues of concern. Figure 1.1 displays some of the major anthropogenic changes over the past century across the world. Rapid changes in population, carbon dioxide (CO_2) and temperature are clear indicators of the changes that humans are bringing to the environment. We are constantly reminded of the harm that we are doing to the planet and the looming threat of rising natural disasters as a result of climate change. Drought is only one element of this scenario. Increased flooding, heat waves, cold snaps, hurricanes, rising sea levels, melting glaciers and disappearing species are all becoming signs of global change. The rising population, and the need to improve agricultural output, to reduce global poverty and to further development, all conspire to make these problems worse, or to make adaptation very difficult. Drought is, however, increasingly being recognized as a key indicator and instigator of the problems that the planet faces, not only because of the costs and harm that it inflicts, but also because it is a manifestation of physical processes of climate, weather and hydrology, which all dictate where and when water is available. As we shall see, drought is inextricably linked with the cycle of water and how this is shaped by weather and climate, and it is in this context that we shall explore how drought has occurred historically and how it might change in the future.

Drought is also a key factor of what may be an impending global water crisis. Although the amount of fresh water globally is sufficient to supply the world's population and agricultural needs – at present we use less than a third of the 12,500km^3 of annually renewed and accessible fresh water (Postel et al, 1996) – it is distributed unevenly. Some sobering statistics exemplify this fact. More than a billion people lack access to clean water, 2.5 billion people do not have water for sanitation and 80 per cent of all disease is borne by dirty water (WWAP, 2009). This is because sufficient renewable water is not located where people live or, in cases where supplies meet demand, the local population does not have the resources to store and redistribute water. In many areas, there are physical limitations to building dams and other water infrastructure.

For example, in sub-Saharan Africa, only 4 per cent of renewable water is stored for later use, in comparison to 60–70 per cent in developed nations (WWAP, 2009). Such regions are often drought-prone, because of the prevailing climate, but more importantly are also vulnerable to drought. In densely populated regions, non-renewable resources may be the major source of water and so overuse is commonplace. Expanding cities and populations are increasing the demand for local water. In developing countries, which are generally water-rich, lifestyles are becoming more demanding of water and crises are emerging.

Agriculture is by far the largest water-consuming sector globally and is exerting increasing pressure on water resources. This is particularly the case in agriculturally intensive regions, which are often the most marginal in terms of water resources and thus rely heavily on the storage and redistribution of water for irrigation. At the same time, waters are being polluted at increasing rates and traditionally reliable sources of water are becoming unusable. The bottom panels in Figure 1.1 summarize how changes in water use, from population increase and per capita demand, has increased at an alarming rate over the 20th century.

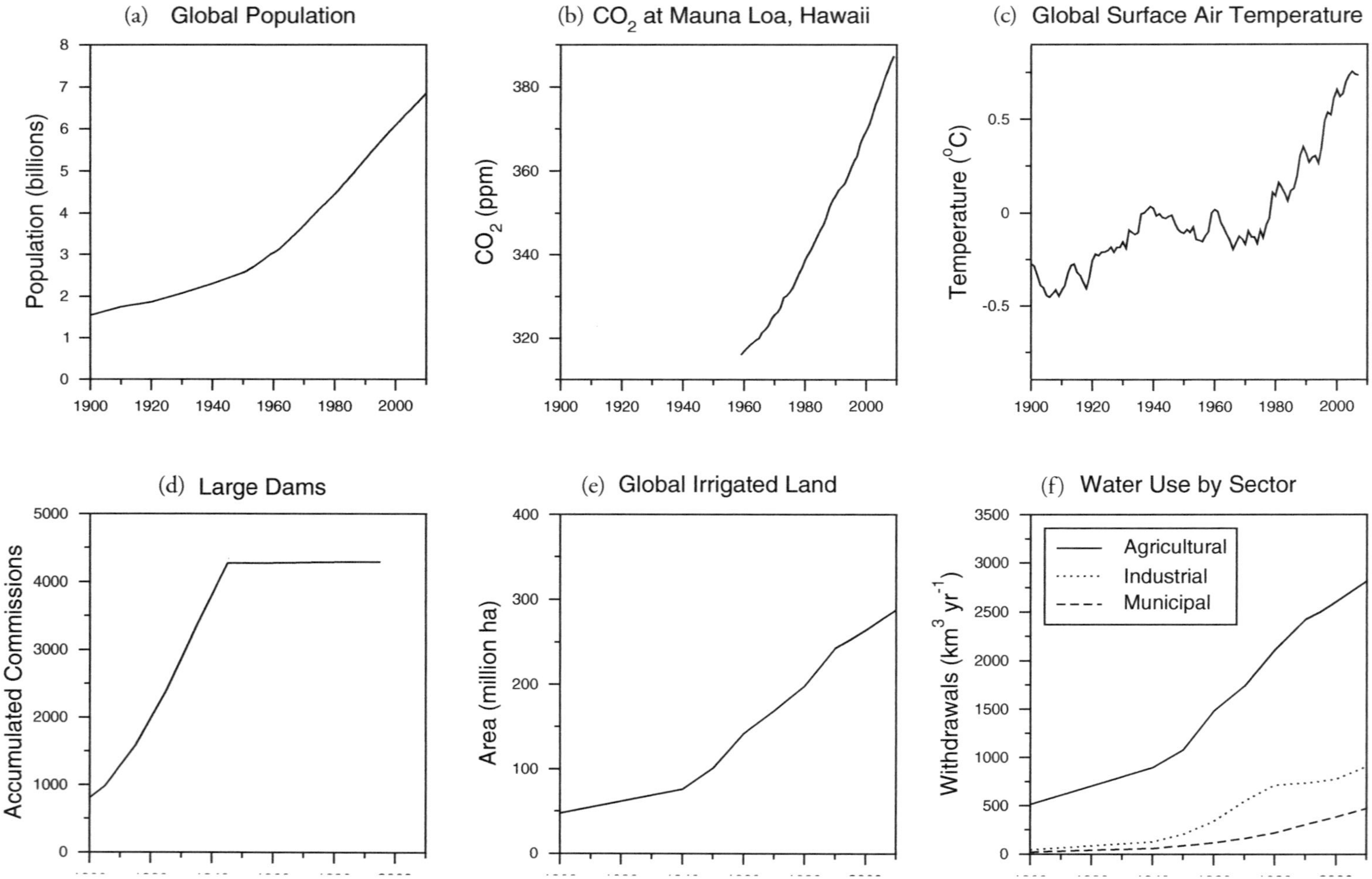

Figure 1.1 *Time series of anthropogenic changes over the 20th century*

Note: a) global population, b) atmospheric CO_2 concentration, c) global near-surface air temperature, d) commissions of large dams, e) area of land under irrigation, f) water usage in terms of agricultural, industrial and municipal withdrawals.

Source: Population data from the CIA World Factbook (The World Factbook 2009. Washington, DC: Central Intelligence Agency, 2009); CO_2 data from the NOAA Earth System Research Laboratory Global Monitoring Division (Dr Pieter Tans, NOAA/ESRL, www.esrl.noaa.gov/gmd/ccgg/trends); temperature data from the CRUTEM3 dataset of Brohan et al (2006); large dam commissions from World Commission on Dams, ICOLD Register of Dams (www.dams.org); irrigated area and water withdrawals from Shiklomanov (2009).

Physical changes to the climate and hydrological systems are not the only current and impending problems that the world is facing. Over the past years, global crises in food supply and price have dominated the news media, with increasing prices causing inconvenience in the developed world (10 per cent of income is spent on food in the US) and potentially lethal impacts in poorer regions, where more than 80 per cent of income is spent on food by the poorest 10 per cent of the population. The recent energy crisis caused oil prices to soar to record levels at the end of 2008, in the light of doubts about reserves and future supply, Middle East tensions and global speculation. Most recently, the global financial crisis has led to increased unemployment, housing loss and decreased levels of economic activity.

In a globally connected world, these crises have far-reaching impacts, and may compound local poverty and degraded environmental conditions, with the potential to lead to political instability and conflict (Kundzewicz and Kowalczak, 2009), although the latter has not been recorded in modern history (Barnaby, 2009). Solving current and future water crises may significantly improve the resilience of people and nations in the face of a gamut of potential natural and economic disasters. History has demonstrated that there is a strong correlation between economic development and poverty alleviation, and access to safe and reliable water supplies. Some 340 million Africans are still living without adequate access to safe drinking water (WWAP, 2009) and the number of people existing in absolute poverty is the same as it was 25 years ago. Nevertheless, competition between meeting our basic needs, achieving a better standard of living and sustaining the environment make this one of greatest challenges of the 21st century. In the face of potential increases in drought severity and frequency, driven by population increase, land use and climate change (Kabat et al, 2003), addressing this challenge is ever more difficult and pressing.

The Motivation for this Book

This book offers an overview of global drought. Given the pervasive nature of drought and the magnitude of its impacts, there is need for greater understanding in terms of its occurrence and potential environmental changes. In the following chapters we describe where and when drought has occurred over past millennia, and in particular during the 20th century, why these droughts happen and what may arise in the future. Drought is a recurring natural hazard that affects every part of the world, with sometimes devastating consequences. The reduction of the impacts of drought therefore has obvious and direct benefits in human, economic and environmental terms. As discussed, the resilience that water and food security brings to populations in the light of adverse conditions, offers further incentive to understanding the consequences of drought. Ultimately, it is the goal of any study of drought, and any other natural disaster, to reduce these impacts. This book aims to make progress towards alleviating the consequences of drought by establishing the context in which it can be understood.

As we shall see, this calls for better observations of drought, a clearer understanding of why droughts occur and recede, and improved monitoring of current conditions and forecasting of future conditions. To make progress, we need to develop the tools for operational monitoring and forecasting over seasonal and longer timescales, and in turn improve the documentation of the occurrence of drought, and further the understanding of the mechanisms that drive its initiation, persistence and recovery.

Despite its omnipresent nature, our knowledge of how drought develops is lacking. We have an insufficient understanding of the hydrological and weather conditions, and mechanisms that lead to drought and its recovery. This hampers not only our ability to monitor drought, but also to predict its occurrence, and to evaluate climate model projections of long-term future changes. Part of the reason for this is the lack of detailed data about its spatial and temporal variability across large scales and the impact on the environment and society. The generally poor predictive ability of climate models to forecast drought a few months in advance indicates this lack of understanding. Nevertheless, our ability to provide reasonable estimates of water cycles is approaching a stage of maturity that warrants the development of global estimates of long-term drought variability and its underlying mechanisms, while acknowledging that much scientific research is needed to improve and understand potential limitations.

This book describes some of the latest research on quantifying the global terrestrial water cycle and its application to the analysis of large-scale drought. It is hoped that it will provide the reader with a sense of the importance of drought, improve understanding of why it occurs and indicate how it fits into the broader picture of hydrology and climate change. Understanding the potential changes and risks of drought under future global warming may be regarded as prequisite information for stakeholders, governments and international agencies that need to assess the potential hazard and take appropriate steps to reduce losses. Long-term recent droughts in the US Midwest and southeast Australia, for example, hint that drought severity is only likely to get worse, given projected climate warming and changes in precipitation for the 21st century. From a scientific standpoint, improved quantification of long-term variability in the hydrological cycle and its extremes, falls under the umbrella of long-standing research into the variability of the global climate, and as such intersects with many other research paths in the geosciences. At a more general level, quantifying the variability in the terrestrial water cycle is central to achieving an understanding of the interaction of the hydrological cycle with human activities and identifying the signs of climate change. Improved estimates of seasonal to interannual variability are vital for determining the requirements of water resource operations, quantifying the risks of flooding and droughts, and understanding the effect of climate change on water resources and environmental well-being (Oki and Kanae, 2006).

An Overview of the Book

The following eight chapters follow a loose timeline, from the early Holocene (the last 12,000 years), through the 20th century to the present day, and into the future to the end of the 21st century. Along the way we describe the occurrence of global drought events, their characteristics, causes and impacts. Interspersed with this is a description of the physical processes of drought development and the science behind drought reconstruction, observation and modelling. We also describe how the occurrence of drought fits into the broader picture of global change.

The first part (Chapters 2–4) provides some background into the hazard of drought and its impacts (Chapter 2), the scientific concepts behind the processes that lead to drought (Chapter 3) and the tools used to quantify, monitor and predict drought (Chapter 4). This information will offer the reader a basic understanding of climate and hydrology, establishing a foundation for the technical concepts behind the later chapters. We have striven to present information in a non-technical style where possible, so it is feasible to read the later chapters first.

Chapter 2 provides an overview of the impacts of drought in the past few decades, which have been collected from various agency reports and disaster databases. The focus of this book rests on the physical side of drought and the impacts as driven by the severity and duration of an event. However, the impacts are as much, if not more, dictated by the vulnerability of a population, which we briefly discuss but do not explore in great detail. For more information on the socio-economics of drought, we refer the reader to more focused texts on disaster vulnerability and resilience.

In Chapter 3, we discuss the science behind drought, beginning with a broad overview of the global hydrological cycle, and its variation in space and time. This includes a discussion of the processes that bring variations in drought across the world and some of the major climate variations that drive the occurrence of drought. We then focus on the propagation of drought through the hydrological system, from low levels of precipitation to drying soils and falling reservoir levels. We look at the feedbacks between the land and atmosphere, which cause drought to persist, and the means of eventual recovery.

Much of our knowledge of drought has advanced in recent years because of the development of scientific and computational tools for gathering and analysing drought-related data, and for monitoring and predicting drought. These tools are described in Chapter 4. As well as observing drought using in-situ measurement devices, such as rain gauges, much of our knowledge of drought occurrence is derived from models of climate and hydrology – tools which are described here. An emerging tool in drought observation is the use of remote-sensing techniques, particularly from satellites that provide global views of, for example, the state of forests, lakes and agriculture, and the hydrological cycle in general. Remote sensing from satellites may prove to be the best way to monitor large-scale drought in the future. We also introduce some of the techniques for reconstructing drought records before instrumental records, using palaeoclimate data, for example based on tree-ring chronologies, cave formations or lake and ocean sediments.

Of particular importance is the use of computer models of climate and hydrology to reconstruct past climates and drought events, and their role in deciphering the physical processes. The merging of hydrometeorological measurements and output from computer modelling forms the basis for the description of historic drought occurrence in the second half of the book. Computer modelling is arguably the only tool available for projecting what might happen to drought in the future under global warming.

The rest of Chapter 4 introduces the methods for observing and assessing historic drought at large scales. This involves taking a series of hydrological measurements or outputs from a computer model and calculating a drought index. There are many different types of indices for different applications, including operational monitoring, which is a reflection of the difficulty in using a single index for all applications. We discuss some of the most popular indices and their advantages and disadvantages.

The second part of the book documents the occurrence of drought across the world in the historic record, the current state of global drought and the possibilities for change in the future. As we obviously know more about what happened in the recent past, compared to thousands of years ago, much of the content is focused on the occurrence of drought in the 20th century during the instrumental record. However, this is viewed in the context of longer-term climate variation since the last ice age in Chapter 5, which gives a broad overview of the swings in climate over the past 12,000 years and how this relates to drought. We concentrate on recent millennia because of the availability of higher resolution and more accurate palaeoclimate data and, in later years, the presence of human documentation that provides direct written evidence of past climates or indirect evidence, such as harvest dates or the frequency of agriculturally related religious festivals. The chapter describes some of the major drought periods and their impacts, and discusses the evidence that major drought episodes have been linked to the downfall of some of the great civilizations of world history.

Chapter 6 moves forwards to the 20th century and a more detailed discussion of variations in climate that have led to variations in hydrology and drought. The analysis is based on computer simulations of the global hydrological cycle, driven by the relative abundance of observations of precipitation and other meteorology, complemented by direct observations of hydrological data, such as river flows, lake and reservoir levels, and data from satellites. The level of detail in time and space, and the availability of data from all parts of the hydrological cycle (including soil moisture, streamflow, snow and lakes) enable us to drill down into the characteristics of drought and how they vary across space and time.

For example, we identify which parts of the world are most prone to long-term or short-term drought, and which periods of the 20th century endured the most frequent occurrences of drought. We make links between drought and climate, in which we explore the mechanisms that cause drought to initiate, develop and recede. In particular we look at what drives the hydrological cycle and its extremes, of which drought is only one facet. We also consider the difference between local and remote causes of drought. The former are concerned with the hydrological cycle in the local region, whereas the latter are concerned with climate phenomenon known as teleconnections, which generally act over large distances to modify the occurrence and severity of drought. This covers large-scale climate variability and oscillations such as the El Niño Southern Oscillation (ENSO) and other modes of climate variability.

We remain in the 20th century in Chapter 7 but shift the focus to documenting the individual major drought events of the century. The discussion includes an analysis of the spatial and temporal characteristics of these events, their causes and their impacts, as seen through the eyes of the monitoring and modelling tools. The debate draws from the large amount of written documentation, including media reports and official databases, as well as human recollections of past events and their impacts. For example, one of the most famous droughts of the 20th century was the Dustbowl drought of the 1930s in the US Midwest. This had huge impacts on the farmers of the region, but also had far-reaching effects nationwide because of abandonment and migration westwards, and its contribution to the continuation of the financial problems of the Great Depression. Elsewhere, the droughts in the Sahel Region of western Africa during the 1970s and 1980s, caused widespread famine and the deaths of hundreds of thousands of people, hampering development of many of the nations of the region. These are but two of the major events that struck in the 20th century. We cover a number of events from every inhabited continent, including those that are less well known and not so well documented. This chapter focuses on a comparison of these events, in terms of their spatial and temporal characteristics, and provides a quantitative assessment of the severest droughts of the 20th century.

Chapter 8 brings us to the 21st century, and a discussion of the prospect of change in drought occurrence and severity under recent and future climate change. We begin by highlighting how drought has been changing in recent years. There has been much interest lately in the impacts of climate change on extreme weather and natural hazards, especially given the continuous media coverage of natural disasters and the devastating impacts to local populations.

As we saw at the beginning of this introduction, drought was pervasive across all continents during 2009, impacting all sectors and causing problems, ranging from water restrictions to ongoing agricultural failure, which led to famine and aggravation of conflict. Long-term droughts in East Africa, southwest Australia, the US West, Argentina and elsewhere in recent years are hinting that global warming is playing a role in increasing the severity and duration of drought across the world. The purpose of the chapter is to discuss the latest research on the relationships between recent drought occurrence and other environmental changes, especially global warming, and whether the latter is exacerbating the former.

As we look to the future, current projections for climate change over the 21st century include rising global temperatures and decreasing precipitation regionally. This will have direct impacts on drought, with worrisome prospects for agriculture and water supply. Chapter 8 explores what the latest climate model projections are for the 21st century and how this will affect the occurrence and severity of drought. We give an overview of future projections in terms of climate (e.g. temperature and precipitation); the mechanisms of change, including local thermodynamic and large-scale circulation processes; projected changes in drought and the potential impacts to water resources, agriculture, ecology and society.

Finally, Chapter 9 summarizes the previous chapters and places the work in the context of broader issues of global change. We conclude with some recommendations for achieving a better understanding of drought and establishing a strategy of predicting drought, with the aim of reducing the impacts through early warning of impending droughts.

References

Barnaby, W. (2009) 'Do nations go to war over water?', in *Nature*, vol 458, pp282–283, doi:10.1038/458282a

Brohan, P., Kennedy, J. J., Harris, I., Tett, S. F. B. and Jones, P. D. (2006) 'Uncertainty estimates in regional and global observed temperature changes: A new dataset from 1850', in *Journal of Geophysical Research*, vol 111, D12106, p21, doi:10.1029/2005JD006548

FEWS-NET (2010) 'East Africa Regional Food Security Update', Famine Early Warning Systems Network (FEWS-NET)', www.fews.net/docs/publications

Kabat, P., Schulze, R. E., Hellmuth, M. E., Veraart, J. A. and Lenton, R. (2003) 'Climate variability and change and freshwater management', in *International Review for Environmental Strategies*, vol 3, pp294–302

Kundzewicz, Z. W. and Kowalczak, P. (2009) 'The potential for water conflict is on the increase', in *Nature*, vol 459, p31, doi:10.1038/459031a

Oki, T. and Kanae, S. (2006) 'Global hydrological cycles and world water resources', in *Science*, vol 313, no 5790, pp1068–1072, doi:10.1126/science.1128845

Postel, S. L., Gretchen, C. D. and Ehrlich, P. R. (1996) 'Human appropriation of renewable fresh water', *Science*, vol 271, no 5250, pp785–788, doi:10.1126/science.271.5250.785

Shiklomanov, I. A. (2009) *World Water Resources and their Use*, database on CD Rom, UNESCO, Paris

WWAP (World Water Assessment Programme) (2009) *The United Nations World Water Development Report 3: Water in a Changing World*, UNESCO, Paris, and Earthscan, London

2
What is Drought?

The Hazard of Drought

Drought is a silent and pervasive disaster that creeps up over weeks and months, often without warning, to impact on human activity on a global scale. Unlike floods, tornados and hurricanes, there are no headline-grabbing photographs, just the aftermath of failing crops, dying livestock and dried-up riverbeds. Drought can cover extensive areas and last for many years, with devastating impacts on agriculture, water supply and the environment. Historically, drought has persistently affected human activity (e.g. Stine, 1994; Hodell et al, 1995) and impacts on every part of the globe in which habitation is possible. Economic and social development have historically been strongly linked to the development of water resources, and the breakdown of this connection might have contributed to the fall of great civilizations.

Ultimately, the impact of drought may be an accumulation of losses over multiple sectors, extending to infrastructure, industry and tourism. Food exports may be reduced or reversed. Food imports may be increased. Economic development may be severely restricted. As such, drought is considered to be one of the most damaging natural hazards in terms of economic cost (Wilhite, 2000). Lives may be disrupted and loss of life is not uncommon. In extreme cases, there is evidence to suggest that drought may be a precursor to civil conflict and war (Gleick, 1998). One of the reasons for this is the large spatial extent and lengthy duration of droughts, sometimes reaching continental scales and lasting for many years. The intensity, duration and timing of droughts share characteristics with other natural hazards, such as floods, earthquakes and tornados. But in comparison to these hazards, droughts tend to start slowly, last for a long time and cover vast areas of the Earth.

Globally, droughts are one of the most devastating of natural hazards, affecting millions of people and inflicting widespread economic consequences (Wilhite et al, 2007). In the US, drought had previously been the costliest weather-related hazard (Ross and Lott, 2003) until it was surpassed by the deadly hurricane season of 2005, during which Hurricane Katrina caused an astonishing $125 billion in damages alone (Lott et al, 2008). Droughts in the US result in estimated annual damages of US$6–8 billion (FEMA, 1995; Wilhite, 2000). Losses from other major natural disasters, such as floods (US$2.4 billion per year) and hurricanes (US$1.2–4.8 billion per year) (Lott et al, 2008) are relatively small by comparison (Table 2.1). The estimated loss from the 1988 central US drought was US$40 billion in total direct and indirect costs (AMS, 1997) and the estimated loss for the state of Texas alone from the 1996 drought was US$6 billion (Wilhite, 2000).

In Europe, the latest figures indicate that the cost of droughts over the past 30 years is about €100 billion or US$135 billion (EC, 2007). It is estimated that on average 15 per cent of the European Union (EU) land area and 17 per cent of the population suffered from the impact of drought each year during 2000–2006 (EC, 2006). The drought in Europe in 2003 is estimated to have cost a total of more than €11 billion or US$15 billion (EurAqua, 2004). In China, the drought in 2001 caused a reduction in water supply for 33 million people. Meanwhile, 22 million head of livestock and agricultural losses were estimated to reach US$6.4 billion (Shen et al, 2007). Globally, there have been an estimated 7000 major disasters of any kind since 1970, causing at least US$2 trillion in damage and killing at least 2.5 million people (UN, 2008). Table 2.2 lists the total estimated losses from natural disasters over recent decades, combined and adjusted from several disaster databases. The table shows that drought is one of the major causes of economic loss.

On a global scale, the direct economic costs of drought are estimated to have been almost US$80 billion over most of the 20th century (EM-DAT, 2009). Although this is less than earthquake, flood and storm (including hurricanes) disasters, these estimates are for direct costs only. Estimating direct and indirect costs for drought is notoriously difficult, given its often subtle and protracted development, and especially in developing countries, where the frequency and quality of reporting is uneven. Compared to other natural disasters, a greater proportion of the population can be affected, especially in less developed regions. In Africa, droughts account for less than 20 per cent of disasters, but account for more than 80 per cent of the affected population (UN/ISDR, 2007). The estimated impact of the floods of 1997–1998 and drought between 1998 and 2008 in Kenya is US$4.8 billion, which is equivalent to a reduction in GDP of 16 per cent. In Ethiopia, GDP is estimated to decline by 38 per cent over 2003–2015, with an increase in poverty of 25 per cent. These numbers may be underestimated because of the high proportion of indirect losses (ECLAC, 2003; Below et al, 2007) beyond the direct impacts to the environment.

As well as economic deficit, loss of life is commonplace, either directly or indirectly. Determining the loss of life due to drought is difficult because of the indirect effects on livelihoods through, for example, lack of food and water, poor health and disease (de Waal, 1989). But it is estimated that more than half of all deaths associated with natural disasters are attributable to drought (Table 2.2). Over the 20th century, this amounts to almost 12 million people having died directly as a result of drought. Almost 1.9 billion people have been affected in some way and only floods are ranked higher in terms of the number of people affected (Below et al, 2007). The semi-arid parts of Africa have particularly suffered (Glantz, 1987; Sheffield et al, 2009). It is estimated that 300,000 people died in Ethiopia alone during the height of the Sahel drought in 1984 (EM-DAT, 2009), due to crop failure, lack of drinking water and disease. Across Africa, half a million people are estimated to have died because of drought-related impacts during the 1980s (Kallis, 2008). The deaths of more than 30,000 are associated with the European severe drought and accompanying heat wave in 2003.

Estimates of drought impacts in recent years indicate that losses are increasing, although it is difficult to know this with certainty because of the short periods of many records. Furthermore, if losses are increasing, it is difficult to isolate the impacts of climate change from changes in vulnerability and wealth. The annual economic costs increased from US$30 billion to $70 billion between 1990 and 1999. Globally, combined losses from drought and flooding increased tenfold over the second half of the 20th century, to US$300billion in the 1990s (Kunkel et al, 1999). The rise in losses is partly due to increased exposure and vulnerability. Economic development increases the exposure of population, assets and wealth. Populations may also

Table 2.1 *Summary of US weather-related disasters, 1980–2009*

Disaster type	Number of events	Frequency (%)	Damages (US$ billion)	Damages (%)
Tropical storms/hurricanes	27	28.1	367.3	51.1
Non-tropical floods	20	20.8	38.4	5.3
Heat waves and droughts	15	15.6	185.2	25.7
Severe weather	13	13.6	70.5	9.8
Fires	10	10.4	19.2	2.7
Freezing weather	6	6.3	18.6	2.6
Blizzards	2	2.1	11.9	1.7
Ice storms	2	2.1	5.9	0.8
East coast storms (Nor'easters)	1	1.0	2.2	0.3
Total	96		719.2	

Source: Adapted from NCDC, www.ncdc.noaa.gov/oa/reports/billionz.html

Table 2.2 *Losses and affected population for drought and other natural hazards globally, 1900–2004*

Disaster Type	Deaths	Affected Population	Economic Losses (US$1000)
Drought	11,707,968	1,885,890,567	80,453,813
Earthquake	2,140,134	99,735,332	340,281,331
Heat waves	101,577	11,500,287	31,691,419
Flood	6,886,838	2,748,389,208	362,498,105
Landslides	56,730	10,752,494	7,127,374
Volcanoes	95,951	4,197,447	2,890,348
Storms	1,215,094	649,097,058	446,320,656
Wildfires	2989	4,032,748	33,871,601
Total	22,207,281	5,413,595,141	1,305,134,647

Source: EM-DAT, 2009

become more vulnerable. For example, more people may live in lowland regions and river floodplains, and more people may occupy areas with marginal access to water supply and agricultural produce. Escalating costs may also be due to an increase in the number and severity of events attributed to climate change and to anthropogenic changes, such as deforestation.

Although these estimates are uncertain, subject to many biases (Gall et al, 2009) and often contradictory, they highlight the magnitude of the impact of drought as a global problem with substantial direct losses and indirect effects. The risk of drought impacts is a product of exposure and vulnerability (UN/ISDR, 2007). In general, loss of life is greater in developing regions and economic impacts are higher in the developed world. The impacts of drought are intimately linked to the vulnerability of a population to adverse conditions (UNDP, 2004) and how society responds within the constraints of changing economies.

From 1991 to 2000, more than 665,000 people died in 2557 natural disasters. Of these, 90 per cent were water-related and 97 per cent of the victims were from developing countries. Vulnerability depends on many factors, most directly related to poverty and demography, but is also linked to economic and political conditions, water supply, infrastructure, management and policy, and influenced by cultural responses to lack of water. The merging of the indirect impacts of drought with other external factors and the complexities of vulnerability ensure that it is extremely difficult to isolate the indirect impacts of drought. Often vulnerability is mapped using a weighted average of a set of vulnerability indicators. For example, these may be poverty, population density, water accessibility and institutional capacity (the ability of the local or national government to respond to a disaster), although these may not be applicable in all cases and can often be misleading in more developed regions.

Defining Drought

So what exactly is drought? The simplest definition is a deficit of water relative to normal conditions. This is governed by the hydrological cycle, which describes the movement of water through the atmosphere, land and oceans. Chapter 3 describes the hydrological cycle in detail and how it is driven by climate and human influences. The major components of the hydrological cycle are precipitation, evaporation, run-off and snow, soil and groundwater storage. Human activities also play a role in modifying the hydrological cycle, by removing, adding or storing water at a location. For drought, we are most interested in the land (terrestrial) and atmospheric components, and the linkages between them, most notably precipitation and evaporation.

When there is no precipitation for a while, perhaps for weeks, months or even years, the stores of water on the land surface (in snowpack, glaciers, ice sheets, lakes, wetlands and rivers) and in the ground (in soil water and aquifers) reduce and there is less water available for use. These stores diminish either through evaporation back into the atmosphere, drainage to lower soil layers, recharge to aquifers, export through rivers or abstraction by humans. There is also a set of feedback mechanisms that may have to be overcome for drought to occur. For example, the evaporation of water back into the atmosphere is a process of recycling water that may fall back again as precipitation. To break this cycle, either the available water on the land for evaporation would have to be depleted, or the evaporated water in the atmosphere would have to be transported from the region by the wind. These feedbacks may also work to prolong a drought. As a drought intensifies, less water is available to be recycled, thus intensifying the drought further. There are many other factors that influence the development of drought, including land cover changes (e.g. deforestation) and disturbances (e.g. fire), which are discussed in more detail in Chapter 3.

Drought may then be defined more specifically as a low amount of water in one or a combination of these stores (river, lake, reservoir, snowpack, soil water and groundwater) or fluxes (precipitation, evapotranspiration and run-off). This definition may be further qualified by adding that a drought occurs when the lack of water is sustained and spatially extensive, and is a deficit below a threshold that has adverse impacts. Depending upon the nature of the water deficit and the objectives of its use, several definitions of drought have come into existence (e.g. Palmer, 1965; Dracup et al, 1980; Rasmussen et al, 1993; AMS, 1997; Heim, 2002).

In the scientific literature, droughts are typically classified into four major types:

1 *meteorological drought*, a significant negative deviation from mean precipitation;
2 *hydrological drought*, a deficit in the supply of surface and subsurface water;
3 *soil moisture* or *agricultural drought*, a deficit in soil moisture, driven by meteorological and hydrological drought, reducing the supply of moisture for vegetation;
4 *socio-economic drought*, a combination of the above three types leading to undesirable social and economic impacts.

These classifications of drought are not rigid, since the definitions incorporate many different physical, biological and socio-economic variables. Further definitions may apply, based on environmental impacts. For example, ecological drought relates to impacts on ecosystems. In fact, definitions associated with impacts are sometimes preferred, because they imply a real impact rather than simply a physical lack of water.

These different types of drought loosely follow the propagation of drought through the hydrologic system and into ecological and economic systems (Figure 2.1). Starting with little or no precipitation for some period of time, meteorological drought begins. This filters through into the soil and can become soil moisture drought, for which the store of water in the upper layers of soil is not replenished by precipitation infiltrating from the surface and existing soil water drains into streams or recharges into groundwater below. This can be exacerbated by high evaporation if temperatures are high or winds are strong. If this occurs in an agricultural region and the soil moisture is not sufficient for crop growth, then this is an agricultural drought. Irrigation may be able to offset this, but the supply of irrigation water may also be affected by drought.

Hydrologic drought in surface waters such as rivers, lakes and reservoirs can develop as supply from the soil moisture diminishes. It can also develop or be intensified by a lack of direct precipitation, and as run-off and groundwater discharge diminishes. Hydrological droughts are important for a variety of reasons. Water supply for drinking, irrigation, hydropower, cooling and industrial use can be affected. Low streamflow that

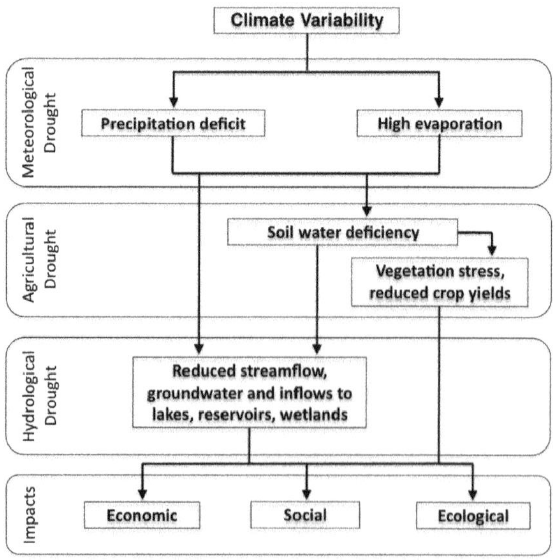

Figure 2.1 *Types of drought, their connections and impacts*

Source: Adapted from the US National Drought Mitigation Center (NDMC)

occurs every year in the dry season (or in the cold season when a river ices over) is an integral part of the flow regime of rivers. However, the maintenance of low flows above minimum thresholds is essential for environmental sustainability and the many human activities that rely on adequate and clean water supply. Low flows are especially important for water quality management, particularly for the dilution of pollution during dry periods and the level of saltwater intrusion in near-ocean river reaches. In terms of water quantity management, low flow levels have a large influence on water-use policy (Kroll et al, 2004). At the same time, low flows are critical time periods for aquatic ecology, which are sensitive to low dissolved oxygen, high water temperatures and high pollution concentrations. More often than not, these various stream-related activities are in competition, but must be assessed and managed in an integrated manner. Low-flow conditions can create biological responses and changes in habitat, such as reduced population size of aquatic species and shifts in the quantity of species type.

Recharge from the soil water store to the underlying groundwater store can be reduced and reflected in lowered well levels and also in discharge to rivers. This is groundwater drought and again water supply can be impacted with the possibility of overexploitation of groundwater. Drought in groundwater is a frequently overlooked aspect of drought. This is somewhat problematic because groundwater accounts for almost 20 per cent of water use globally and provides 45 per cent of drinking water supply (WWAP, 2009). With major issues of sustainability in some areas of the world, groundwater drought is growing in importance.

Any one of these drought types can have ecological and/or economic impacts, either directly or indirectly. Generally the most important are the effects on agricultural production, including crops and livestock, because the impacts are direct and can affect large swathes of agricultural land, with the potential for famine in regions of subsistence agriculture. Other sectors that may be impacted in the short term are hydropower and industrial use, river navigation and recreation, and water quality. Long-term drought may give rise to permanent damages, including the loss of habitat and species diversity, drainage of aquifers and land degradation and desertification. Herein lies one of the problems of drought, in that no single definition of drought is satisfactory because a drought may mean very different things to different sectors and populations. For example, a farmer is interested in the amount of water in the soil that is available to crops. A farmer in an irrigation district may only be interested in reservoir levels as this is the only source of water. Villagers in India who rely on well water are only concerned with the dwindling supplies at the bottom of the well.

Droughts also have characteristics that make them very difficult to define, in comparison to other geophysical hazards such as floods (WMO, 2006). They are slow-onset hazards, which may also recede as slowly. Thus it is difficult to establish the exact timing of its start and finish. Furthermore, the societal impacts may follow suit and accumulate slowly, and then endure for months after the drought has physically waned. Often a drought has been going on for several months before government institutions declare an official drought. Spatially, droughts are also problematic because they vary in complex ways that also change in time, so that defining their spatial extent becomes difficult. Since a drought is ignorant of political boundaries, drought-monitoring activities are challenging because they are often carried out on the basis of political regions. These challenges have important implications for drought relief because funds are usually only distributed when and where an official drought is called. Thus droughts are difficult to identify and characterize, they impact in non-structural and indirect ways and may linger for months or years over large regions. Nevertheless, progress is being made towards characterizing droughts in objective ways, which is providing the impetus for a better understanding of drought development that will feed into progress towards better drought management.

References

AMS (American Meteorological Society) (1997) 'Meteorological drought-policy statement', in *Bulletin of the American Meteorological Society*, vol 78, pp847–849

Below, R., Grover-Kopec, E. and Dilley, M. (2007) 'Documenting drought-related disasters: A global reassessment', *The Journal of Environment & Development*, vol 16, no 3, pp328–344

de Waal, A. (1989) *Famine That Kills: Darfur, Sudan, 1984–5*, Oxford University Press, Oxford, UK, p272

Dracup, J. A., Lee, K. S. and Paulson E. G., Jr. (1980) 'On the statistical characteristics of drought events', in *Water Resources Research*, vol 16, pp289–296

EC (European Commission) (2006) 'Water scarcity and drought', First Interim Report, http://ec.europa.eu/environment/water/pdf/1st_report.pdf

EC (2007) 'Communication addressing the challenge of water scarcity and droughts in the European Union (COM 2007 414)', Commission of the European Communities website, http://ec.europa.eu/environment/water/quantity/scarcity_en.htm

ECLAC (2003) 'Handbook for estimating the socio-economic and environmental effects of disasters', Economic Commission for Latin America and the Caribbean, Mexico City, p111

EM-DAT (2009) 'The OFDA/CRED International Disaster Database', Université Catholique de Louvain, Brussels, Belgium, website www.emdat.net

EurAqua (2004) 'Towards a European drought policy', www.euraqua.org/download/18.4a4d22a41128e56161b800015566/Drought_brochure_992kb.pdf

FEMA (1995) 'National Mitigation Strategy', in *Partnerships for Building Safer Communities*, Mitigation Directorate, Federal Emergency Management Agency, Washington, DC

Gall, M., Borden, K. A. and Cutter, S. L. (2009) 'When do losses count?' in *Bulletin of the American Meteorological Society*, vol 90, no 6, pp799–809, doi:10.1175/2008BAMS2721.1

Glantz, M. H. (1987) *Drought and hunger in Africa: Denying famine a future*, Cambridge University Press, Cambridge

Gleick, P. H. (1998) 'Water and conflict' (see Chronologies A and B), in P. H. Gleick, *The World's Water 1998–1999*, Island Press, Washington, DC, pp105–135

Heim, R. R., Jr. (2002) 'A review of twentieth century drought indices used in the United States', in *Bulletin of the American Meteorological Society*, vol 83, no 8, pp1149–1165

Hodell, D. A., Curtis, J. H. and Brenner, M. (1995) 'Possible role of climate in the collapse of the Classic Maya civilization', in *Nature*, vol 375, pp391–394

Kallis, G (2008) 'Droughts', in *Annual Review of Environment & Resources*, vol 33, pp85–118

Kroll, C., Luz, J., Allen, B. and Vogel, R. M. (2004) 'Developing a watershed characteristics database to improve low streamflow prediction', in *Journal of Hydrologic Engineering*, vol 9, no 2, pp116–125

Kunkel, K. E., Pielke, R. A. and Changnon, S. A. (1999) 'Temporal fluctuations in weather and climate extremes that cause economic and human health impacts', in *Bulletin of the American Meteorological Society*, vol 80, pp1077–1098

Lott, N., Ross, T., Houston, T. and Smith, A. (2008) 'Billion dollar US weather disasters, 1980–2008' (fact sheet), National Climatic Data Center, Asheville, North Carolina

Palmer, W. (1965) 'Meteorological drought' (Research Paper no 45), US Department of Commerce, Washington, DC

Rasmussen, E. M., Dickinson, R. E., Kutzbach, J. E. and Cleaveland, M. K. (1993) 'Climatology', in D. R. Maidment (ed) *Handbook of Hydrology*, McGraw-Hill, New York, pp2.1–2.44

Ross, T. F., and Lott, J. N. (2003) 'A climatology of 1980–2003 extreme weather and climate events', Tech. Rep. 2003–1, National Climatic Data Center, Asheville, North Carolina

Sheffield, J., Andreadis, K. M., Wood, E. F. and Lettenmaier, D. P. (2009) 'Global and continental drought in the second half of the 20th century: Severity-area-duration analysis and temporal variability of large-scale events', in *Journal of Climate*, vol 22, no 8, pp1962–1981

Shen, C., Wang, W.-C., Hao, Z. and Gong, W. (2007) 'Exceptional drought events over eastern China during the last five centuries', in *Climatic Change*, vol 85, pp453–471, doi:10.1007/s10584-007-9283-y

Stine, S. (1994) 'Extreme and persistent drought in California and Patagonia during medieval time' in *Nature*, vol 369, pp546–549

UNDP (2004) 'Reducing disaster risk. A challenge for development', Bureau for Crisis Prevention and Recovery/United Nations Development Programme website, www.undp.org/bcpr/disred/rdr.htm

UN/ISDR (2007) 'Drought risk reduction framework and practices: Contributing to the implementation of the Hyogo Framework for Action', United Nations Secretariat of the International Strategy for Disaster Reduction, Geneva, Switzerland

UN (United Nations) (2008) 'World economic and social survey 2008: Overcoming economic insecurity', Department of Economic and Social Affairs, United Nations, www.un.org/esa/policy/wess/wesp2008files/wesp2008.pdf

Wilhite, D. A. (2000) 'Drought as a natural hazard: Concepts and definitions', in D. A. Wilhite (ed) *Drought: A Global Assessment*, Routledge, London, pp3–18

Wilhite, D. A., Svoboda, M. D. and Hayes, M. J. (2007) 'Understanding the complex impacts of drought: A key to enhancing drought mitigation and preparedness', in *Water Resources Management*, vol 21, pp763–74

WMO (2006) *Drought Monitoring and Early Warning: Concepts, Progress and Future Challenges*, World Meteorological Organization, Geneva, Switzerland, no 1006

WWAP (World Water Assessment Programme) (2009), *United Nations World Water Development Report 3, Water in a Changing World*, Paris, UNESCO, and London, Earthscan

3

The Science of Drought

Introduction

Chapter 1 highlighted the large number of drought events that were impacting across the world in recent years. These droughts did not start and end at the same time, or reach their peak concurrently. They also inflicted different impacts, partly because of their different severities and extents, but also because of where and when they occurred. While these droughts were happening, floods were occurring in other places and normal conditions were prevalent elsewhere. These patterns highlight the complex variation in water across time and space that governs when and where hazards such as droughts occur. This chapter focuses on the science of drought, describing these variations and their causes, and how they combine to form conditions of drought. To begin with, we shall consider the fundamental role of the hydrological cycle: the underlying mechanism that governs the movement of water across and within the land surface. The hydrological cycle is itself driven by natural climate variations and man-made influences, such as changes in land use and groundwater pumping. The following description considers these natural and human drivers of the variation of drought in time and space.

The Hydrological Cycle

Drought is an extreme state of the hydrological cycle. This cycle describes the continuous movement of water between and within the atmosphere, ocean and land, and through its three phases: gas (water vapour), liquid (rain, soil moisture, streams and lakes) and solid (snow and ice). The cycle can be viewed as the Earth's thermostat, keeping the climate warm enough to stop it from freezing over and cool enough to stop all the water from being vaporized. In this section, a brief overview of the hydrological cycle is given, with an emphasis on the land component, which is where droughts develop and impact. The land hydrological cycle is closely linked with the atmospheric water cycle and to a lesser degree with the oceanic water cycle, although the ocean exerts a very large influence on climate. The atmospheric and oceanic cycles, and how they interact to drive the occurrence of drought, are described below.

Figure 3.1 shows a simplified version of the hydrological cycle. This is a generalized picture and all parts of the cycle may or may not be present in any part of the world. For instance, snow and ice are absent in tropical regions, except at high elevations, and groundwater may not be a significant component in steep terrain. Precipitation, in the form of rain, snow, or some other frozen variety, is ultimately derived from evaporation from the ocean. Evaporated water vapour is transported (advected) by wind to the land, where through various processes it is condensed as precipitation, which falls either directly onto the land surface or a water body (river, lake, wetland). When precipitation falls as snow, it will accumulate where the temperature is cold enough.

Rainfall will be partitioned into interception that is collected on vegetated canopies and surfaces, infiltration that soaks into the ground and run-off that flows directly into streams and channels. Melting snow and ice will also contribute to run-off and infiltration. The infiltrated water then percolates down through the soil column to either recharge groundwater stores, or to be discharged into springs, rivers, lakes, or even directly into the sea. Water is also sent back to the atmosphere via evaporation and transpiration, directly from the vegetation canopy and soil surfaces, or as transpiration from plants. Evaporation is the process of conversion of a substance (in this case water) from a liquid into a gas. Transpiration is the evaporation of water from the soil by plants via their stomata, which are small openings on the leaves. The combined process of evaporation and transpiration is called evapotranspiration and requires energy from the sun, atmosphere or land surface. This evaporated water may then either fall again as precipitation or be advected to another location. Run-off, soil and groundwater discharge eventually accumulate in rivers that lead to the oceans to complete the hydrological cycle.

The hydrological cycle is a series of inflows and outflows that are balanced by a set of stores – the hydrological water balance. For example, soil moisture is a store of water that is replenished by infiltrating water from above (an inflow) and is diminished by drainage from below, and by evapotranspiration (both outflows). Over time, if the inflows become less than the outflows, the store will decrease, resulting in a lack of water or a drought. Human activity can also play a major role in the hydrological cycle, by changing the rate of in- and outflows, or the size of the stores. For instance, reservoirs provide a store of water for later use, and can govern downstream river flows to mitigate flooding. Groundwater pumping may extract water for purposes of irrigation that is then redistributed over a wider area, which will drain away as run-off or infiltration into the soil, where it will be transpired by plants or will percolate back into the groundwater.

Spatial and Temporal Patterns of Hydrology

The distribution of water in the form of these inflows, outflows and stores varies widely in time and space. At any one time and location, different sets of hydrological processes can be in operation, which are shaped by the driving weather and climate, and the underlying characteristics of the land.

Spatial patterns

The spatial variation of water, from local to global scales, is driven by the patterns of climate and the underlying land-surface characteristics of elevation, slope, vegetation, land use and water bodies. Some of these factors are not mutually exclusive, such as climate and vegetation, whereas others, such as land use, may

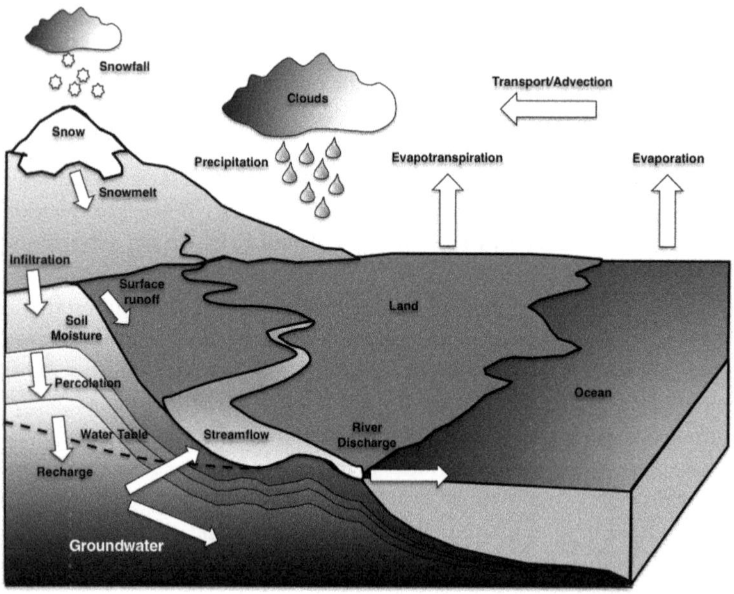

Figure 3.1 *The hydrological cycle*

be independent of climate in the case of irrigated agricultural land. At large scales, the climate is the dominant driver of the spatial variation in water.

Plate 3.1 shows a global map of climate zones based on the Köppen-Geiger classification (Peel et al, 2007). This classification uses information on monthly precipitation and temperature to assign each point on Earth to a climate zone. The major climate zones are (A) tropical, (B) dry, (C) temperate mid-latitude, (D) cold mid-latitude, (E) polar and (H) highland. These classes are divided further into a number of subzones, listed in Table 3.1. The largest climate zone is (B), which includes deserts, and warm and cold steppe regions, and covers 30 per cent of the land. This consists mainly of the Sahara and other desert regions. The second largest is climate zone (D) (cold, mid-latitude climates), which

Table 3.1 *Climate zones of the world based on the Köppen-Geiger classification (percentage area of the major zones are given in parentheses)*

Climate Class	Description	Features
A (19%)	**Tropical**	High temperature and precipitation. Temperature of the coldest month exceeds 18°C.
Af	Tropical rain forest	No dry season. The driest month has at least 60mm of rain. Rainfall is generally evenly distributed throughout the year.
Am	Tropical monsoonal	Pronounced wet season. Short dry season. One or more months with less than 60mm of rain. Highest annual temperature occurs just prior to the rainy season.
Aw	Tropical savanna	Winter dry season. There are more than two months with less than 60mm of rain.
B (30.2%)	**Arid/Semi-arid**	Potential evaporation exceeds precipitation.
BSh	Subtropical steppe	Low-latitude dry. Evaporation exceeds precipitation on average, but is less than potential evaporation. Average temperature is more than 18°C.
BSk	Mid-latitude steppe	Mid-latitude dry. Evaporation exceeds precipitation on average, but is less than potential evaporation. Average temperature is less than 18°C.
BWh	Subtropical desert	Low-latitude desert. Evaporation exceeds precipitation on average, but is less than half the potential level of evaporation. Average temperature is more than 18°C. Frost is absent or infrequent.
BWk	Mid-latitude desert	Mid-latitude desert. Evaporation exceeds precipitation on average, but is less than half the potential level of evaporation. Average temperature is less than 18°C. Winter temperatures are below freezing.
C (13.4%)	**Temperate**	Average temperature of the coldest month is less than18°C, but above –3°C. Average temperature of the warmest month is more than 10°C.
Cfa	Humid subtropical	Mild with no dry season and a hot summer. Average temperature of warmest month is more than 22°C. Average temperature of coldest month is less than 18°C. Year-round rainfall, but highly variable.
Cfb	Maritime temperate	Mild with no dry season and a warm summer. Average temperature of all months is lower than 22°C. At least four months with average temperatures of more than 10°C. Rainfall is equally spread around the year.
Cfc	Maritime subarctic	Mild with no dry season and a cool summer. Average temperature of all months is lower than 22°C. There are one to three months with average temperatures of more than 10°C. Rainfall is equally spread around the year.

Table 3.1 *Climate zones of the world based on the Köppen-Geiger classification* (Cont'd)

Csa	Mediterranean	Mild with dry, hot summer. Warmest month has average temperature of more than 22°C. At least four months with average temperatures of over 10°C. Frost danger in winter. At least three times as much precipitation falls during the wettest winter months as in the driest summer month.
Csb	Mediterranean maritime	Mild with dry, cool summer. On average, the temperature of the warmest month does not exceed 22°C. At least four months with average temperatures of more than 10°C. Frost danger in winter. At least three times as much precipitation falls during the wettest winter months as in the driest summer month.
Cwa	Humid subtropical	Mild with dry winter, hot summer. On average, the temperature of the warmest month exceeds 22°C. Frost danger in winter. At least ten times as much precipitation falls during the wettest summer month as in the driest winter month.
D (24.6%)	**Cold**	Average temperature of the warmest month is over 10°C and that of the coldest month is less than –3°C.
Dfa	Humid continental	Humid with hot summer. At least 30mm of rain falls in the driest month. Average temperature of the warmest month exceeds 22°C.
Dfb	Humid continental	Humid with severe winter, no dry season, warm summer. At least 30mm of rain falls in the driest month. Average temperature of the warmest month exceeds 22°C.
Dfc	Subarctic	Severe winter, no dry season, cool summer. At least 30mm of rain falls in the driest month. Less than four months have an average temperature of below 10°C.
Dfd	Subarctic	Severe, very cold winter, no dry season, cool summer. At least 30mm of rain falls in the driest month, average temperature of coldest month is below –38°C.
Dwa	Humid continental	Humid with severe, dry winter, hot summer. At least 10 times as much precipitation falls in the wettest month of summer as in the driest month of winter.
Dwb	Humid continental	Humid with severe, dry winter, warm summer. At least 10 times as much precipitation in the wettest month of summer as in the driest month of winter.
Dwc	Subarctic	Severe, dry winter, cool summer. At least 10 times as much precipitation falls in the wettest month of summer as in the driest month of winter
Dwd	Subarctic	Severe, very cold and dry winter, cool summer. At least 10 times as much precipitation falls in the wettest month of summer as in the driest month of winter.
E (12.8%)	**Polar**	Temperatures are below freezing for most of the year.
ET	Tundra	Tundra, average temperature of warmest month exceeds 0°C.
EF	Ice cap	No month with temperature of above 10°C.
H	**Mountainous**	Complex zone that can encompass any of the above classes because of its mountainous terrain.

covers about 25 per cent of the land, but has many subclasses defined by differences in the wetness and temperature of the seasons.

The basic distribution of climate is a result of the general circulation of the atmosphere, which represents the major flows of water and energy around the world. The circulation is a result of the differential heating of the Earth by the sun, the rotation and tilt of the Earth, and the distribution of the land masses. Over the year, the global incoming radiation from the sun (solar radiation) is balanced by the outgoing energy emitted by the Earth (terrestrial radiation). However, the sun's energy is most intense in the tropics and lowest at the poles (because of the angle of incidence of the sun's rays), where outgoing radiation is greater than incoming solar radiation. This creates a surplus of energy near the equator and a deficit of energy near the poles (Figure 3.2). The climate system tries to maintain a balance and so the surplus of energy is dissipated towards the poles by various atmospheric and ocean flows.

The intense solar radiation in the tropics causes the heating and rising of moist air, which cools and condenses to form a region of intense clouds and precipitation, called the Intertropical Convergence Zone (ITCZ). This is where the tropical rain forests are located. This region is also known as the doldrums, where the pressure gradient is very small, creating little wind and thus making it difficult for sailing boats to cross the equator. This also means that there is very little transfer of weather systems between the northern and southern hemispheres. The rising moist air in the ITCZ spreads polewards when it hits the stable air of the stratosphere at the tropopause (the boundary between the troposphere and stratosphere). In the absence of other factors (the tilt and rotation of the Earth, land masses), this spreading would continue to the poles. However, this air movement is restricted to a band about 30° north and 30° south of the equator because of the rotation of the Earth and the height of the tropopause (Held and Hou, 1980; Frierson et al, 2007). At these latitudes, air descends back to the surface to form a band of high pressure called the subtropical high. Having lost most of its moisture in the rising part of the circulation in the ITCZ, the descending branch of the circulation is relatively dry. Coupled with the warming process of compression as the air subsides, this results in hot and dry conditions that are relatively free of convection and thunderstorms. Many of the world's major deserts are located in the region of the subtropical highs, such as the Sahara

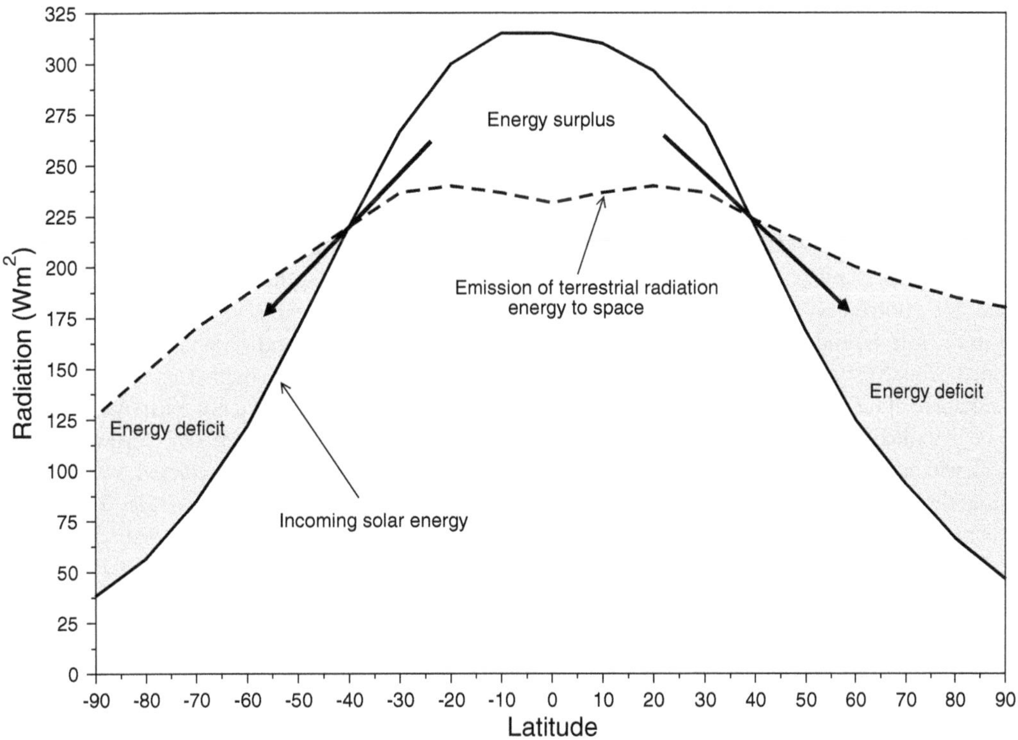

Figure 3.2 *The global energy balance*

Note: Around the equator, incoming radiation from the sun (solar radiation) that is not reflected by clouds, the ocean and land surfaces, is greater than the outgoing radiation emitted by the Earth (terrestrial radiation). This creates a surplus of energy. Towards the poles, the solar radiation is less than the outgoing radiation, creating a deficit. The imbalance of energy surplus and deficit between the equator and the poles causes a transport of energy, which drives the general circulation of the atmosphere.

Desert and the Kalahari Desert in Africa, the Gobi Desert in Asia, the Southwest Desert in the US, and the Australian Desert in the southern hemisphere.

Part of the subtropical dry surface air travels back to the lower pressure at the equator to replace the moist rising tropical air and complete the circulation. The Coriolis force, induced by the spin of the Earth, deflects this air movement from east to west for the equator-bound air to form the Trade Winds or Tropical Easterlies that blow steadily from the northeast in the northern hemisphere and southeast in the southern hemisphere. The other portion of dry air moves towards the poles, and is deflected from west to east to produce the Westerlies. This circulation between the equator and the subtropics is called a Hadley cell, first proposed by the English meteorologist George Hadley. The Hadley cell is not centred exactly on the equator but follows the thermal equator, which is the point directly below the sun (or zenith). This undulates north and southwards with the seasons because of the tilt of the Earth and gives rise to the movement of the ITCZ, which tracks north in the northern hemisphere summer and south in the southern hemisphere summer. The width of the Hadley cell is related to global temperatures and the temperature gradient between the equator and the poles (Frierson et al, 2007), and this is one mechanism for the expansion of the dry subtropics under global warming that is discussed in Chapter 8 (Drought in the 21st Century).

Beyond the subtropics, other general circulation cells exist in the mid-latitudes and polar regions, which are explained primarily on the basis of convection and the Coriolis force. The average atmospheric flows can therefore be described by a three-cell circulation model (Figure 3.3), which consists of the Hadley cell straddling the equator, the mid-latitude Ferrel cell between 30° and 60° (consisting of westerly winds at the surface and the opposite flow aloft) and the polar cell at higher latitudes (with easterly flows at the surface). At the boundaries of these cells, jet streams will form, which are narrow, fast-moving air currents near the tropopause, which flow westwards. Warm air moving polewards from the subtropical high meets with cold polar easterly air. As the angular momentum is perpetuated, the air increases speed to form the polar jet stream. Similarly, at the subtropical high, the weaker subtropical jet stream gathers. Jet streams are not stationary in latitude but meander greatly, as well as moving north and south

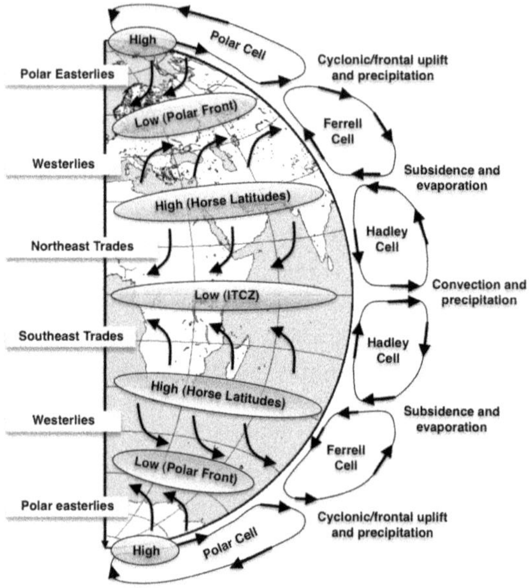

Figure 3.3 *General circulation of the atmosphere*

with the seasons. The subtropical and polar jet streams can even join together. These meanders are called Rossby waves, which are embedded in the flow but move much more slowly than the actual wind speed. The meander of the polar jet stream has a large impact on mid-latitude weather systems and thus drought.

In reality, the general circulation of the atmosphere is far more complex and variable than that predicted by the three-cell model, although it explains the distribution of climate adequately for our purposes. For instance, observations and theory are not consistent with wind and energy transport as predicted by the three-cell model, and other factors come into play. The distribution of land masses is one feature that alters the wind patterns: for example, the Andes mountain range interrupts the latitudinal subtropical high-pressure belt. Furthermore, the general atmospheric circulation is driven to a large degree by the motion of large eddies – cyclonic and anticyclonic – in the mid-latitudes.

Temporal variation

As well as varying spatially, the climate varies across multiple timescales, from hourly, to daily, to seasonal, to centuries and longer spans. The climate responds naturally to external forcings, such as solar radiation (Christensen and Lassen, 1991) and atmospheric composition (Robock and Mao, 1995). Furthermore, it varies as a result of internal interactions between components of the coupled climate system (Trenberth and Hurrell, 1994), including feedbacks with the oceans and land surface (e.g. Jiang et al, 2006; McCabe and Palecki, 2006). These variations occur as both regular and semi-regular cycles, and as chaotic fluctuations. At daily to weekly timescales, the chaotic fluctuations of the weather systems tend to dominate. These are the most unpredictable variations in climate, but in general bring extreme weather to extra-tropical and polar regions, causing flooding, ocean storm surges, droughts and heat waves. Furthermore, climate variation in the past century has increasingly been influenced by human activity.

Solar forcing

The main external driver of climate variability is the intensity of the sun's energy reaching the Earth. This is caused by the daily rotation of the Earth, the tilt of the Earth's orbit around the sun on an annual cycle, and the changes in these factors on millennial timescales – which give rise to major events such as the ice ages. There are also other quasi-periodic changes in the intensity of the sun's energy itself, whose variations have been linked to changes in climate and global temperatures in particular (Crowley, 2000). At daily or diurnal timescales, the rotation of the Earth causes variation in the intensity of the sun's energy reaching the Earth. This heats and cools the atmosphere and land surface, giving rise to many daily weather phenomena, such as the temperature change between night and day, land–sea breezes and afternoon convective storms.

The annual orbit of the Earth around the sun drives the seasonal changes in climate. The largest changes are brought by the monsoons. These are shifts in wind directions caused by the north–south migration of the ITCZ. For instance, the south Asian monsoon brings copious amounts of rainfall to the Indian subcontinent during June to September as the heating of the land lowers the pressure and draws moisture-laden air from the Indian Ocean, which hits the Himalayan mountain range, rises and falls as precipitation. Other monsoons in Southeast Asia, northern Australia and North America develop similarly. Monsoons can be very important to these regions, since most of the annual rainfall can occur in the monsoon period, making agriculture highly dependent on the consistency of the amount and timing of the monsoon rains.

The location and strength of the polar and subtropical jet streams also varies with the seasons, moving polewards in the summer and back in the winter. In general, jet stream winds blow from west to east, but they often meander north and southwards, following the boundaries between cold and hot air, diverging to form eddies and even disappearing and reappearing elsewhere. This is a result of many complex interactions between high- and low-pressure systems, hot and cold air masses, and seasonal influences. Winds tend to be strongest in the winter when temperature gradients are strongest.

Ocean–atmosphere internal variations

As mentioned before, the climate results from a blend of external forcings (natural and anthropogenic), and natural internal interactions. The latter are mostly a result of interactions of the ocean–atmosphere system. The oceans play a huge role in shaping the climate on all timescales, ranging from days to millennia, mainly because of their large heat-storage capacity. Of particular importance are variations at interannual to decadal timescales, which play a very important role in determining the variability of the Earth's climate. They can be a direct cause of variability, in the absence of any significant change in the forcing, through interactions between various elements of the coupled ocean–atmosphere–land system. Internal variations in the coupled system may conspire to generate extremes of climate. Anomalies at the ocean surface interact with the overlying atmosphere to alter the transport of moisture around the globe by intensifying local convection, shifting storm tracks, and changing the frequency of hurricanes, to bring flooding or drought to different regions of the globe. The relationship between climatic variation in one part of the ocean and its impacts at remote locations is called a teleconnection. These are very important for understanding the somewhat ordered nature of the coupled ocean–atmosphere system and the lag between the dynamics of the ocean and the atmosphere. These linkages and their remote impacts, which occur at a later time, offer the potential for climate prediction.

Much of the large-scale dynamics of the oceans and overlying atmosphere follow quasi-periodic cycles or oscillations (see Box 3.1). The most important and well known of these is the El Niño Southern Oscillation (ENSO), which oscillates about every two to seven years and impacts on the climate of the tropics and many regions in the mid-latitudes (Ropelewski and Halpert, 1987). Other climate oscillations and modes of large-scale variability, such as the North Atlantic Oscillation (NAO), the Pacific Decadal Oscillation (PDO), and the Atlantic Multidecadal Oscillation (AMO), act on other timescales and interact with the ENSO, or are the primary drivers of regional variations in climate. For example, the NAO is known to affect the climate in eastern North America and Europe (Hurrell and van Loon, 1997), as well as North Africa (Wang, 2003). The PDO (Mantua et al, 1997) is a primary driver of climate around the Pacific basin and interacts with the ENSO, modifying the climate on a global scale (Newman et al, 2003; Verdon and Franks, 2006). The AMO (Kerr, 2005) impacts on the North Atlantic, especially on the North American climate (Enfield et al, 2001; McCabe et al, 2004) and the European climate (Sutton and Hodson, 2005), and is a potential modulating force of the ENSO (Dong et al, 2006).

Box 3.1 Climate oscillations and teleconnections

El Niño Southern Oscillation (ENSO)

The ENSO is the most important driver of global climate on interannual timescales. It manifests in the tropical Pacific Ocean, in the form of anomalies in sea surface pressure and sea surface temperature (SST). The major circulation pattern in this region is the Walker circulation, which is a zonal (east–west) overturning circulation, driven by the trade winds. These set up differences in temperature between the east and west, via a shallower thermocline (boundary between the well-mixed upper layer and the stable lower layer) and thus increased upwelling and cooling in the east. The SST differences force air to ascend in the west and to subside in the east, with the circulation being completed by westward flows aloft. The Walker circulation amplifies the zonal easterly winds, but oscillates irregularly.

The Southern Oscillation Index (SOI) is a measure of the strength of this, in terms of the pressure difference between the east and west tropical Pacific (specifically at Tahiti and Darwin, Australia). This is intimately related to temperature as well, providing an oscillation between warm and cool SSTs. When the SOI is in its negative phase, the difference between the east and west Pacific is smaller and so the easterly winds decrease, resulting in reduced upwelling and warmer SSTs in the eastern tropical Pacific. This is referred to as the El Niño (meaning in Spanish 'boy' or 'Christ child', since it tends to occur near Christmas). In its positive phase, the SOI characterizes colder temperatures in the eastern Pacific as the thermocline tilts even further upwards towards the east. This phase is known as La Niña ('girl' in Spanish). The tight coupling between oscillations in surface pressure, as characterized by the SOI, and that of SSTs, as typified by El Niño/La Niña, has given rise to the more popular use of the term El Niño Southern Oscillation (ENSO) to describe this phenomenon. The positive feedback is eventually overturned by a negative feedback related to the adjustment of the thermocline in the ocean, but at a delayed timescale, thus inducing irregular fluctuations in the ENSO.

Figure 3.4 shows the regions of the world affected by the ENSO, indicating that these stretch well beyond the local area of the tropical Pacific. Warmer SSTs during El Niño force convection to migrate eastwards, bringing wetter conditions to the central Pacific and drier conditions to Indonesia and northern Australia. Further afield, El Niño brings reduced rainfall in southwestern Africa, but wetter conditions in parts of the US, and in East Asia weakens the monsoon, causing warmer conditions (Wang et al, 2000).

North Atlantic Oscillation (NAO) and Arctic Oscillation (AO)

In the northern hemisphere, the North Atlantic Oscillation (NAO) is characterized by a year-to-year seesaw in the difference in pressure between the Icelandic Low and the Azores High pressure regions. These modes are particularly

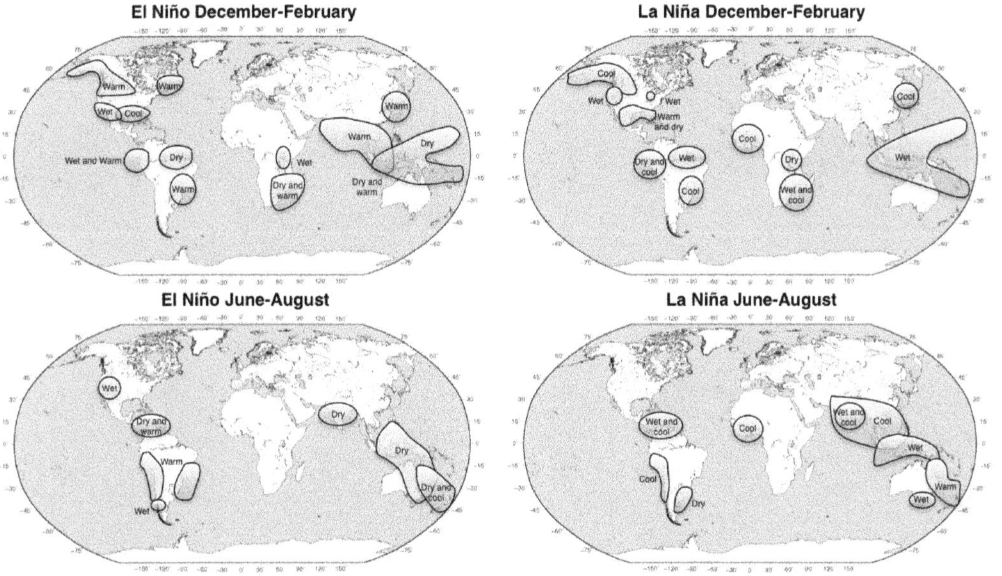

Figure 3.4 *Map of regions impacted by the ENSO*

Source: Adapted from US National Weather Service, website www.srh.weather.gov/srh/jetstream/tropics/enso_impacts.html

strong in the northern hemisphere winter, although summertime manifestations are apparent (Folland et al, 2009). The variations are actually part of an extended pattern throughout the northern hemisphere and are better known as the Arctic Oscillation (AO), or Northern Annular Mode (NAM; Thompson and Wallace, 1998). The AO manifests as anomalies of different size between the Arctic and mid-latitudes, at around 40°N. The phases of the AO and NAO deflect the jet stream further north or south in the Atlantic Ocean, changing the strength of westerly winds and storm tracks, and affecting the climates of North America and Europe. The positive phase of the AO (and NAO), in which arctic sea level pressure is below normal, brings warmer and wetter weather to northern Europe and into northern Eurasia, cooler and drier weather to the Mediterranean and stronger trade winds at lower latitudes (Osborn, 2000) (Figure 3.5). The negative phase is characterized by higher than normal sea level pressures in the Arctic, colder winters in North America and Europe, warm and wet conditions in the Mediterranean and weaker trade winds. Interestingly, the recent long-term trend in the weather pattern of the NAO may be related to northern hemispheric winter warming (Hurrell, 1996).

Pacific Decadal Oscillation (PDO)

The PDO is similar to the ENSO but manifests in the northern Pacific Ocean and varies across much longer timescales of 20 to 30 years (Mantua et al, 1997). It is quantified by the variability in sea surface temperatures and sea level pressure in the north Pacific. Different phases of the PDO have persisted for several decades, most notably the 'cold' or 'negative' phase of 1890–1924 and 1947–1976, and the 'warm' or 'positive' phase of 1925–1946, and from 1977 until recent times. The impacts of the PDO are somewhat similar to those of the ENSO. Over North America, for example, the positive phase of the PDO is associated with wetter conditions in the southwestern US, and warmer winter and spring temperatures in the northwest, accompanied by lower precipitation extending to the Great Lakes region.

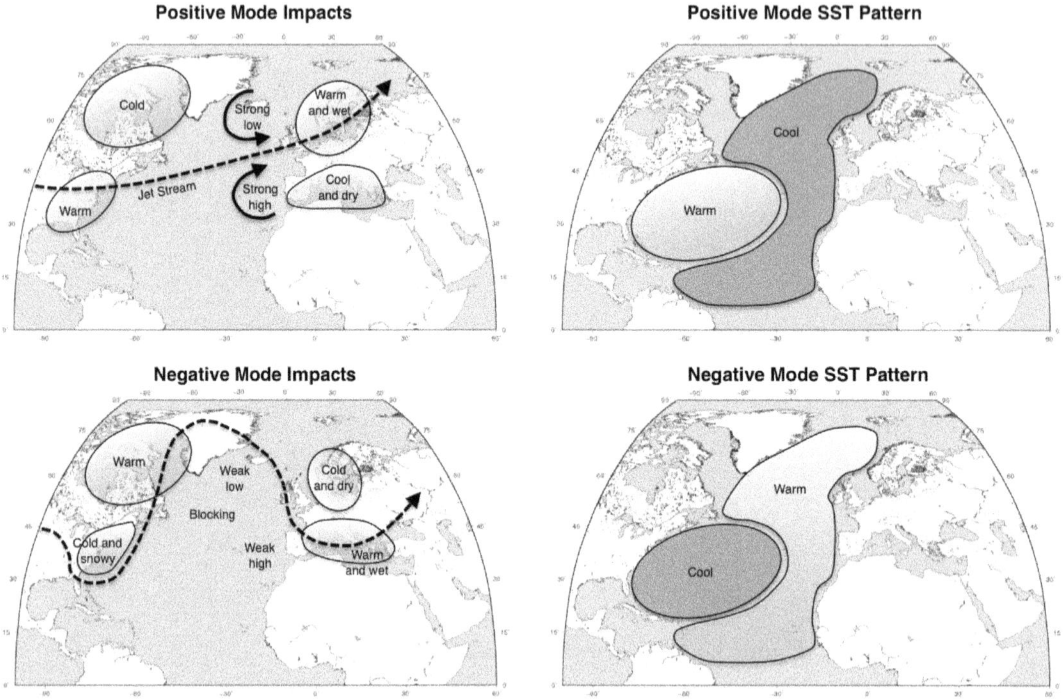

Figure 3.5 *Map of regions impacted by the NAO*

Source: Adapted from the website http://airmap.unh.edu/background/nao.html

The negative phase brings persistent dry conditions to the southern US and northern Mexico during the winter and spring, and more moisture to the northwest. The PDO is a strong modulator of snowpack and streamflow in the western US, affecting the water supply (Webb et al, 2004). The PDO exerts a modulating influence on the ENSO, therefore changing the strength and persistence of the ENSO's effects around the world. For example, the impact of the ENSO on the East Asian winter monsoon is only robust when the PDO is in its cold phase (Wang et al, 2008). Over North America, the impacts of the ENSO are strongest only during the positive PDO phase (Gershunov and Barnett, 1998), and only in the negative phase over Australia (Power et al, 1999).

Atlantic Multidecadal Oscillation (AMO)

On the other side of the world, the Atlantic Multidecadal Oscillation (AMO) describes changes in surface temperatures in the North Atlantic over periods of several decades (Kerr, 2005). It is thought to be a factor in the frequency of Atlantic hurricanes and may be influenced by anthropogenic warming (Enfield and Cid-Serrano, 2010). The AMO is related to temperature and precipitation fluctuations over the northern hemisphere. Figure 3.6 shows the regions that are impacted, in particular North America and Europe. The warm phase of the AMO is associated with dry conditions in the Midwest and southwest of the US, whereas wetter conditions prevail in the southeast and northwest (Enfield et al, 2001). There is also evidence from climate models and palaeoclimate data that the AMO is associated with climate over North Africa, India and northeastern Brazil (Zhang and Delworth, 2006; Shanahan et al, 2009), discussed further in Chapter 5.

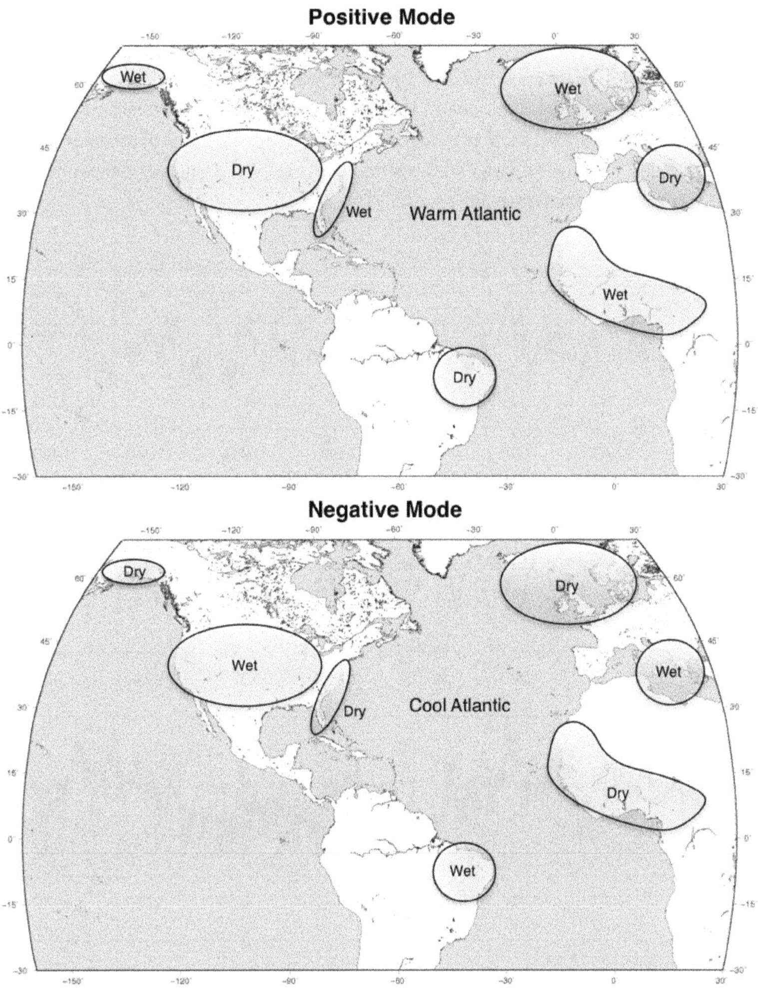

Figure 3.6 *Map of regions impacted by the AMO*

Source: Adapted from website www.aoml.noaa.gov/phod/faq/amo_faq.php

Indian Ocean Dipole (IOD)

In the Indian Ocean the periodic cooling of the southeast and warming of the western equatorial ocean is called the Indian Ocean Dipole (IOD). This causes the centre of convection over the warm east Indian Ocean to migrate towards the west, bringing high precipitation to East Africa and dry conditions (droughts and forest fires) across Indonesia. Further afield, the IOD also appears to be an important driver of the climate over Australia (Ummenhofer et al, 2009), and may override the usually wetter conditions brought about by a La Niña event.

Southern Annular Mode (SAM)

The southern hemisphere equivalent of the NAM, the Southern Annular Mode (SAM; Limpasuvan and Hartmann, 2000) characterizes the atmospheric dynamics of the middle latitudes and describes shifts in pressure between the

South Pole and the mid-latitudes. The SAM varies on weekly to annual timescales and dictates much of the short-term variability in weather in these regions. The SAM is linked to temperature variations across Antarctica, the Southern Ocean and the distribution of sea ice, and also to storm tracks in the mid-latitudes. The SAM is also known as the Antarctic Oscillation (AAO) and has been linked with drought in Australia (Verdon-Kidd and Kiem, 2009).

Figure 3.7 shows time series of some of these major modes of climate variability over the 20th century and illustrates the differences in their timescales. The NAO varies from weekly to annual timescales, whereas the AMO and PDO act on decadal to multidecadal timescales.

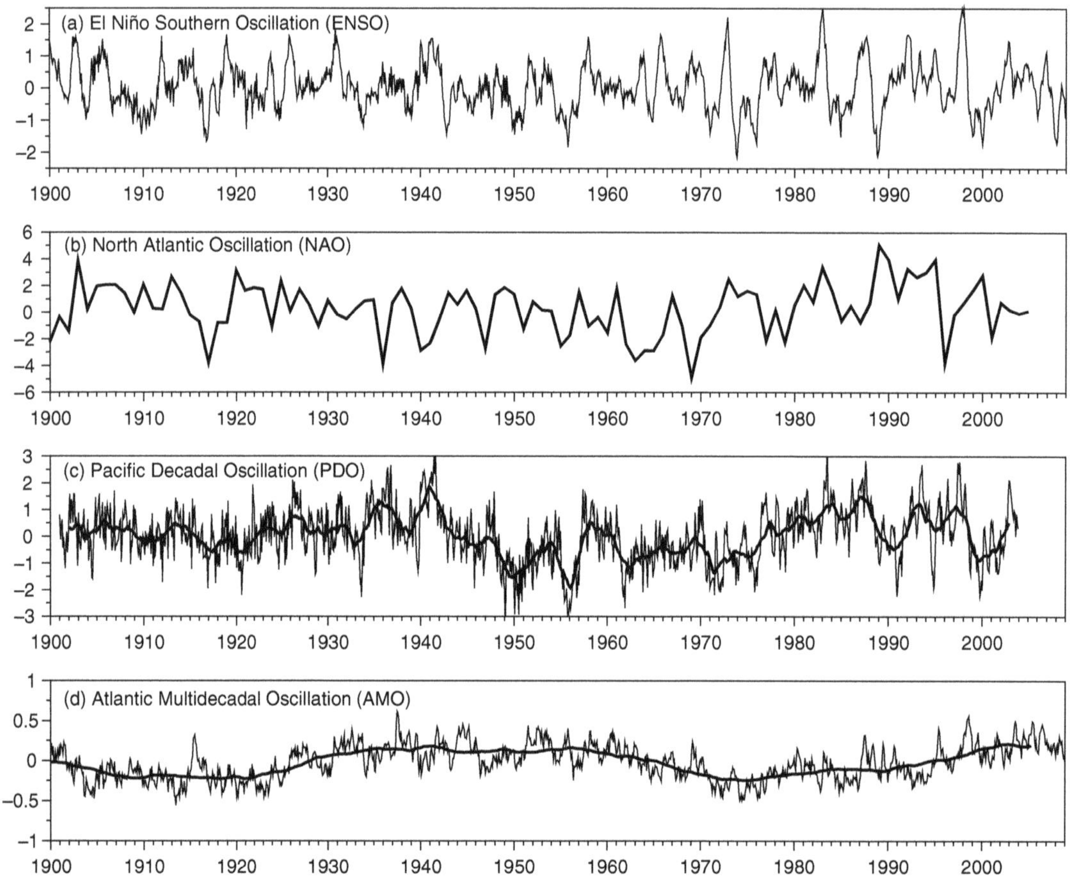

Figure 3.7 *Time series of major modes of climate variability*

(a) El Niño Southern Oscillation (ENSO) Niño3.4 SST anomalies based on HadSST1 data

Source: www.esrl.noaa.gov/psd/enso

(b) The North Atlantic Oscillation (NAO) wintertime index, based on the difference of normalized sea level pressures (SLP) between the Azores and Iceland

Source: www.cgd.ucar.edu/cas/jhurrell/indices.html

(c) The Pacific Decadal Oscillation (PDO) (Mantua et al, 1997). The data are calculated as the principal component of North Pacific (north of 20°N) monthly SST variability

Source: http://jisao.washington.edu/pdo

(d) The Atlantic Multidecadal Oscillation (AMO) (Enfield et al, 2001), which is the detrended, area-weighted average of SST data over the North Atlantic (0–70°N)

Source: www.cdc.noaa.gov/Correlation

Volcanoes, dust storms and aerosols

Climate can also be appreciably impacted by the presence of aerosols in the atmosphere. Aerosols are tiny particles suspended in the air that occur naturally or are human-made. Naturally occurring aerosols originate from volcanoes, dust storms, forest fires and sea spray. The impact of aerosols is most obvious when they are large enough to scatter sunlight and reduce visibility, which can, for example, cause red sunrises and sunsets. More importantly, aerosols have a large impact on climate, although this is not currently well understood and is a subject of intense scientific research. Aerosols have a direct cooling effect because most types reflect sunlight back into space, depending on the size and composition of the particles. This mechanism may have partially offset the impact of global warming induced by greenhouse gases between the 1940s and 1960s.

Aerosols also have an indirect effect on climate by changing the properties and behaviour of clouds. Clouds would not form without aerosols, as small particles are required for cloud droplets to condense upon (known as cloud condensation nuclei). For small dust aerosols, as their concentration increases, cloud water is spread more thinly over the large number of particles, changing the character of the clouds, their thickness and optical properties. This impacts the absorption and reflection of sunlight, and the amount of precipitation. In the case of ocean salt aerosols, their relatively larger size tends to enhance precipitation directly. Aerosols can also be important as sites for chemical reactions, such as those involved in the destruction of the ozone layer in the stratosphere by polar stratospheric clouds.

Volcanoes are a major source of aerosols, and can have significant impacts on climate. This is because aerosols of volcanic origin are generally ejected into the stratosphere, where they are spread to almost every part of the globe and can remain there for up to two years. Here they reflect sunlight, cooling the lower atmosphere and land surface. The massive Mount Pinatubo eruption in 1991 in the Philippines caused a subsequent global cooling of about 0.5°C and damage to the ozone layer, after ejecting nearly 20 million tonnes of sulphur dioxide.

Dust storms also produce aerosols. For example, these can be seen in satellite images as giant plumes blowing from the Sahara Desert, across the North Atlantic Ocean, and even reaching the Americas. Similar storms originate from the dry interior of China and blow towards coastal cities. Australia experienced a massive dust storm in 2009, which spread from the interior to cover the city of Sydney in a cloak of red dust for several days. Such storms can have impacts on cloud formation and the absorption of sunlight. The Dust Bowl drought in the US Midwest in the 1930s was exacerbated by poor land management practices, including overgrazing, which caused erosion and dust storms that further contributed to the drought conditions (Cook et al, 2004). The Sahel drought in the 1970s and 1980s was similarly intensified by vegetation degradation and erosion due to overgrazing, poor management and erosion.

Human influences

Greenhouse gases such as water vapour, CO_2, methane and nitrous oxide are naturally present in the atmosphere. The presence of these gases keeps the Earth about 33°C warmer, because they absorb radiation emitted from the Earth's surface, which would otherwise escape back into space. Without this greenhouse effect, life on Earth would not be possible. However, because of human activities such as fossil fuel burning, land-use change and other industrial activities, concentrations of these gases have been rising over the past 150 years, since the start of the industrial revolution (although there is some evidence to suggest that humans have had an impact through land-use change and wood burning for 8000 years; Ruddiman, 2003). The result has contributed to a rise in global temperatures of about 0.74°C since the beginning of the 20th century, and about 0.13°C per decade since 1955 (IPCC, 2007). The impact that this has had on drought over the 20th and the early 21st centuries is discussed in later chapters.

Agricultural and industrial activity adds large amounts of aerosols to the atmosphere. These include black soot (black carbon) and smoke from forest and agricultural fires, wood fuel and diesel burning, and sulphates from the burning of fossil fuels. The concentration of anthropogenic sulphate aerosols has increased dramatically since the start of the industrial revolution and is currently estimated to be more than half of all sulphate aerosols (Kinne et al, 2006). Because of industrial activity, northern hemisphere concentrations are much higher. Although sulphates only survive for between three and five days in the atmosphere, they can have dramatic impacts on climate through the reflection of sunlight and the alteration of cloud properties. This tends to be more localized than, say, the impacts of greenhouse gases, being more important downwind of industrial activity and forest fires.

Land use not only has an impact on the amount of carbon emitted or absorbed from the atmosphere, but it can also have a direct impact on the weather and climate (Diffenbaugh, 2009). This occurs mainly through changes to the energy budget of the Earth's surface, as different land-cover types reflect sunlight in different ways due to their reflective properties or *albedos*. For instance, grain crops are generally lighter in colour than forests and so will reflect more radiation.

Different vegetation types also partition rainfall into evapotranspiration and run-off in different ways, and interact with soil moisture differently through plant-specific canopy configurations, rooting depth distributions and water-use strategies. These characteristics not only give rise to different feedbacks to the atmosphere, in terms of energy and moisture, which then affect local temperature and precipitation, but may also have downstream impacts in remote regions. At the ground surface, land-use change can also influence the local hydrology, which we discuss in more detail later in this chapter, in relation to drought. For example, replacing natural forest cover with agricultural land generally increases the rate of run-off and sediment loss. Increasing the area of impervious surfaces (asphalt, concrete) in urban areas can increase run-off and, in some cases, has led to increased flooding (Leopold, 1968).

What Causes Drought?

Drought occurs as a result of extremes in climate. It may also be exacerbated or dampened by anthropogenic influences. For instance, over-pumping groundwater supplies, diverting rivers and intensively cultivating agricultural land may all reduce water supplies, thereby either causing a drought or exaggerating problems in the case of a climate-driven drought. Other human influences such as urbanization, overgrazing and deforestation may alter the local hydrology, thus indirectly reducing available water.

Of course, most human activity in relation to water has the goal of (1) making the most of available resources and (2) guarding against potential drought by storing water in surface or groundwater reservoirs. Sometimes, however, these two goals are not met at the same time, and the overexploitation of resources can lead to problems during periods of climate-induced drought. It is worthwhile noting here that drought is a temporary aberration of climate and not a permanent phenomenon. This distinguishes it from arid climates, which are dry all the time in absolute terms, but can still suffer from drought when local conditions are drier than normal.

The scientific body of research into the processes that cause drought to develop and persist is growing rapidly. This research has highlighted a number of factors that may potentially impact on drought occurrence, including large-scale atmospheric mechanisms that are associated with modes of climate variability characterized by pressure or SST anomalies (Schubert et al, 2004; Seager et al, 2005), and evidence that land–atmosphere feedbacks play a role in their persistence (Long et al, 2000; Wang and Eltahir, 2000). Droughts have been studied in detail regionally and globally in the past, and much more so in the past decade, in terms of their occurrence and forcing mechanisms. This has been particularly so for the US (e.g. Namias, 1982; Chang and Wallace, 1987; Trenberth and Branstator, 1992; Lyon and Dole, 1995; Beljaars et al, 1996; Trenberth and Guillemot, 1996; Mo et al, 1997) and the Sahel (e.g. Nicholson, 2000; Giannini et al, 2003; Held et al, 2005; Bell and Lamb, 2006; Biasutti and Giannini, 2006; Lau et al, 2006; Herceg et al, 2007). This research has arisen from the large impacts inflicted by droughts in these regions. Recent droughts and heat waves, such as that in Europe in 2003, have also received considerable attention (e.g. Fischer et al, 2007). The long-term drought in southeast Australia has prompted a number of studies investigating the causes and feedbacks of climate change (Karoly et al, 2003; Nicholls, 2004; Lockart et al, 2009).

Drought and climate

The variation in climate over a spectrum of temporal and spatial scales is the main driver of drought. The previous section described how the climate varies naturally and as a result of human activity. The extremes of the natural variations affect the terrestrial water cycle in general and impact on human activities in terms of bringing changes to the availability and absence of water, e.g. in the case of flooding or drought. Typically, droughts are initiated by atmospheric circulation and weather systems that conspire to cause lower precipitation and/or higher evaporation than normal in a region. For example, high-pressure weather systems (anticyclones) are associated with dry weather, and low-pressure systems (cyclones) give rise to wet weather. Droughts occur when the usual pattern of weather changes its seasonal timing, its location or persists for longer than normal. These weather systems in turn may be related to short-lived atmospheric anomalies that arise from the chaos of the atmosphere or oceanic temperature anomalies, which can persist for decades (such as different phases of the ENSO or AMO). Thus there is a chain of processes, ranging from multiple-year timescales and large space scales, with the ocean being the main driver, to regional and local scales, where seasonal and daily weather processes and land features play major roles.

In mid-latitude temperate climate zones, where the weather is more variable, drought can occur at any time. Figure 3.8 shows time series of soil moisture (from a hydrological model) in two locations in the US. The first lies in the southeast US, where the climate is highly variable in the short term. This fact is reflected by the high-frequency changes in the values of soil moisture, which indicate that there is potential for drought at any time, but only for short durations. Nevertheless, during the summer, when temperatures are high and atmospheric demand for evapotranspiration is at its peak, the potential for drought is also at its highest. The high variation in drought risk is governed by the location of the jet stream, which meanders not only with the interaction of cool polar air and warm tropical air, but also with the seasonal radiation-forced north–south migration, bringing alternate dry and wet spells, and shifting circulation features.

In these regions, and especially in eastern North America and Europe, a frequent cause of drought is atmospheric blocking. This describes the persistence of stationary anticyclones (circulation around regions of high pressure), which deflect rain-bearing features around them, such as small-scale weather systems and cyclones. The NAO manifests in shifts of the North Atlantic jet stream, pushing it further north in its positive mode (and bringing warmer weather to areas of the eastern US and northern Europe), and forcing it southwards in its negative mode (which covers the eastern US and northern Europe in cold arctic air).

Marginal climates are particularly exposed to drought. These are climates where precipitation is infrequent and/or unreliable, causing the productivity of water-dependent activities, such as agriculture, to be

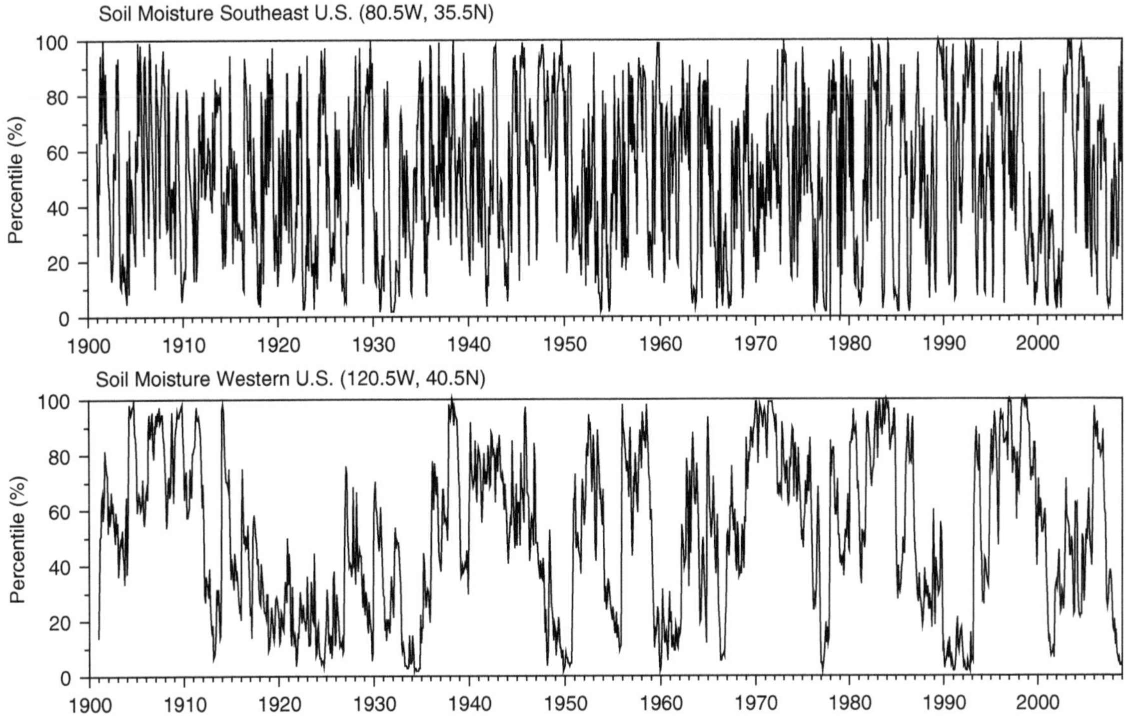

Figure 3.8 *Differences in hydrological regime between the (top) southeast and (bottom) western US in terms of soil moisture*

highly variable. These typically occur in climate transition zones between temperate and dry, and tropical and dry climates. The bottom time series in Figure 3.8 gives an example for a generally dry climate in the western US. The longer timescale variation in climate can be seen in long periods of drought (and wet spells), which may last for decades.

In tropical transition zones, the location and strength of the ITCZ varies each year and can therefore be a major factor in the occurrence of drought. For example, in the Sahel and East Africa, the northern migration of the ITCZ governs the reliability of the rains, leading to drought when the migration stalls southwards, and floods when it makes more northerly progress than usual. In monsoon regions, a number of factors including the strength of the ITCZ, as well as land–ocean interactions, control the onset of the rainy season, its penetration onto the land mass and ultimately how much rain will occur. If the monsoon fails, drought is a likely consequence, given that many of these regions rely on the monsoon for the majority of their annual rainfall. In marginal regions, drought may persist for decades, as seen in the Sahel during the 1970s and 1980s.

Drought across the globe is primarily controlled by the circulation and temperature of the oceans. The oceans are the generator of the world's weather and climate, and so they provide the impetus for large-scale changes in weather, such as those shifts characterized by the ENSO and NAO, which can lead to drought (and flooding in other places) around the world. A large body of research related to climate processes has indicated that the ocean plays a dominant role in cases of drought. Analysis of observations (e.g. Long et al, 2000; Barlow et al, 2001; Cassou et al, 2005; Hoerling et al, 2006; Bronnimann, 2007; Garcia-Herrara et al, 2007) and modelling studies (Schubert et al, 2004; Held et al, 2005) suggest that large-scale climate variability and SSTs are key factors in causing precipitation variability and drought occurrence. Indeed, some studies have indicated global patterns of drought occurrence (Sheffield et al, 2009), including hemispheric symmetry (Herweijer and Seager, 2008).

Climate effects on drought are compounded or negated by feedbacks with the land surface, and their delays and propagation through the hydrological system. For example, long-term drought conditions in the Sahel, during the 1970s and 1980s, had devastating social and environmental consequences. However, the forcing mechanism of this drought has been and continues to be the subject of debate. Previously, it was thought that overuse, in the form of overgrazing, deforestation and poor land management, was responsible. Yet recent studies have shown that a combination of factors were to blame, including land–atmosphere interactions, ocean temperatures and anthropogenic forcing (Nicholson, 2000; Giannini et al, 2003; Held et al, 2005; Biasutti and Giannini, 2006; Lau et al, 2006; Los et al, 2006). Nevertheless, these research findings were mostly based on modelling studies, and ensemble assessments of coupled climate model simulations of the 20th century indicate that not all models can replicate observed trends in rainfall.

Although, feedbacks with the land may not be the primary cause of droughts, there is much evidence to suggest that the land surface plays a role in prolonging droughts (Long et al, 2000; Wang and Eltahir, 2000; Zampieri et al, 2009). In particular, observational (Mo and Juang, 2003; Koster and Suarez, 2004; D'Odorico and Porporato, 2004; Raddatz, 2005) and modelling studies (Koster et al, 2004; Chen et al, 2005; Wu et al, 2007) indicate that soil moisture anomalies impact on seasonal precipitation and contribute to the maintenance of drought conditions through precipitation recycling (Trenberth, 1999; D'Odorico and Porporato, 2004; Teuling et al, 2005; Bierkens and van den Hurk, 2007; Dekker et al, 2007).

In general, precipitation variability and drought occurrence are driven by a combination of the variability of advected atmospheric moisture and the strength of local recycling. For instance, analysis of the central US drought in 1988 and the Oklahoma–Texas drought in 1998 has shown that a combination of anomalous SSTs and atmospheric patterns were responsible for their development, and land surface interactions were instrumental in their persistence (Sud et al, 2003; Hong and Kalnay, 2000). Plate 3.2 illustrates how the process of precipitation recycling can affect hydrological conditions and lead to enhanced drought locally or downstream (Sheffield et al, 2010). In this example, the source of moisture for the Ohio River basin is from the south and the west, as depicted by lines of atmospheric flow. The southerly moisture source is mainly derived from oceanic evaporation that is at its height during the summer months, when warm moist air brings plentiful precipitation to the region. The moisture from the west is derived not only from both oceanic sources (the Pacific Ocean and Gulf of California), but also from recycled precipitation falling on the western and central US, which evaporates and travels via the prevailing winds eastwards, where it may fall as precipitation again. It is estimated that this mechanism accounts for the fact that 60 per cent of the precipitation falling on the Ohio River basin comes from the evaporation of water elsewhere, mainly in the western US (Sheffield et al, 2010). This effect can be quantified by a recycling ratio, which is the ratio of regional precipitation derived from land evaporation to the total precipitation (from land and ocean evaporation), and is generally lowest in coastal regions and highest in continental interiors. In times of drought, this mechanism can be broken because dry soil conditions reduce evaporation and thus cause recycled precipitation to decrease. This can lead to the persistence of dry conditions within the same region and/or to dry conditions elsewhere, in cases in which moisture would usually be transported further downwind. For example, the drought in 2002 in the Midwest US was enhanced by dry conditions upwind in the Arizona–New Mexico region (Dominguez et al, 2009). Similar mechanisms can be linked to the initiation of drought conditions in southeast US in 2006–2007 by the preceding drought in Texas in 2006 (Sheffield et al, 2010).

As well as soil moisture, feedbacks between the land and the atmosphere can be controlled by the presence of snow, which has a high albedo value, and therefore reflects more incoming radiation with impacts on the state of the atmosphere. Some studies suggest that large-scale anomalies in snow cover can influence atmospheric circulation (Barnett et al, 1988). For example, anomalies in Eurasian snow cover can impact on the Asian monsoon (Barnett et al, 1989), amplify the NAO (Saito and Cohen, 2003) and influence Canadian hydrology (Déry et al, 2005). Similar mechanisms can occur through changes in albedo from black carbon lying on snow, desertification that converts dark vegetation to highly reflective dry soils, as well as the direct injection of atmospheric aerosols from the land via dust storms, wildfires and smoke.

Drought and Hydrology

Although droughts are generally driven by a period of low precipitation, there are other factors that modulate their occurrence, depending on the characteristics of the region and how drought propagates through the hydrological system (van Lanen et al, 2004). These characteristics include the underlying climate, soil type and stratigraphy, elevation and slope, vegetation cover and physiology, the groundwater system and its connections with surface hydrology and neighbouring regions. Drought generally propagates as follows. The initial stage is meteorological drought, driven by low precipitation, but can also be initiated by atmospheric demand through increased evapotranspiration (Figure 3.9). Low precipitation will impact in different ways, depending on the season, antecedent conditions and characteristics of the region. This is linked to the storage of water in the soil column. A decrease in precipitation will lead to a decrease in soil moisture. This will reduce evapotranspiration, with

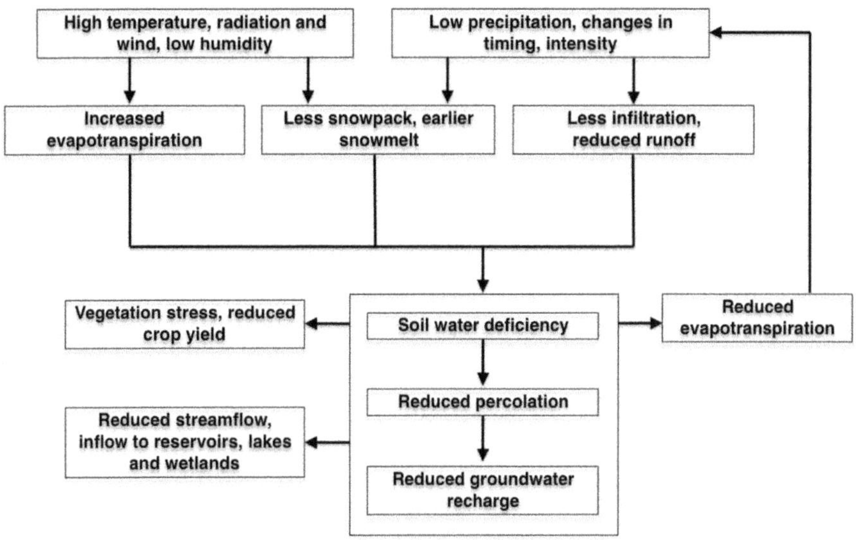

Figure 3.9 *Propagation of drought through the hydrological system and the feedbacks to the atmosphere*

Source: Adapted from NDMC

subsequent impacts on the vegetation, leading to ecological drought. If this occurs during the dry season, then recharge to groundwater will be small and hydrological drought will not emerge. However, if the depletion of soil moisture is high then it will need to be restored in the wet season, thereby reducing subsurface flow and recharge, giving rise to hydrological drought. In the wet season, hydrological drought can occur because evapotranspiration continues as normal, depleting soil moisture and thus reducing subsurface flow and recharge. The subsequent dry season may then be affected (van Lanen et al, 2004).

High potential evapotranspiration can also cause a reduction in soil moisture. Potential evapotranspiration is the maximum possible level of evapotranspiration that can be reached, given unlimited water supply and current atmospheric conditions. It is enhanced by higher temperature and radiation, and stronger and drier winds. If the climate is humid, the reduction may not be significant for plants, but may eventually reach wilting point (the amount of water below which a plant starts to wilt). In dry regions, the recharge is low anyway and the reduction in soil moisture does not matter. The reduction in soil moisture is not indefinite, as the soil moisture store will eventually deplete. There are also feedback mechanisms with the vegetation that may shut down transpiration if soil moisture becomes too low, although the vegetation may not be affected in the long term. If low soil moisture persists near the wilting point for a period of time, then the vegetation may wither and die. This depends on the type of vegetation and its strategy for coping with drought. Many tree species are able to cope with multiyear droughts because of their deep roots and ability to search for other sources of water (Nepstad et al, 2002).

Figure 3.10 provides an example of how hydrological anomalies (wet and dry events) propagate through the

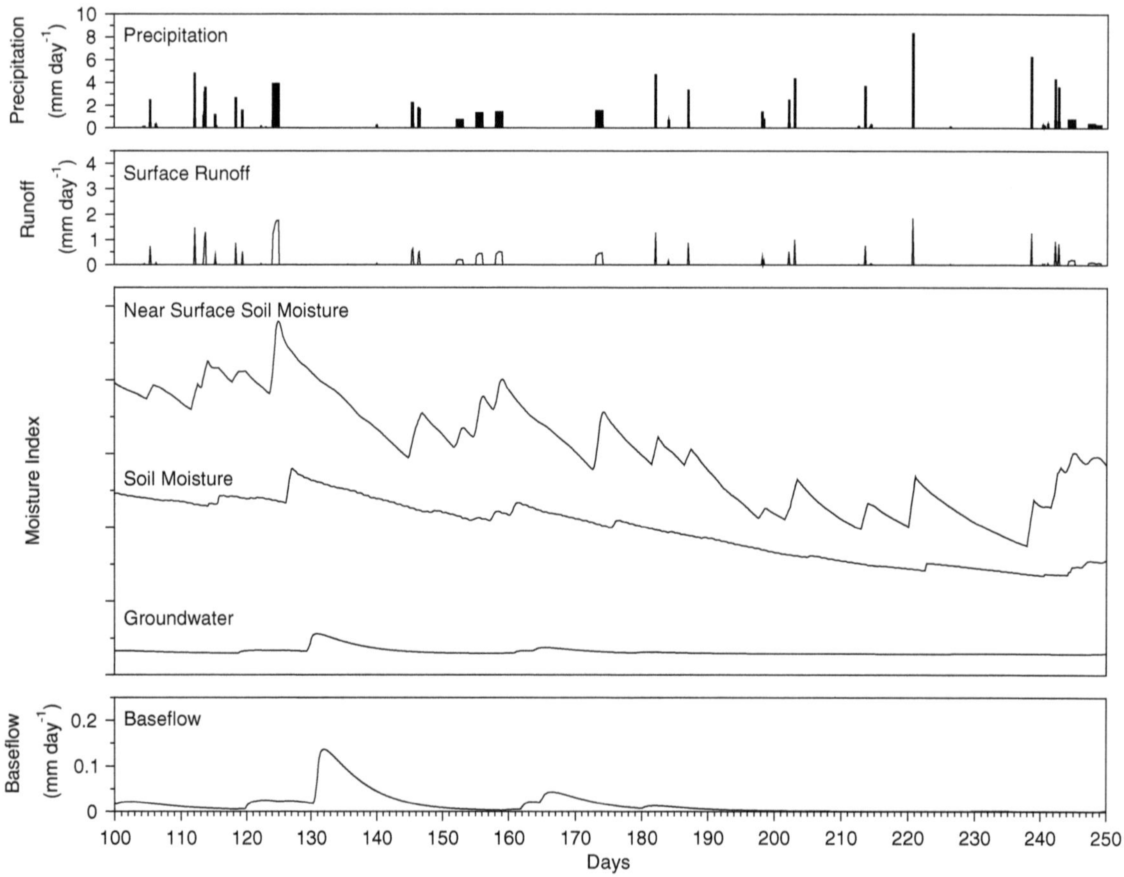

Figure 3.10 *The propagation of wet and dry events through the hydrological system*

hydrological system for a temperate humid region. In this case, precipitation events are reflected immediately in the surface run-off. Moisture in the near surface of the soil also responds quickly. However, the propagation of the pulse of water is gradually dampened with depth until it barely registers in the groundwater. As the region dries down in a drought, soil moisture in the near surface dries down rapidly over days to weeks, whereas moisture in the top few metres takes weeks to months. Groundwater is not affected and it may require months to years to deplete these deep water storages. The speed of propagation of drought to different parts of the hydrological cycle is dependent on the local and regional characteristics (Eltahir and Yeh, 1999; Keyantash and Dracup, 2004; Laaha and Blöschl, 2007). Regionally, different flow generation processes come into play (Poff, 1996).

In humid regions, low flows in perennial streams may be dominated by baseflow derived from groundwater or interflow; in dry regions, streamflow may be intermittent or seasonal, and low flows follow very different patterns of distribution. Even within a region, the characteristics of hydrology and low flows may be heterogeneous (e.g. Vogel and Kroll, 1992). At the river catchment level, the hydrology is controlled by a wide variety of processes and combinations of these, depending on climate, surface characteristics and flow paths, and underlying geology: surface run-off or overland flow; summer evapotranspiration; wintertime snow accumulation and the timing and speed of spring melt; and groundwater (which can vary greatly in its extent and properties). One of the most important factors in drought propagation is the capacity of the region or catchment to store water in, for example, lake, aquifers, snowpack, wetlands, groundwater and the upper soil layers. These create delays in the propagation of drought and, in some cases, can cause a drought signal to be completely attenuated, whereas in other instances, can make droughts persist for many years.

Figure 3.11 presents streamflow time series for four rivers across the US, with very different hydrological regimes. The San Marcos in Texas is fed by the Edwards aquifer, and exhibits a steady baseflow, upon which streamflow peaks are superimposed. The Augusta Creek in Missouri is typical of a small river in a temperate climate, showing large variability from day to day and month to month. The Colorado is mainly fed by snowmelt and so is dominated by a large peak when the region warms in late spring. Lastly, Cave Creek in Arizona is in a dry climate, where baseflow is zero and streamflow is flashy after infrequent short but intense storms.

In baseflow-dominated catchments (i.e. where most streamflow is derived from groundwater, lake and glacier sources), drought develops slowly, because these stores are generally large and extensive, and dominate the sources of water (Eltahir and Yeh, 1999). In dry periods, water is still available for a considerable amount of time and maintains a steady flow. Once a drought starts, they also tend to persist for longer with little or no response to individual rainfall events and diminish slowly. In surface-flow dominated regions, the response of drought is much more rapid, developing quickly and possibly diminishing even faster after a single storm. Precipitation tends to bypass the soil and groundwater, and flows directly towards streams, especially where the slopes are high and/or the topsoil is shallow and underlain by impermeable bedrock (Poff, 1996). However, it is possible to have highly permeable soils and sloping groundwater systems that deliver storm water quickly to streams. This can be influenced by the type of land use in the catchment area: forested land restricts overland flow and encourages infiltration, whereas ploughed fields and urban impermeable land have low infiltration rates and high overland flow. In flat regions that are dominated by wetlands and multiple flow channels, the water table can be very high and there is no available storage. Here the response can be very quick as well, draining rapidly after a period of no precipitation and becoming saturated after a single storm (van Lanen et al, 2004). As the size of a region grows, the mixture of sources of flow (from quick and base) expands, but generally becomes increasingly dominated by baseflow as the maintainer of water levels.

In cold regions, the propagation is linked to the existing store of water as snowpack and ice, which plays a role through the storage and delay of water flows. Drought can also be a result of low snowpack, caused by low snowfall and/or high temperatures, leading the snow to melt early and run off quickly. This can give rise to drought later in the spring or summer. For example, water resources in California and other regions rely on the winter snowpack in the mountains to provide water through the spring and summer. The impact of higher levels of potential evapotranspiration

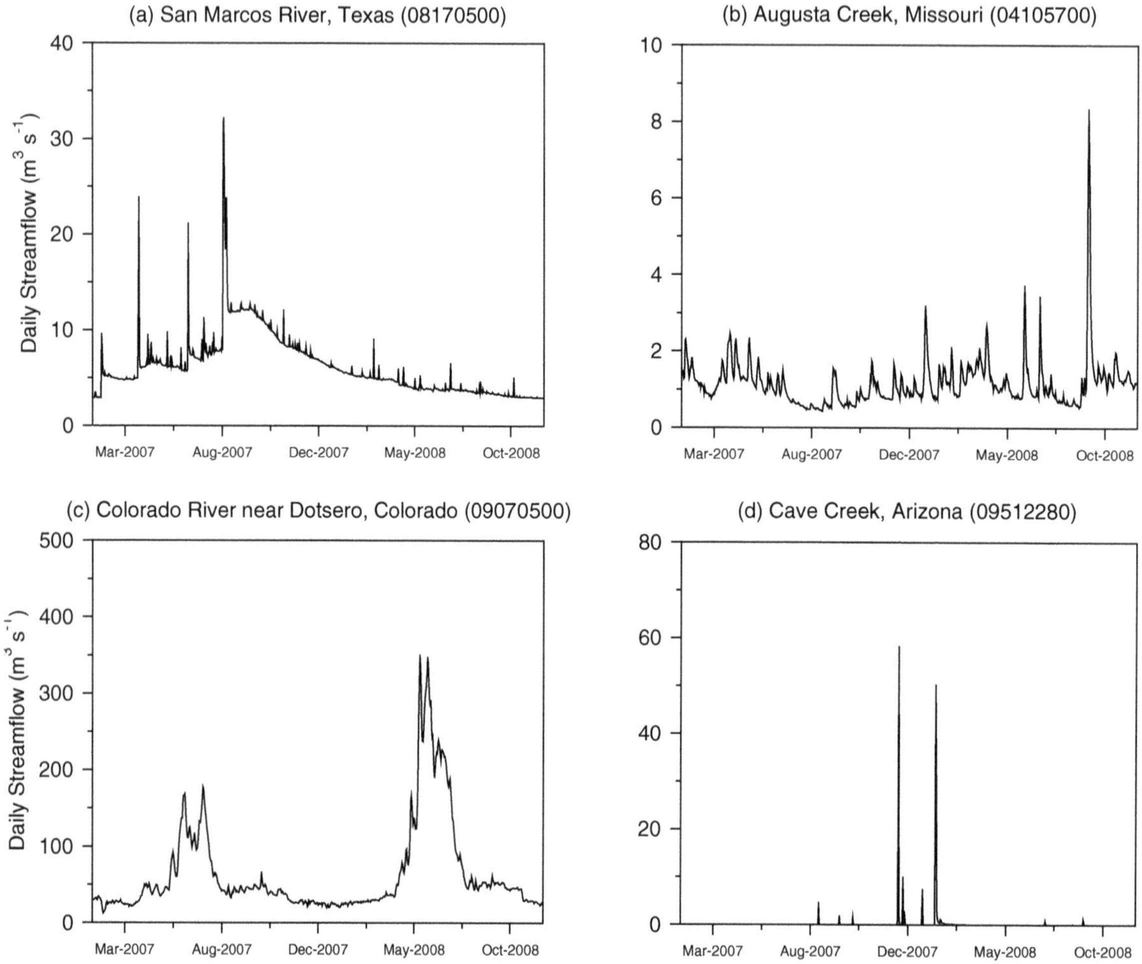

Figure 3.11 *Different hydrological regimes as depicted by daily streamflow time series for four rivers across the US*

Note: gauge numbers are given above each panel.

Source: US Geological Survey (USGS)

can also manifest in cold seasons as sublimation (evaporation) of the snowpack or canopy snow storage. Although cold season evaporation is generally low or even negligible, these sublimation events can reduce snowpack significantly overnight. Conversely, cooler temperatures can delay the spring melt, giving rise to winter drought as levels of soil moisture, streamflow and recharge fall, and this can be exacerbated by freezing of rivers and soil moisture.

Drought and human activity

Humans play a major role in governing the amount and flow of water in a region, and heightening the occurrence of drought. Throughout history, humans have endeavoured to tame water flows to reduce hazards, enable navigation and transport, and store water for regular supply. Some examples of human interference are: land-use change, irrigation of farmland, reservoir building, river diversion, wetland drainage and groundwater abstraction. The purpose of these activities may be to reduce the occurrence of drought, as in the case of reservoirs, to safeguard against future dry periods, but they may also inadvertently increase the frequency of drought. Land-use change has been significant in many parts of the world, especially for the purpose of cultivating agriculture, driven by the demands of a growing population, and more recently the expansion of urban areas (Foley et al, 2005). The deforestation of rain forests, afforestation, conversion of grasslands to agriculture, urbanization, and so on, have all changed the landscape hydrologically. The impact of the changes are direct, in terms of the changes to the interception of precipitation by the vegetation itself, and the related impacts on throughfall to the ground, infiltration and overland flow. There are also associated indirect impacts, such as changes to the soil surface (compaction, crusting) and subsurface (macropores in roots), which can change the hydrology, and feedbacks to the atmosphere through evapotranspiration (Diffenbaugh, 2009).

Human activity may have played a role in exacerbating the droughts in the Sahel during the 1970s and 1980s through overgrazing and feedbacks to the atmosphere driven by the reduced vegetation cover (Giannini et al, 2008). The Dustbowl drought in the central US in the 1930s was exacerbated by land management practices (Cook et al, 2009), borne out by the subsequent drought of the 1950s, whose heightened severity was managed far better as a result of the lessons learnt from the Dustbowl (Cook et al, 2007). Forest burning for land clearance, as in the case of the Amazonian rain forest, results in large hydrological effects and feedbacks into the atmosphere, which may affect drought development. It can immediately change levels of interception, infiltration and overland flow, because of the loss of vegetative cover. There may also be a hydrophobicity impact from forest fires, whereby chemical reactions in the soil surface result in the formation of a water repellent layer (Lukey et al, 1995; Burch et al, 2006).

Urbanization describes the process of changing the natural earth into a landscape of impervious surfaces and managed storm flow. This can result in reduced evapotranspiration, increased overland flow and quick flow through the storm-water drainage network. Many of these changes lead to higher run-off rates and flooding (Villarini et al, 2009), but at the same time reduce infiltration into the soil and recharge to groundwater, which increases the potential for drought.

Increased pressure on water availability may also come from growing populations. Over the 20th century, the global urban population increased from 220 million to 2.8 billion, and is expected to double in Africa and Asia by 2030 (WWAP, 2009). Population-driven urbanization can alter recharge in many different ways. On the one hand, the development of impervious surfaces and drainage channelling may remove water from the area. On the other hand, the use of more water in urban and suburban areas can lead to more recharge, for instance through the maintenance of parks and greenery. The removal of vegetation can also reduce evapotranspiration, leading to the greater availability of water for discharge and recharge. Over-pumping groundwater resources can have negative impacts on drought. As water tables are reduced, groundwater flow to streams and springs also decreases. There is also evidence to suggest that groundwater drought may positively feedback to precipitation (Bierkens and van den Hurk, 2007; Anyah et al, 2008). Irrigation may also impact on precipitation downwind through increased evaporation (DeAngelis et al, 2010).

The largest direct human impact on drought most likely stems from the engineering of water resources, whereby a multitude of human activities impose upon surface-water flows and storage. These include diversions to river courses, drainage of lakes and wetlands, construction of reservoirs and dams, pumping of groundwater and irrigation of dry lands. The purpose of these changes is manifold: the alleviation of low flows; the control of floods; the provision of water supply for irrigation, public and industrial consumption and cooling for power plants; navigation systems, ecosystem habitats and recreational facilities. Although these changes are often implemented to relieve drought and provide year-round access to water, they can have a considerable impact on remote downstream drought regimes, especially where no consideration is given to ecosystems or other countries. For instance, the Aral Sea in Central Asia was once among the largest lakes in the world, but has virtually disappeared due to appropriation of its upstream sources for irrigation. This has devastated the local ecology and economy, and has probably had an effect on the local climate (Glantz, 2005). The human influence on climate change is also an increasingly important factor in drought, and this will be discussed further in Chapter 8. It is sufficient to say here that the prospect of rising temperatures, coupled with decreasing regional precipitation, or more intense and variable precipitation in some regions, will lead to increased and more severe drought.

Summary

This chapter has covered some of the essential elements of the climate and hydrological systems that are associated with drought. Drought is fundamentally driven by the variations in time and space of climate and weather, which occur on multiple scales. At the longest timescales, ranging from 10,000 to 100,000 years, variations in the Earth's orbit are associated with regional shifts in humid and dry climates, seen in the transformation of forest to desert. Overlapping these are changes in solar output, which occur on timescales of between decades and centuries, or across even longer spans. Finally there are changes that occur over years to decades, forced by volcanic activity and the internal variability of the coupled climate system. The most important of these is the ENSO, which controls much of the variability in climate in the tropics and parts of the extra-tropics on interannual timescales, and hence the occurrence and intensity of drought.

Drought is then augmented or diminished by various interactions at the land surface, where impacts are felt most keenly. At first, the drought is propagated through the hydrological system, which may attenuate or delay the drought signal as it moves from meteorological, to agricultural, to hydrological drought. The drought's propagation is influenced by the properties of the land surface and hydrological system, including the type of vegetation and land use, local topography, underlying geology, and the connectivity of the river network between various hydrological stores. Second, the drought is influenced by human activity: in general, this is aimed at reducing hydrological extremes, to ensure there is sufficient stored water in times of drought and enough unused storage in times of flood. However, there are many examples of how human activity can cause and exaggerate drought, including pollution, the removal of vegetation, over-pumping of groundwater and construction of reservoirs.

Hence there are two major factors that cause drought to happen: climate variation and human activity. The first is the primary cause and is essentially beyond our control. Fortunately, the second cause is relatively minor in its direct effects. However, indirect impacts, such as the effect of anthropogenic greenhouse gas emissions, may turn out to be a dominant factor in global climate and drought in the future (see Chapter 8). The next chapter discusses the data sources and methods that are used to quantify and analyse drought over large scales.

References

Anyah, R., Weaver, C. P., Miguez-Macho, G., Fan, Y. and Robock, A. (2008) 'Incorporating water table dynamics in climate modelling: 3. Simulated groundwater influence on coupled land-atmosphere variability', in *Journal of Geophysical Research*, vol 113, D07103, p15, doi:10.1029/2007JD009087

Barlow, M., Nigam, S. and Berbery, E. H. (2001) 'ENSO, Pacific decadal variability, and US summertime precipitation, drought and streamflow', in *Journal of Climate*, vol 14, pp2105–2128

Barnett, T. P., Dumenil, L., Schlese, U. and Roeckner, E. (1988) 'The effect of Eurasian snow cover on global climate', in *Science*, vol 239, pp504–507

Barnett, T. P., Dumenil, L., Schlese, U., Roeckner, E. and Latif, M. (1989) 'The effect of Eurasian snow cover on regional and global climate variations', in *Journal of the Atmospheric Sciences*, vol 46, pp661–685

Bell, M. A. and Lamb, P. J. (2006) 'Integration of weather system variability to multidecadal regional climate change: The West African Sudan-Sahel zone, 1951–98', in *Journal of Climate*, vol 19, no 20, pp5343–5365

Beljaars, A., Viterbo, P., Miller, M. and Betts, A. (1996) 'The anomalous rainfall over the US during July 1993 – Sensitivity to land-surface parameterization and soil moisture', in *Monthly Weather Review*, vol 124, pp362–383

Biasutti, M. and Giannini, A. (2006) 'Robust Sahel drying in response to late 20th century forcings', in *Geophysical Research Letters*, vol 33, L11706, p4

Bierkens, M. F. P. and van den Hurk, B. J. J. M. (2007) 'Groundwater convergence as a possible mechanism for multiyear persistence in rainfall', in *Geophysical Research Letters*, vol 34, L02402, p5, doi:10.1029/2006GL028396

Bronnimann, S. (2007) 'Impact of El Niño Southern Oscillation on European climate', in *Reviews of Geophysics*, vol 45, p2

Burch, G. J., Moore, I. D. and Burns, J. (2006) 'Soil hydrophobic effects on infiltration and catchment run-off', in *Hydrological Processes*, vol 3, no 3, pp211–222

Cassou, C., Terray, L. and Phillips, A. S. (2005) 'Tropical Atlantic influence on European heat waves', in *Journal of Climate*, vol 18, no 15, pp2805–2811

Chang, F. C. and Wallace, J. M. (1987) 'Meteorological conditions during heat waves and droughts in the United States Great Plains', in *Monthly Weather Review*, vol 115, pp1253–1269

Chen, M., Pollard, D. and Barron, E. J. (2005) 'Hydrological processes in China and their association with summer precipitation anomalies', in *Journal of Hydrology*, vol 301, pp14–28

Christensen, E. and Lassen, K. (1991) 'Length of the solar cycle: An indicator of solar activity closely associated with climate', in *Science*, vol 254, pp698–700

Cook, E. R., Woodhouse, C. A., Eakin, C. M., Meko, D. M. and Stahle, D. W. (2004) 'Long-term aridity changes in the Western United States', in *Science*, vol 306, pp1015–1018

Cook, E. R., Seager, R., Cane, M. A. and Stable, D. W. (2007) 'North American droughts: Reconstructions, causes and consequences', in *Earth-Science Reviews*, vol 81, pp93–134

Cook, B. I., Miller, R. L. and Seager, R. (2009) 'Amplification of the North American "Dust Bowl" drought through human-induced land degradation', in *Proceedings of the National Academy of Sciences*, vol 106, pp4997–5001

Crowley, T. J. (2000) 'Causes of climate change over the past 1000 years', in *Science*, vol 289, pp270–277

DeAngelis, A., Dominguez, F., Fan, Y., Robock, A., Deniz Kustu, M. and Robinson, D. (2010) 'Observational evidence of enhanced precipitation due to irrigation over the Great Plains of the United States', in *Journal of Geophysical Research*, vol 115, D15115, p115, doi:10.1029/2010JD013892

Dekker, S. C., Rietkerk, M. and Bierkens, M. F. P. (2007) 'Coupling microscale vegetation-soil water and macroscale vegetation-precipitation feedbacks in semiarid ecosystems', in *Global Change Biology*, vol 13, no 3, pp671–678

Déry, S. J., Sheffield, J. and Wood, E. F. (2005) 'Connectivity between Eurasian snow cover extent and Canadian snow water equivalent and river discharge', in *Journal of Geophysical Research*, vol 110, D23106, p14

Diffenbaugh, N. S. (2009) 'Influence of modern land cover on the climate of the United States', in *Climate Dynamics*, vol 33, no 7–8, pp945–958, doi:10.1007/s00382-009-0566-z

D'Odorico, P. and Porporato, A. (2004) 'Preferential states in soil moisture and climate dynamics', in *Proceedings of the National Academy of Sciences of the United States of America*, vol 101, pp8848–8851

Dominguez, F., Villegas, J. C. and Breshears, D. D. (2009) 'Spatial extent of the North American Monsoon: Increased cross-regional linkages via atmospheric pathways', in *Geophysical Research Letters*, vol 36, L07401, p5, doi:10.1029/2008GL037012

Dong, B. W., Sutton, R. T. and Scaife, A. A. (2006) 'Multidecadal modulation of El Nino-Southern Oscillation (ENSO) variance by Atlantic Ocean sea surface temperatures', in *Geophysical Research Letters*, vol 33, no 8, p4

Eltahir, E. A. B. and Yeh, P. (1999) 'On the asymmetric response of aquifer water level to floods and droughts in Illinois', in *Water Resources Research*, vol 35, no 4, pp1199–1217

Enfield, D. B. and Cid-Serrano, L. (2009) 'Secular and multidecadal warmings in the North Atlantic and their relationships with major hurricane activity', in *International Journal of Climatology*, vol 30, no 2, pp174–184, doi:10.1002/joc.1881

Enfield, D. B., Mestas-Nunez, A. M. and Trimble, P. J. (2001) 'The Atlantic Multidecadal Oscillation and its relationship to rainfall and river flows in the continental US', in *Geophysical Research Letters*, vol 28, pp2077–2080

Fischer, E., Seneviratne, S. I., Lüthi, D. and Schär, C. (2007) 'The contribution of land-atmosphere coupling to recent European summer heat waves', in *Geophysical Research Letters*, vol 34, L06707, p6, doi:10.1029/2006GL029068

Foley, J. A. et al (2005) 'Global consequences of land use', in *Science*, vol 309, no 5734, pp570–574, doi:10.1126/science.1111772

Folland, C. K., Knight, J., Linderholm, H. W., Fereday, D., Ineson, S. and Hurrell, J. W. (2009) 'The summer North Atlantic Oscillation: Past, present, and future', in *Journal of Climate*, vol 22, pp1082–1103

Frierson, D. M. W., Lu, J. and Chen, G. (2007) 'Width of the Hadley cell in simple and comprehensive general circulation models', in *Geophysical Research Letters*, vol 34, L18804, p5, doi:10.1029/2007GL031115

Garcia-Herrera, R., Paredes, D., Trigo, R. M., Trigo, I. F., Hernandez, E., Barriopedro, D. and Mendes, M. A. (2007) 'The outstanding 2004/05 drought in the Iberian Peninsula: Associated atmospheric circulation', in *Journal of Hydrometeorology*, vol 8, no 3, pp483–498

Gershunov, A. and Barnett, T. P. (1998) 'Interdecadal modulation of ENSO teleconnections', in *Bulletin of the American Meteorological Society*, vol 79, no 12, pp2715–2725

Giannini, A., Saravanan, R. and Chang, P. (2003) 'Oceanic forcing of Sahel rainfall on interannual to interdecadal time scales', in *Science*, vol 302, pp1027–1030

Giannini, A., Biasutti, M. and Verstraete, M. M. (2008) 'A climate model-based review of drought in the Sahel: Desertification, the re-greening and climate change', in *Global and Planetary Change*, vol 64, pp119–128, doi:10.1016/j.gloplacha.2008.05.004

Glantz, M. H. (2005) 'Water, climate, and development issues in the Amu Darya basin', in *Mitigation and Adaptation Strategies for Global Change,* vol 10, no 1, pp23–50

Held, I. M. and Hou, A. Y. (1980) 'Nonlinear axially symmetric circulations in a nearly inviscid atmosphere', in *Journal of the Atmospheric Sciences,* vol 37, pp515–533

Held, I. M., Delworth, J., Lu, K., Findell, L. and Knutson, T. R. (2005) 'Simulation of Sahel drought in the 20th and 21st centuries', in *Proceedings of the National Academy of Sciences of the United States of America,* vol 102, pp17,891–17,896

Herceg, D., Sobel, A. H. and Sun, L. Q. (2007) 'Regional modelling of decadal rainfall variability over the Sahel', in *Climate Dynamics,* vol 29, pp89–99

Herweijer, C. and Seager, R. (2008) 'The global footprint of persistent extra-tropical drought in the instrumental era', in *The International Journal of Climatology,* vol 28, no 13, pp1761–1774, doi:10.1002/joc.1590

Hoerling, M., Hurrell, J., Eischeid, J. and Phillips, A. (2006) 'Detection and attribution of twentieth century northern and southern African rainfall change', in *Journal of Climate,* vol 19, pp3989–4008

Hong, S. Y. and Kalnay, E. (2000) 'Role of sea surface temperature and soil-moisture feedback in the 1998 Oklahoma-Texas drought', in *Nature,* vol 408, no 6814, pp842–844

Hurrell, J. W. (1996) 'Influence of variations in extratropical wintertime teleconnections on Northern Hemisphere temperature', in *Geophysical Research Letters,* vol 23, pp665–668

Hurrell, J. W. and van Loon, H. (1997) 'Decadal variations in climate associated with the North Atlantic Oscillation', in *Climatic Change,* vol 36, nos 3–4, pp301–326

IPCC (2007) S. Solomon, D. Qin, M. Manning, Z. Chen, M. Marquis, K. B. Averyt, M. Tignor and H. L. Miller (eds) *Climate Change 2007: The Physical Science Basis. Contribution of Working Group I to the Fourth Assessment Report of the Intergovernmental Panel on Climate Change,* Cambridge University Press, Cambridge, UK

Jiang, T., Qiang, Z., Deming, Z. and Yijin, W. (2006) 'Yangtze floods and droughts (China) and teleconnections with ENSO activities (1470–2003)', in *Quaternary International,* vol 144, pp29–37

Karoly, D. J., Risbey, J. and Reynolds, A. (2003) 'Global warming contributes to Australia's worst drought, research report', WWF, Sydney, Australia

Kerr, R. A. (2005) 'Climate change: Atlantic climate pacemaker for millennia past, decades hence?' in *Science,* vol 309, no 5731, pp41–43

Keyantash, J. A. and Dracup, J. A. (2004) 'An aggregate drought index: Assessing drought severity based on fluctuations in the hydrological cycle and surface water storage', in *Water Resources Research,* vol 40, no 9, p14

Kinne, S., Schulz, M., Textor, C., Guibert, S., Balkanski, Y., Bauer, S. E., Berntsen, T., Berglen, T., Boucher, O., Chin, M., Collins, W., Dentener, F., Diehl, T., Easter, R., Feichter, H., Fillmore, D., Ghan, S., Ginoux, P., Gong, S., Grini, A., Hendricks, J., Herzog, M., Horrowitz, L., Isaksen, I., Iversen, T., Kirkevg, A., Kloster, S., Koch, D., Kristjansson, J. E., Krol, M., Lauer, A., Lamarque, J. F., Lesins, G., Liu, X., Lohmann, U., Montanaro, V., Myhre, G., Penner, J., Pitari, G., Reddy, S., Seland, O., Stier, P., Takemura, T. and Tie, X. (2006) 'An AeroCom initial assessment optical properties in aerosol component modules of global models', in *Atmospheric Chemistry and Physics,* vol 6, pp1815–1834, doi:10.5194/acp-6-1815-2006

Koster, R. D. and Suarez, M. J. (2004) 'Suggestions in the observational record of land-atmosphere feedback operating at seasonal time scales', in *Journal of Hydrometeorology,* vol 5, no 3, pp567–572

Koster, R. D. et al (2004) 'Regions of strong coupling between soil moisture and precipitation', in *Science,* vol 305, no 5687, pp1138–1140

Laaha, G. and Blöschl, G. (2007) 'A national low flow estimation procedure for Austria', in *Hydrological Sciences Journal–Journal Des Sciences Hydrologiques,* vol 52, no 4, pp625–644

Lau, K. M., Shen, S. S. P., Kim, K. M. and Wang, H. (2006) 'A multimodel study of the twentieth century simulations of Sahel drought from the 1970s to 1990s', in *Journal of Geophysical Research-Atmospheres,* vol 111, D07111, p9

Leopold, L. B. (1968) 'Hydrology for urban planning: A guidebook on the hydrological of urban land use', Professional paper no 669, US Geological Survey

Limpasuvan, V. and Hartmann, D. L. (2000) 'Wave-maintained annular modes of climate variability', in *Journal of Climate,* vol 13, pp4414–4429

Lockart, N., Kavetski, D. and Franks, S. W. (2009) 'On the recent warming in the Murray-Darling Basin: Land surface interactions misunderstood', in *Geophysical Research Letters,* vol 36, L24405, p6, doi:10.1029/2009GL040598

Long, M., Entekhabi, D. and Nicholson, S. E. (2000) 'Interannual variability in rainfall, water vapor flux, and vertical motion over West Africa', in *Journal of Climate,* vol 13, pp3827–3841

Los, S. O., Weedon, G. P., North, P. R. J., Kaduk, J. D., Taylor, C. M. and Cox, P. M. (2006) 'An observation-based estimate of the strength of rainfall-vegetation interactions in the Sahel', in *Geophysical Research Letters,* vol 33, L16402, p5

Lukey, B. T., Sheffield, J., Bathurst, J. C., Lavabre, J., Mathys, N. and Martin, C. (1995) 'Simulating the effect of vegetation cover on the sediment yield of Mediterranean catchments using SHETRAN', in *Physics and Chemistry of the Earth*, vol 20, nos 3–4, pp427–432

Lyon, B. and Dole, R. (1995) 'A diagnostic comparison of the 1980 and 1988 US summer heat wave-droughts', in *Journal of Climate*, vol 8, pp1658–1676

Mantua, N. J., Hare, S. R., Zhang, Y., Wallace, J. M. and Francis, R. C. (1997) 'A Pacific interdecadal climate oscillation with impacts on salmon production', *Bulletin of the American Meteorological Society*, vol 78, no 6, pp1069–1079

McCabe, G. J. and Palecki, M. A. (2006) 'Multidecadal climate variability of global lands and oceans', in *International Journal of Climatology*, vol 26, pp849–865

McCabe, G. J., Palecki, M. A. and Betancourt, J. L. (2004) 'Pacific and Atlantic Ocean influences on multidecadal drought frequency in the United States', in *Proceedings of the National Academy of Sciences of the United States of America*, vol 101, pp4136–4141

Mo, K. C. and Juang, H. M. H. (2003) 'Relationships between soil moisture and summer precipitation over the Great Plains and the Southwest', in *Journal of Geophysical Research-Atmospheres*, vol 108, D19, p17

Mo, K., Nogues-Paegle, C. J. and Higgins, R. W. (1997) 'Atmospheric processes associated with summer floods and droughts in the central United States', in *Journal of Climate*, vol 10, pp3028–3046

Namias, J. (1982) 'Anatomy of Great Plains protracted heat waves (especially the 1980 US summer drought)', in *Monthly Weather Review*, vol 110, pp824–838

Nepstad, D. C., De Carvalho, C. R., Davidson, E. A., Jipp, P. H., Lefebvre, P. A., Negreiros, G. H., Da Silva, E. D., Stone, T. A., Trumbore, S. E. and Vieira, S. (2002) 'The role of deep roots in the hydrological and carbon cycles of Amazonian forests and pastures', in *Nature*, vol 372, pp666–669, doi:10.1038/372666a0

Newman, M., Compo, G. P. and Alexander, M. A. (2003) 'ENSO-forced variability of the Pacific decadal oscillation', in *Journal of Climate*, vol 16, pp3853–3857

Nicholls, N. (2004) 'The changing nature of Australian droughts', in *Climate Change*, vol 63, pp323–336, doi:10.1023/B:CLIM.0000018515.46344.6d

Nicholson, S. (2000) 'Land surface processes and Sahel climate', in *Reviews of Geophysics*, vol 38, pp117–139

Osborn, T. (2000) 'North Atlantic Oscillation', Information Sheet 11, Climate Research Unit, University of East Anglia, Norwich, UK, www.cru.uea.ac.uk/cru/info/nao

Peel, M. C., Finlayson, B. L. and McMahon, T. A. (2007) 'Updated world map of the Köppen-Geiger climate classification', in *Hydrological Earth Systems Science*, vol 11, pp1633–1644

Poff, N. L. (1996) 'A hydrogeography of unregulated streams in the United States and an examination of scale-dependence in some hydrological descriptors', in *Freshwater Biology*, vol 36, pp71–91

Power, S., Casey, T., Folland, C., Colman, A. and Mehta, V. (1999) 'Interdecadal modulation of the impact of ENSO on Australia', in *Climate Dynamics*, vol 15, no 5, pp319–324

Raddatz, R. L. (2005) 'Moisture recycling on the Canadian prairies for summer droughts and pluvials from 1997 to 2003', in *Agricultural and Forest Meteorology*, vol 131, pp13–26

Robock, A. and Mao, J. (1995) 'The volcanic signal in surface temperature observations', in *Journal of Climate*, vol 8, pp1086–1103

Ropelewski, C. F. and Halpert, M. S. (1987) 'Global and regional scale precipitation patterns associated with the El Niño/Southern Oscillation', in *Monthly Weather Review*, vol 115, pp1606–1626

Ruddiman, W. F. (2003) 'The anthropogenic greenhouse era began thousands of years ago', in *Climatic Change*, vol 61, pp261–293

Saito, K. and Cohen, J. (2003) 'The potential role of snow cover in forcing interannual variability of the major Northern Hemisphere mode', in *Geophysical Research Letters*, vol 30, no 6, p1302, doi:10.1029/2002GL016341

Schubert, S. D., Suarez, M. J., Pegion, P. J., Koster, R. D. and Bacmeister, J. T. (2004) 'Causes of long-term drought in the US Great Plains', in *Journal of Climate*, vol 17, pp485–503

Seager, R., Kushnir, Y., Herweijer, C., Naik, N. and Miller, J. (2005) 'Modelling of tropical forcing of persistent droughts and pluvials over western North America: 1856–2000', in *Journal of Climate*, vol 18, pp4065–4088

Shanahan, T. M. et al (2009) 'Atlantic forcing of persistent drought in West Africa', in *Science*, vol 324, no 5925, pp377–380, doi:10.1126/science.1166352

Sheffield, J., K., Andreadis, M., Wood, E. F. and Lettenmaier, D. P. (2009) 'Global and continental drought in the second half of the 20th century: Severity-area-duration analysis and temporal variability of large-scale events', in *Journal of Climate*, vol 22, no 8, pp1962–1981

Sheffield, J., Luo, L. and Wood, E. F. (2010) 'Exploring mechanisms and predictability of US drought using remote sensing, hydrological modeling, reanalysis and the NCEP climate forecast system' (Abstract 5.1), 24th Conference on Hydrology, AMS Annual Meeting, Atlanta, GA, January 2010

Sud, Y. C., Mocko, D. M., Lau, K.-M. and Atlas, R. (2003) 'Simulating the Midwestern US drought of 1988 with a GCM', in *Journal of Climate*, vol 16, no 23, pp3946–3965

Sutton, R. T. and Hodson, L. R. (2005) 'Atlantic forcing of North American and European summer climate', in *Science*, vol 309, pp115–118

Teuling, A. J., Uijlenhoet, R. and Troch, P. A. (2005) 'On bimodality in warm season soil moisture observations', in *Geophysical Research Letters*, vol 32, L13402, p4 doi:10.1029/2005GL023223

Thompson, D. W. J. and Wallace, J. M. (1998) 'The Arctic Oscillation signature in the wintertime geopotential height and temperature fields', in *Geophysical Research Letters*, vol 25, pp1297–1300

Trenberth, K. E. (1999) 'Atmospheric moisture recycling: Role of advection and local evaporation', in *Journal of Climate*, vol 12, no 5, pp1368–1381

Trenberth, K. E. and Branstator, G. W. (1992) 'Issues in establishing causes of the 1988 drought over North America', in *Journal of Climate*, vol 5, pp159–172

Trenberth, K. E. and Guillemot, C. J. (1996) 'Physical processes involved in the 1988 drought and 1993 floods in North America', in *Journal of Climate*, vol 9, pp1288–1298

Trenberth, K. E. and Hurrell, J. W. (1994) 'Decadal atmosphere-ocean variations in the Pacific', in *Climate Dynamics*, vol 9, p303

Ummenhofer, C. C., England, M. H., McIntosh, P. C., Meyers, G. A., Pook, M. J., Risbey, J. S., Gupta, A. S. and Taschetto, A. S. (2009) 'What causes southeast Australia's worst droughts?', in *Geophysical Research Letters*, vol 36, L04706, doi:10.1029/2008GL036801

van Lanen, H. A. J., Fendeková, M., Kupczyk, E., Kasprzyk, A. and Pokojski, W. (2004) 'Flow generating processes', in L. M. Tallaksen and H. A. J. van Lanen (eds) *Hydrological Drought: Processes and Estimation Methods for Streamflow and Groundwater*, Developments in Water Science, Elsevier Science B.V., vol 48, pp53–96

Verdon, D. C. and Franks, S. W. (2006) 'Long-term behaviour of ENSO: Interactions with the PDO over the past 400 years inferred from paleoclimate records', in *Geophysical Research Letters*, vol 33, no 6, p5

Verdon-Kidd, D. C. and Kiem, A. S. (2009) 'Nature and causes of protracted droughts in southeast Australia: Comparison between the Federation, WWII, and Big Dry droughts', in *Geophysical Research Letters*, vol 36, L22707, p6, doi:10.1029/2009GL041067.12

Villarini, G., Smith, J. A., Serinaldi, F., Bales, J., Bates, P. D. and Krajewski, W. F. (2009) 'Flood frequency analysis for nonstationary annual peak records in an urban drainage basin,' in *Advances in Water Resources*, vol 32, pp1255–1266

Vogel, R. M. and Kroll, C. N. (1992) 'Regional geohydrologic-geomorphic relationships for the estimation of low-flow statistics', in *Water Resources Research*, vol 28, no 9, pp2451–2458

Wang, B., Wu, R. and Fu, X. (2000) 'Pacific-east Asian teleconnection: How does ENSO affect east Asian climate?', in *Journal of Climate*, vol 13, no 9, pp1517–1536

Wang, G. L. (2003) 'Reassessing the impact of North Atlantic Oscillation on the sub-Saharan vegetation productivity', in *Global Change Biology*, vol 9, no 4, pp493–499

Wang, G. L. and Eltahir, E. A. B. (2000) 'Role of vegetation dynamics in enhancing the low-frequency variability of the Sahel rainfall', in *Water Resources Research*, vol 36, pp1013–1021

Wang, L., Chen, W. and Huang, R. (2008) 'Interdecadal modulation of PDO on the impact of ENSO on the east Asian winter monsoon', in *Geophysical Research Letters*, vol 35, L20702, p4, doi:10.1029/2008GL035287

Webb, R. H., Hereford, R. and McCabe, G. J. (2004) 'Climatic fluctuations, drought, and flow in the Colorado River', USGS Fact Sheet 2004-3062, version 2

WWAP (World Water Assessment Programme) (2009) *The United Nations World Water Development Report 3: Water in a Changing World*, UNESCO, Paris, and Earthscan, London

Wu, W. R., Dickinson, R. E., Wang, H., Liu, Y. Q. and Shaikh, M. (2007) 'Covariabilities of spring soil moisture and summertime United States precipitation in a climate simulation', in *International Journal of Climatology*, vol 27, pp429–438

Zampieri, M., D'Andrea, F., Vautard, R., Ciais, P., de Noblet-Ducoudré, N. and Yiou, P. (2009) 'Hot European summers and the role of soil moisture in the propagation of Mediterranean drought', in *Journal of Climate*, vol 22, no 18, p4747

Zhang, R. and Delworth, T. L. (2006) 'Impact of Atlantic multidecadal oscillations on India/Sahel rainfall and Atlantic hurricanes', in *Geophysical Research Letters*, vol 33, L17712, doi:10.1029/2006GL026267

4

Quantifying Drought

Introduction

One of the main goals of scientific research on climate and hydrology is to reduce the impact of natural hazards and disasters such as hurricanes, floods, heat waves and droughts, by improving the monitoring of their development and predictions of their future occurrence. Knowing where a disaster is happening, how severe it is and how it is impacting is crucial for adequately responding to the needs of populations and maintaining the health of ecosystems. Furthermore, with ample forewarning, governments, agencies and the public can prepare for the possibility of such a disaster. In the case of short-term, fast-acting hazards such as hurricanes, for which a maximum warning of only a few days is possible, the best response is often to evacuate an area with public safety as the main goal. At the other extreme, a long-term and slowly evolving drought will usually require long-term measures to be put into place. The widespread nature of droughts means that huge areas and numbers of people may be affected. In developing regions, food security may be at risk and warning of drought several months ahead is necessary to put food aid and relief measures into place. The monitoring and prediction of drought are therefore essential tools for the reduction of impacts.

Progress in using such tools and improving their accuracy and capabilities is being made across the world, yet they are generally still in their infancy, especially in terms of prediction. Common to both, nevertheless, is the need to characterize drought in quantitative terms, for example by measuring the current spatial extent of drought, its severity and whether it is manifesting as meteorological drought or has propagated into the soil, streams or reservoirs. These types of data are needed to make assessments of current drought conditions and make decisions on which actions should be taken, which in turn depend on the severity of the drought and what sectors are affected. In general, quantifying current conditions should be done with respect to historic conditions, because this allows one to make comparisons and objectively assess whether the current drought is more or less severe than previous droughts, of which we have experience and knowledge of the impacts. Thus long series of drought data are required to put current conditions into context. Furthermore, improving the prediction of drought relies on attaining a better understanding of the physical mechanisms that lead to drought, its propagation through the hydrological, ecological and human systems, and its impacts. Again, this calls for quantitative data on climate, weather and hydrology, which can be analysed to reveal these mechanisms and form the basis for developing new predictive tools and for the assessment and improvement of existing ones.

This chapter discusses the various tools that are used to observe, quantify and model the hydrological cycle and drought, and the various ways that drought is identified and characterized. In subsequent chapters, these different types of data are used to reconstruct and analyse drought over past millennia, and particularly during the 20th century, and for future projections over the 21st century. These include direct (in-situ) instrumental measurements of hydrological data, output from computer models of the terrestrial hydrological cycle and climate systems, and palaeoclimate proxy data, which are used to reconstruct climate and hydrology for time periods before instrumental measurements became available.

Drought Data

To understand how drought varies (and other characteristics of the large-scale hydrological cycle), long-term observations of the hydrological cycle are required. The terrestrial hydrological cycle is composed of the fluxes of precipitation (rain and snowfall), evapotranspiration (soil and canopy water evaporation, plant transpiration and snow sublimation) and run-off (surface and subsurface flow), together with storage on the land surface (snowpack, vegetation canopy, lakes, wetlands, rivers) and subsurface (soil moisture, groundwater). These components are related through the water budget (Equation 1), which states that the fluxes of precipitation (P), evapotranspiration (E) and run-off (Q) are balanced by the change in water storage (S) at the Earth's surface:

$$\Delta S = P - E - Q \tag{1}$$

In terms of drought, relevant observations can be of precipitation, streamflow, soil moisture, groundwater levels or any other physical quantity related to water. Precipitation data can be used to determine the state of meteorological drought. Soil moisture and evapotranspiration data can represent agricultural drought. Streamflow, lake and reservoir levels, and groundwater levels are used to quantify hydrological drought. In cold regions, the state of the snowpack can be indicative of current and future hydrological drought. In mountainous regions such as western China and Chile, measurements of glacier extent and melt, which are major sources of water supply in these regions, can also be indicators of drought conditions. Soil moisture, in particular, is a good index of drought, reflecting recent precipitation and antecedent conditions, and indicating agricultural potential and available water storage (Keyantash and Dracup, 2004). To some extent, soil moisture captures the aggregate balance of all hydrological processes and represents available water. The soil column acts as a filter between incoming precipitation and throughfall (direct precipitation plus canopy drainage) and the processes that remove water from the hydrological system, such as evapotranspiration and subsurface drainage (Rawls et al, 1993; Entekhabi et al 1996). The amount of water in the top layers of the soil is correlated with short-term precipitation and so also gives a measure of meteorological drought. The soil moisture in the root zone (the region of influence of vegetation roots) is a governing factor of the state of vegetative growth through the availability of water for transpiration and thus is a direct indicator of agricultural drought (Keyantash and Dracup, 2004). In deeper soil layers, the soil moisture levels and depth to groundwater reflect long-term or hydrological drought, because it is this water that is available for recharge to aquifers and rivers.

In general, hydrological data are used for a wide variety of monitoring and management activities, including the forecasting and warning of natural hazards. They can also be used for planning and design, for example, to estimate the probability of a major flood or drought and thereby to devise flood defences or reservoirs to handle such events. In the context of environmental change, hydrological data are used to track trends and shifts in processes, such as those induced by global warming or the impacts of land-use change. More generally, huge amounts of data are collected by various national and regional agencies across the world to aid hydrological and climate monitoring and research. For drought in particular, high quality data are required to understand how drought develops and why and how it interacts with the broader hydrological and climate systems. These data can be measurements at a location which, when taken over multiple days, months or years, form a time series. Measurements taken at multiple points can form a map over a region that gives a view of the spatial variation in a quantity and allows it to be related to other quantities, such as climate, land use or elevation.

Observations of hydrological variables and processes are, however, scarce over the large spatial scales that are of interest for the monitoring and analysis of drought. Global datasets of these variables are generally lacking at high spatial resolution or are available only for limited time periods, notably from satellite remote sensing. Long records of data are useful for uncovering links with climate variability that acts on interannual to decadal timescales, such as the El Niño Southern Oscillation (ENSO) and the Pacific Decadal Oscillation (PDO) (see Box 3.1), and long-term trends and shifts due to climate change and anthropogenic influences. Streamflow is probably the best-measured hydrological variable over large regions, given that a measurement at a gauging station encapsulates the hydrology of the upstream basin, although it does not provide information about the spatial distribution of hydrological processes within the basin. Precipitation is also one of the best-observed variables, with long-term gauges (back to the mid-1800s) in many regions of Europe, North America, Australia and South Africa, among others, although observations are generally limited in spatial and temporal coverage, especially over sparsely populated regions and areas of complex terrain. Measurements of lake levels go back at least a century for some regions. Soil moisture and evapotranspiration are among the least-observed components of the hydrological cycle in terms of long-term, large-scale measurements. Only for a few measurement locations in the US, Russia, China and the Ukraine are there multidecadal observations of soil moisture.

Monitoring Drought from Space

Obtaining hydrological data at large scales from ground-based measurements alone remains a challenge, in terms of both accuracy and in achieving water budget closure (i.e. satisfying Equation 1). This is particularly true at larger scales and in less developed regions, where the paucity of gauges and measurement stations, and their inherent inaccuracies, prevents the development of a consistent picture of the water cycle. To overcome this, data from remote sensing (usually called retrievals) have the potential to provide spatially continuous estimates of components of the terrestrial water cycle over regional to global scales (Alsdorf and Lettenmaier, 2003; Sheffield et al, 2009a).

In recent years, remote sensing has become a major tool for observing climate, weather and the land surface, and has offered the potential to provide data for monitoring drought. Satellite-borne sensors are now capable of taking direct and indirect measurements of most hydrological components, including precipitation, lake and river levels, evaporation, soil moisture, snow and even total water storage (the sum of surface and subsurface water). Remote sensing can also provide useful information on the variation in vegetation state and thus plant productivity and health. Plate 4.1 shows examples of satellite-derived products for vegetation and different components of the hydrological cycle. These are discussed in more detail in later sections. Although some of these methods are in their infancy, and retrieved variables often require models and algorithms that need to be calibrated, the large spatial coverage and high temporal resolution (sub-daily for geostationary and equatorial orbiting satellites) means that they can provide near global information in real time. The relative novelty of many remote-sensing products, however, means that long-term data are not available and so the analysis of historic droughts in a climatic context is difficult.

Remote sensing refers to the observation or measurement of a process from a remote location. This may be from a fixed observing platform, such as a radar that measures rainfall, or a moving platform like an aircraft or a satellite. In general, for monitoring drought, the interest is in large spatial scales, and this is where satellite remote sensing becomes an important tool. It can provide high spatial and temporal resolution data over large areas to give a comprehensive view of the development of a drought. More generally, it is quickly becoming a major source of data for observing the large-scale water and energy cycles on the land, and in the atmosphere and oceans.

The challenge associated with the remote sensing of drought is that there is a disconnect between what is measured by the satellite and what is required to quantify the state of drought in terms of hydrological or vegetative indices. Generally, the satellite sensor measures radiation over one or more bands of the electromagnetic spectrum. These radiances must be converted into surface parameters and quantities (retrievals) that relate to hydrological variables and then provide a quantitative description of drought. This is challenging because of the complex transitions of radiation through the land–atmosphere system and the physical relationships between radiation and water fluxes and states. In practice, this entails relating the reflected, emitted or back-scattered radiation from the Earth's surface (vegetation, soil, water bodies) to the state of the land surface and vegetation. Because of the strong absorption and scattering of radiation by atmospheric constituents, such as water vapour, oxygen, CO_2 and aerosols over many electromagnetic bands, remote-sensing retrievals are restricted to a few atmospheric windows that are transparent and allow the sensor to see the land surface. These wavelength bands are the visible, infrared, short-wave, thermal and microwave.

There exist a multitude of data products from recent and current satellite missions that quantify the components of the water budget, either individually or as an aggregate estimate, at various scales of time and space (Sheffield et al, 2009a). Vegetative health or vigour has been estimated from visible and near-infrared data for several years (e.g. Kogan, 1997). Precipitation is regularly retrieved from multisensor microwave and infrared data, using a variety of techniques (e.g. Joyce et al, 2004; Huffman et al, 2007). Evapotranspiration can be estimated from analyses of the surface energy balance, given remote-sensing inputs of net radiation and surface meteorology (Su et al, 2005) and initial large-scale estimates are becoming available (e.g. Mu et al, 2007). Changes in total surface and subsurface storage can be derived using gravity anomaly measurements (Swenson and Wahr, 2002). Streamflow and surface water storage can be estimated using laser altimetry and interferometric synthetic aperture radiometry technologies (Alsdorf and Lettenmaier, 2003). Based on these systems, advances have been made in recent years in the development of drought-monitoring systems. With the wealth of information coming from the latest generation of Earth Observing System (EOS) satellites, there is now great potential for the regular and high-resolution global monitoring of drought. We next describe in detail the various types of remote-sensing products that can and are being used for drought monitoring.

Vegetative vigour

The state of vegetation is generally representative of environmental water stress, especially in water-limited areas, such as arid and semi-arid regions. Indices for quantifying the state of vegetation are generally based on combinations of measurements of visible and infrared radiation, and also surface temperature. Examples include the Normalized Difference Vegetation Index (NDVI), the Vegetation Condition Index (VCI) and the Temperature Condition Index (TCI). The NDVI has been used extensively to quantify the concentration of green leaf vegetation and thus determine regions of water stress due to lack of water. The signature of vegetative vigour in the form of different intensities of radiation from a plant results from the strong absorption of visible light (VIS) by chlorophyll, as part of the process of photosynthesis and the strong reflection of near-infrared (NIR) radiation by the cell structure of leaves. So if the reflected near-infrared radiation is greater than the visible, then the vegetation is likely to be dense and healthy. Conversely, if there is little difference between the two bands then the vegetation is likely to be sparse or stressed. Mathematically, NDVI is calculated as follows:

$$\text{NDVI} = (\text{NIR} - \text{VIS})/(\text{NIR} + \text{VIS}) \tag{2}$$

The resulting values range from –1 to 1, with higher positive values indicating healthier vegetation. In most regions the deviation of NDVI from the long-term average can be used to characterize the health of vegetation and as a measure of drought. It is also possible that low NDVI may be due to cool temperatures or low radiation because of heavy cloud cover, among other environmental factors. One such satellite-borne instrument is the Advanced Very High Resolution Radiometer (AVHRR), which can discern radiation in the near-infrared and visible wavelengths at a resolution of 1km. More recently launched instruments such as the Moderate Resolution Imaging Spectroradiometer (MODIS) give improved spatial resolution (250m) and increased spectral resolution. Plate 4.1a gives an example from the AVHRR-derived NDVI data for Europe during the 2003 heat wave, showing the signature of drought in the western part of the continent. Some of these instruments are now being used to monitor crops and agricultural productivity.

Precipitation

Precipitation has been retrieved from satellite sensors for several years and the algorithms are quite mature. There are currently two main methods for estimating precipitation, which first evolved in the 1970s with those based on visible and infrared data. VIS/IR retrievals do not measure precipitation directly but are inferred from cloud-top temperatures and are therefore most accurate for deep convective rainfall rather than stratiform rainfall. Microwave sensors alternatively sense precipitation more directly as microwave frequencies interact strongly with precipitation drops and are immune to cloud cover impacts. Precipitation products have been retrieved generally from passive microwave sensors, such as the Special Sensor Microwave Imager (SSM/I), and more recently from the TRMM Microwave Imager (TMI) on board the Tropical Rainfall Measuring Mission (TRMM) satellite, and the Advanced Microwave Scanning Radiometer-EOS (AMSR-E) on the Aqua satellite. The TRMM satellite also hosts the first space-borne precipitation radar (PR), which is an active microwave sensor that can sense the vertical distribution of precipitation. Current research is combining retrievals from the PR with those from the passive TMI sensor. A number of precipitation products are being generated in near real time, which use variations of these technologies enabling environmental monitoring at global scales, including drought monitoring. These products optimally combine data from multiple platforms and sensors. For example, the high temporal and spatial resolution of VIS/IR data are combined with the better accuracy, but lower resolution microwave data to generate high frequency precipitation data. Some examples of these products are the TRMM Multisatellite Precipitation Analysis (TMPA), the Precipitation Estimation from Remotely Sensed Information using Artificial Neural Networks (PERSIANN) and the CPC MORPHing technique (CMORPH). Currently, these products provide data historically for multiple years and in near real time, at resolutions of close to a few kilometres and on a sub-hourly basis. An example of data from the TMPA precipitation product is shown in Plate 4.1b over Southeast Asia and northern Australia during March 2007, highlighting the detailed structure that is possible to achieve from satellite data. Future planned precipitation measurement missions include the US–Japan Global Precipitation Mission (GPM), which will provide next-generation versions of the PR and TMI, but for a constellation of satellites encompassing the Earth, including part of the planned US National Polar-orbiting Operational Environmental Satellite System (NPOESS) fleet. The use of multiple platforms will somewhat overcome the problem of under-sampling precipitation with current microwave sensors.

Evapotranspiration

Evapotranspiration is a key variable that links the terrestrial water, energy and carbon cycles, yet ground observations at useful scales are essentially non-existent (Bastiaanssen et al, 2005). Remote sensing is an emerging potential source of data (Enthekhabi et al, 1999) that is finding real-world applications in the agricultural sector, where high-resolution data are required to quantify water-use efficiency, and for climate research, where long-term, large-scale datasets are needed. While evapotranspiration itself is a simple process, it has proven challenging to accurately retrieve it over natural surfaces using satellite observations. The difficulties include accurately representing the physical mechanisms controlling the process, heterogeneities at the land surface, the interacting influences and feedbacks in the soil, vegetation and meteorological continuum, and imprecise knowledge of variables controlling evapotranspiration. Typical approaches for estimating evapotranspiration from remote sensing are based on empirical regressions that relate evapotranspiration to surface temperature and vegetation indices (e.g. Nemani and Running, 1989; Nishida et al, 2003), predictive equations that approximate the main evaporative processes (e.g. Monteith, 1964; Priestley and Taylor, 1972), or surface energy balance considerations that use radiometric surface temperatures to solve the coupled equations of sensible, latent and available heat energy by assimilation (Caparrini et al, 2004) or directly (Norman et al, 1995; Anderson et al, 1997; Bastiaansen et al, 1998; Su, 2002). For a comprehensive overview of methods for estimating surface heat fluxes using satellite radiometric data, see the work of Kalma et al (2008).

Significant progress has been made in recent years, in particular because of interest in EOS era data to provide satellite remote-sensing inputs at continental to global scales of the required vegetation, radiation and near-surface meteorology data, which are used to retrieve evapotranspiration. However, data from polar orbiting satellites only provide one or two overpasses a day and these approaches therefore require the merging of data from multiple platforms or their combination with ground or modelled data to provide information at sub-daily scales. Geostationary satellites afford an alternative source of data at sub-daily timescales, and can be used to provide evapotranspiration-based drought monitoring at high temporal and spatial resolution. Plate 4.1c shows an example of the Evaporation Stress Index (ESI) of Anderson et al (2007) for the drought in the western US, which peaked in the summer of 2002. The ESI is calculated using thermal data from the Landsat series of satellites to give an index of drought at a 10km resolution. With further processing, retrievals can be achieved at field scales and discern the evaporation signatures of different crop types.

Snow

Wintertime accumulation of snowpack is a major source of water in many regions and careful monitoring is essential for water resource planning. Snow can be characterized by identifying two attributes: (1) snow cover extent (SCE), which is the spatial area covered by snow, irrespective of its depth or density, and (2) snow water equivalent (SWE), which is the equivalent depth of water in the snowpack if it was completely melted. These two quantities can be retrieved from VIS/IR and passive microwave sensors, respectively. Because of the high reflectivity of snow in the VIS/IR electromagnetic bands, it is quite simple to map the presence of snow on the ground, although these types of sensors cannot discern between snow and clouds and are restricted to daytime retrievals. SCE has been retrieved since the late 1960s from VIS/IR sensors, including the AVHRR, and global SCE products are available at 25km and weekly resolution. More recently, the MODIS sensor has been used to retrieve SCE at daily and 500m resolution, and there is potential to use its multispectral bands to differentiate between clouds and snow cover.

Passive microwave radiometers overcome some of the problems with VIS/IR technologies as they can penetrate through clouds and they can be used at night as well. Microwave frequencies are sensitive to snow water equivalent and the microphysical properties of the snowpack, including the snow grain size. Nevertheless, they do have some disadvantages, including a coarser spatial resolution (25km), saturation at higher SWE values and inability to penetrate dense forest canopies and complex topography. The latter can be overcome with active (radar) microwave sensors, which have a 25m resolution, although they suffer from the other microwave restrictions. Progress has recently been made towards generating blended products that optimally use information from VIS/IR and microwave sensors on multiple satellite platforms to provide the best combination of SCE and SWE data.

Soil moisture

Large-scale soil moisture data is extremely useful for monitoring drought, and microwave-based technologies exist to retrieve it from satellites, albeit with restrictions. These use models that simulate the transfer of radiation emitted from the soil through the vegetation canopy and atmosphere to the satellite sensor, based on the sensitivity of the soil dielectric constant to its wetness. The biggest drawback is the penetration depth of the microwave signal, which is dependent on the signal frequency, but is generally restricted to the top centimetre of the soil. In densely vegetated regions, the signal is attenuated and so retrievals are limited to sparsely vegetated regions with a biomass of less than 5kg/m^2. Current passive microwave sensors generally use frequencies of 6GHz or higher, although L-band (1–2GHz) is optimal in terms of penetration through vegetation and into the soil. There are also restrictions because of the spatial resolution of microwave data, which, because of the high spatial variability of soil moisture at the field scale, makes interpretation of data at 25km problematic. Current research is looking to combine data from multiple satellites to improve coverage and resolution, including merging passive and active products. However, these are generally still restricted to higher frequencies in the C-band and as such cannot retrieve results over dense vegetation. The recently launched European Space Agency Soil Moisture Ocean Salinity (SMOS) mission and the future planned NASA Soil Moisture Active and Passive (SMAP) mission will carry L-band instruments, which will improve penetration depth.

Lake and river levels

The level of rivers, lakes and reservoirs can be estimated from remote sensing, using radar altimeters or interferometric synthetic aperture radar (InSAR) (Alsdorf and Lettenmaier, 2003; Alsdorf et al, 2007). These sensors determine the height of the water surface by measuring the time that a radar or lidar pulse takes to travel from the satellite to the surface and back again. The distance is then subtracted from the satellite's position, as measured against an arbitrary datum. Some of the first satellite missions to provide data on the level of large water bodies included the ERS-1 and TOPEX/POSEIDON radar altimeters. Currently

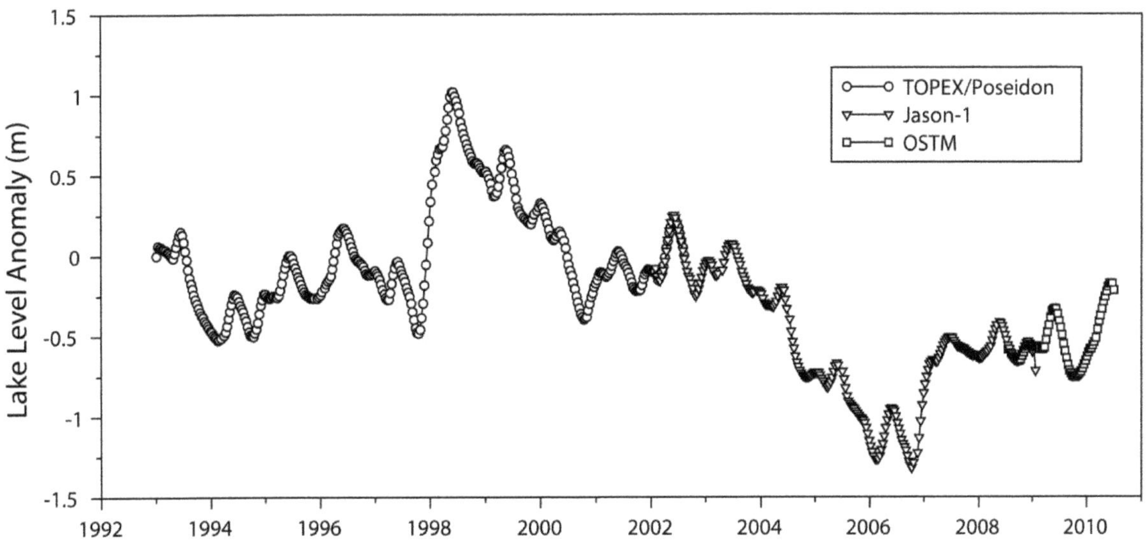

Figure 4.1 *Lake Victoria level variations as calculated from satellite radar altimeter data from the TOPEX/Poseidon, Jason-1 and Jason-2/OSTM satellites*

Source: US Department of Agriculture Global Reservoir and Lake Monitor website, www.pecad.fas.usda.gov/cropexplorer/global_reservoir

flying next-generation instruments include Jason-1, Jason-2, ERS-2 and Envisat, with future planned missions from the Indian, Chinese and European space agencies. A time series of lake levels derived from a combination of these satellites is shown in Figure 4.1 for Lake Victoria in central Africa, which supplies the Upper Nile. The satellite data show a very good match with actual lake level measurements. More importantly, these types of retrievals can be made for almost every large water body on Earth (Berry et al, 2005; Birkett et al, 2009), with huge potential to monitor water resources and drought. A limitation of these types of instruments is, however, their narrow track width and long repeat time, meaning that only a small fraction of the Earth's lakes and rivers are sampled, and with long waits between overpasses. The planned Surface Water Ocean Topography (SWOT) mission will overcome some of these limitations. SWOT will use a Ka band radar, which will give 100km swathes with repeat times of up to three days at high latitudes, thus providing relatively high-resolution data on water bodies down to 1km in size. Combined with a hydrodynamic model, data from SWOT for the slope of water in reasonably large rivers can be used to retrieve streamflow solely from remote-sensing data.

Total water storage

In recent years an exciting new remote-sensing tool that measures changes in the Earth's gravity field has been exploited to determine changes in the total water storage of the land surface, soil and underlying aquifers. The Gravity Recovery and Climate Experiment (GRACE) satellite (Tapley et al, 2004) has been monitoring the Earth's gravity field since its launch in 2002. Variations in the gravity field can be attributed to changes in terrestrial water storage after the removal of atmospheric and ocean bottom pressure changes. GRACE uses global positioning system (GPS) and microwave ranging systems to measure the distance between a pair of satellites and therefore infer the monthly change in the Earth's geoid (idealized reference surface). These shifts can be related to changes in water storage in lakes, reservoirs, wetlands, rivers, snow and ice, soil moisture and groundwater. Although GRACE data have been used for the analysis of ocean circulation and a variety of hydrological studies, including the measurement of changes in ice sheets and water budgets, their application to drought monitoring and water resources assessment is promising. For example, Plate 4.1d shows the GRACE water storage change anomaly (calculated by removing the mean seasonal cycle) during the 2005 drought in the Amazon. The reduction in water centred on the main channel of the Amazon River can be clearly seen. One drawback of the GRACE data is their low spatial resolution of the order of 500–1000km. A follow-on to the GRACE mission is the newly launched European Space Agency (ESA) Gravity field and steady-state Ocean Circulation Explorer (GOCE) (Klees et al, 2000), which is focused on measuring sea-level change.

Modelling Drought

Another source of data is from computer models, which simulate the physical processes of the hydrological and climate systems that govern the distribution and changes of water, energy and momentum. The advantage of models is that they can provide spatially and temporally consistent and continuous fields (maps) of variables at large scales. When given observations as inputs, these models can produce reasonably realistic depictions of observed hydrology and climate, potentially for multiple decades to centuries. The downside is that these models are imperfect and so are subject to biases in their output. They also rely on some form of measured data to drive them. The types of models that are of interest for drought analysis and prediction can be categorized as large-scale hydrological models or land surface models (LSMs), regional climate models (RCMs) and global climate models (more formally known as general circulation models, GCMs). Hydrological models and LSMs differ in their legacy of development and application: the former derive from the hydrological research community and the latter from the development of land surface schemes for climate modelling. Nowadays, there is considerable overlap between them, and they can be considered as identical for our purposes. These types of models simulate the physical processes of the land hydrological cycle and can, when driven by observed surface climate observations (precipitation, temperature, radiation and so on), simulate observed streamflow and other hydrological variables reasonably well (Mitchell et al, 2004; Sheffield and Wood, 2007; Wang et al, 2009). Thus, model simulations of hydrology can potentially provide hydrological variables globally for multiple decades, albeit with some uncertainty in their values.

GCMs simulate atmospheric processes and their interaction with the oceans and land, generally over seasonal to decadal timescales and longer. They can adequately replicate the observed global distribution of climate and its variation across different timescales, when given observed initial and boundary conditions (such as temperature and humidity profiles of the atmosphere, the temperature of the sea surface and time series of external forcings, such as CO_2 concentrations and solar variability). They do, however, suffer from biases and errors because of their generally coarse spatial resolution (about 2° in latitude and longitude, or 150–200km) and use of parameterizations (empirical representations) for finer-scale processes, such as cloud formation and convection. Lately, the resolution of some climate models has increased to about 10–50km, which has allowed the simulation of fine-scale, complex processes such as tropical storms (Knutson et al, 2007). GCMs come in a number of varieties, depending on whether they include only an atmospheric model with specified ocean boundary conditions (AGCM), an ocean model with specified atmospheric conditions (OGCM) or a fully coupled atmosphere ocean model (AOGCM). RCMs are generally higher spatial resolution versions of GCMs, which are applied at regional scales and driven by atmospheric boundary conditions taken from a GCM. The land component of GCMs has until recently often been some form of simple hydrological water budget model. As the importance of the role of the land surface has increased, both as a feedback to the atmosphere and because this is where the impacts of weather and climate are felt, these land models have increased in complexity and accuracy. There is now a move towards Earth system models (ESMs), in recognition of the need to incorporate the complex feedbacks between all components of the natural Earth system, which includes ecosystem dynamics, biospheric feedbacks, atmospheric chemistry and ocean biogeochemistry.

It is also possible to merge observations with models in a process called data assimilation (e.g. Evenson, 2003; Kalnay et al, 2007). This provides the dual advantage of simulating the continuous and consistent fields of atmospheric or hydrological variables, while anchoring the simulation to available observations. An important form of this is model *reanalyses*, which are historic simulations of the global atmosphere–land system over multiple decades using a climate model that ingests observations from weather balloons, satellites and ground observations. Some examples of global reanalyses are the US National Centers for Environmental Prediction/National Center for Atmospheric Research (NCEP/NCAR) Reanalysis (Kalnay et al, 1996), the European Centre for Medium-Range Weather Forecasts (ECMWF) ERA-40 Reanalysis (Uppala et al, 2005) and the NCEP North American Regional Reanalysis (NARR; Mesinger et al, 2006). Operational versions of these models and assimilation systems form the basis for weather forecasting.

Another benefit of models is that they can be used to diagnose the physical mechanisms of the variation in climate and hydrology, including those that lead to drought and its recovery, by analysing the connections between different parts of the coupled physical systems that they represent. For example, AOGCMs can provide insights into the forcing mechanisms of historic drought events, such as the influence of patterns of warm or cool SSTs (Hoerling and Kumar, 2003). Much of the recent research into large-scale drought has favoured model-based approaches, because of the lack of observational data about the occurrence and variability of droughts over large time and space scales, and because they can give insights into the mechanisms that control drought development and persistence. These tools allow for the exploration of the coupled ocean–atmosphere mechanisms that cause drought and the land–atmosphere feedbacks that contribute to its persistence. Models can also be used for prediction at seasonal scales when run in forecast mode, and so can form the basis for seasonal hydrological prediction (Wood and Lettenmaier, 2006) and thus drought forecasting (Luo and Wood, 2007). Similarly, projections of future climate changes over the 21st century and beyond can only be derived from models. Chapter 8 discusses how this can be achieved using climate models and the latest projections of future changes in drought.

In recent years, the development and use of land surface modelling schemes has increased as the role of the land surface in the climate system has gained acceptance (Entekhabi et al, 1996; Koster and Suarez, 2003). State-of-the-art schemes simulate the water and energy cycles at the land surface, and a multitude of physical processes, including evapotranspiration, soil water dynamics, snow and ice, and interactions with rivers, lakes and wetlands. Current research is expanding their capabilities to simulate the impacts of human activities (reservoirs, river diversion, irrigation and land-use change), vegetation dynamics (growth, die-off, species competition and disturbances such as fire, deforestation, overgrazing) and biogeochemistry, including the carbon and nitrogen cycles. They do this by using a variety of advanced techniques, including multilayer soil models with full water and energy accounting, and sub-grid variability of hydrological processes. By modelling the transport and storage of water in detail, these schemes are highly suited to the analysis of the occurrence and intensity of drought, and can do this on time and space scales that are useful for a wide variety of applications.

Until recently, the lack of consistent high-resolution meteorological forcing data required by these models has prevented their full use to this end. With ever-increasing computational power and the advent of observational data (such as precipitation and temperature) from around the world via telemetry and remote sensing, and modelled meteorological data from atmospheric reanalyses and forecast models, it is now feasible to generate land surface hydrological data products over large scales, both in real time and retrospectively, for decades to centuries (e.g. Mitchell et al, 2004). This type of dataset in turn can provide the capability for operational drought monitoring and for quantifying the variation in drought back in time. It could be argued that retrospective model simulations provide our best estimate of the variation of the terrestrial water cycle and its extremes, including drought, over the last 50–100 years, when instrumental data have been available. Box 4.1 provides details of the global land surface hydrological simulations that form the basis for the analyses in Chapters 6 and 7. The use of land surface hydrological modelling also offers the potential for drought forecasting using seasonal climate forecasts from coupled ocean–atmosphere general circulation models.

Box 4.1 Global hydrological modelling

An important tool in the reconstruction and analysis of drought globally are land surface models or large-scale hydrological models. In Chapters 6 and 7 we use data from the Variable Infiltration Capacity (VIC) land surface hydrological model (Liang et al, 1994; Cherkauer et al, 2002) to evaluate the occurrence and characteristics of drought during the 20th and early 21st centuries. The VIC model is characterized as an LSM, and was originally designed as the land component for climate models, but has a legacy rooted in the hydrological research community. It is thus well suited to provide useful predictions of soil moisture and other hydrological variables at large scale and for multiple decades, in the absence of direct measurements. The capabilities of the model are summarized in Figure 4.2. Like many other similar types of model, the VIC simulates the water and energy cycles at the land surface on a grid, where the characteristics of the land surface, such as the elevation, soil type and vegetation cover are specified for each grid cell. The physical processes that constitute the transfer of water and energy are represented by parameterizations, which simulate these processes with varying degrees of complexity, some of which are based on empirical relationships. These include the processes of infiltration, evapotranspiration, water drainage through the soil, run-off generation and subsurface flow, cold season processes such as snow accumulation, sublimation and freeze/thaw, and the functioning of lakes and wetlands. These hydrological processes are described in more detail in Chapter 3.

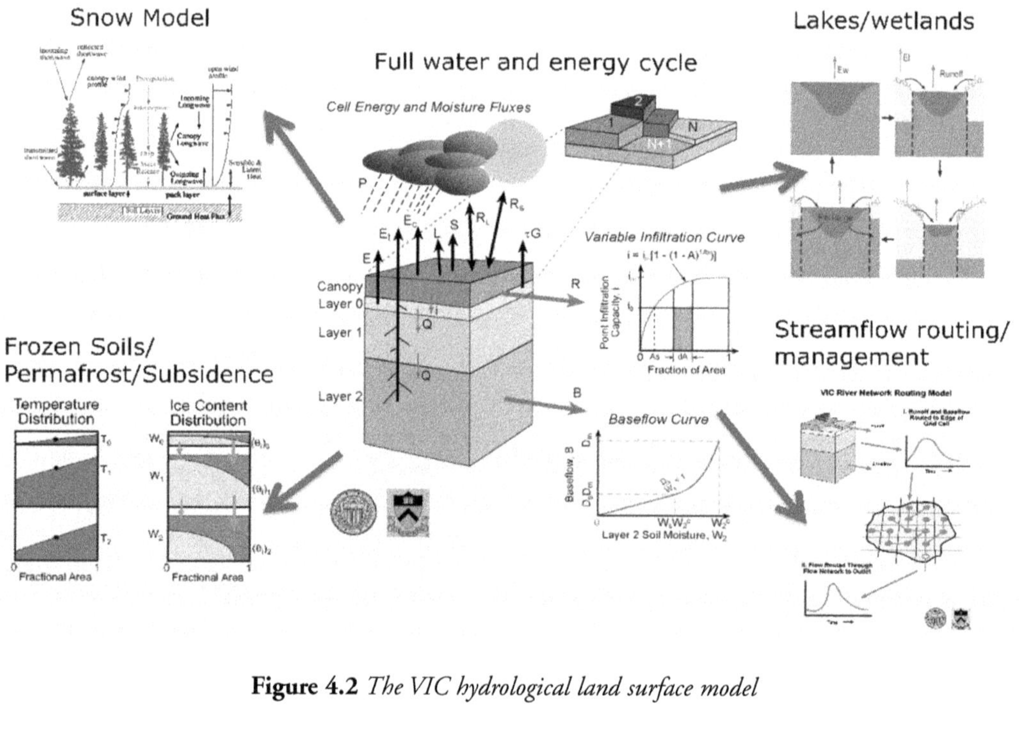

Figure 4.2 *The VIC hydrological land surface model*

Model simulations, input data and evaluations

The VIC model has a rich history of use in numerous applications, from the small-scale modelling of catchment processes (e.g. Cherkauer and Lettenmaier, 1999; Bowling et al, 2003; Bohn et al, 2007), to regional-scale hydroclimate and drought analyses (e.g. Sheffield et al, 2004a; Andreadis et al, 2005; Hamlet et al, 2007; Su and Lettenmaier, 2009), to global studies (e.g. Nijssen et al, 2001; Haddeland et al, 2007; Sheffield and Wood, 2008) and future climate impact assessments (e.g. Christensen et al, 2004; Hayhoe et al, 2007; Maurer et al, 2007). The VIC simulations used here are based on the work of Sheffield and Wood (2007, 2008). The model was run over all global land areas, except for the permanent ice caps of Antarctica and Greenland, at a resolution of 1° latitude and longitude computational grid cells. This is equivalent to grid sizes of about 10,000km^2 at the equator.

The model is driven by a dataset of meteorological data that includes precipitation, air temperature, downward surface short-wave and long-wave radiation, humidity, pressure and wind speed. These data are taken from the global meteorological dataset of Sheffield et al (2006), which combines daily and sub-daily data from atmospheric model reanalysis with monthly observation-based gridded datasets of precipitation, temperature and humidity, and satellite-based datasets of radiation. The reanalysis data are used to downscale the observational datasets in time to sub-daily time steps, which are required to adequately simulate the terrestrial water and energy cycles. At the same time, the observational data are used to bias-correct the reanalysis data at monthly time scales, which suffer from uncertainties in terms of their predictions. Corrections are made to the monthly precipitation totals, mean temperature, diurnal temperature range, storm frequency and intensities, as well as adjustments for elevation effects on meteorological variables.

The VIC model also requires data to represent the land surface. These are elevation, soil type and characteristics, and vegetation type. The elevation data control the topographic dependency of the variation in precipitation, temperature and snow cover within a grid cell. Soil type and soil characteristics provide information on the layering, density and structure of the soil, including its porosity, hydraulic conductivity and surface roughness. Different vegetation types, such as grassland, tropical rain forest, woodlands, deserts and croplands are specified within each grid cell, according to their global distribution. Model parameter values are specified for each vegetation type, such as the leaf area index, stomatal conductance, height and root profile.

Because the model has a number of parameter values which cannot be measured or have no direct physical basis, these must be estimated. Usually this is done through the calibration of the model, so that its output matches a measured hydrological quantity, usually the streamflow, since this is readily available and an aggregate measure of the hydrology of the upstream river basin. The global simulation has been calibrated against measurements of streamflow from a set of large river basins across the globe in different climates, from tropical to arctic. The output from the model has also been evaluated against available observations of terrestrial hydrology, including in-situ measurements of soil moisture and large basin streamflow, remote-sensing-based estimates of snow cover and water equivalent, water storage change and evaporation, and inferred estimates of evaporation from reanalysis atmospheric data.

Uncertainties and caveats in the modelling

The use of this dataset for the evaluation of drought is only as robust as the meteorological data that forces the simulation and the land surface model that is used to derive the soil moisture data. It has been shown that modelled land surface hydrology is sensitive to the forcing dataset that drives it, and especially precipitation (Fekete et al, 2004; Sheffield et al, 2004b; Guo et al, 2006). Fekete et al (2004) showed that the uncertainty in precipitation datasets was of the order of interannual variability and that the impact of precipitation uncertainties on the terrestrial water budget was of at least the same magnitude, especially in semi-arid regions where the hydrological response is highly non-linear. In the second Global Soil Wetness Project (GSWP-2) multimodel comparison, Guo et al (2006) found that uncertainties in the forcings were as large as differences between land surface models. The first-order drivers of drought, monthly precipitation and temperature, are derived in this work from the Climatic Research Unit (CRU) TS3.0 gauge-based dataset (Mitchell and Jones, 2005). For time periods when station observations are limited or non-existent, the CRU dataset relies on climatological values, or so-called 'relaxation to climatology' (Mitchell and Jones, 2005). Additionally, the number of gauge measurements that contributes to a grid cell may force errors in the simulated hydrology for densities of the order of less than 30 gauges per 10^{6}km^2 (Oki et al, 1999). The impact of this is likely to be highest in remote regions, such as the Amazon, Central Africa, Central Asia and high latitudes of northern Canada, Alaska and northern Eurasia, and especially in the earlier part of the record, when few gauges measurements were available.

It is therefore likely that errors in the forcing dataset used here will result in errors in the representation of drought and that the reliability of the time series at the grid scale (1°) may be reduced (Patz et al, 2002), although this can be alleviated through spatial and temporal averaging (Giorgi, 2002). Nevertheless, at larger scales, and especially in data-poor regions, we argue that this is the best estimate we can achieve. However, we highlight in the analysis in later chapters where and for which variables the uncertainties are high. In data-rich regions such as the US and Europe, our estimates may compare less favourably against those obtained when using data from dense station networks. For example, Decharme and Douville (2006) showed that the GSWP-2 forcing dataset drastically overestimated precipitation, in comparison to data from a dense network in France, with resulting impacts on modelled river discharge, and that this overestimation was systematic globally. However, comparisons of drought characteristics derived from the dataset in this book, and datasets based on higher spatial resolution modelling forced with gauge-based observations (Sheffield et al, 2004a; Andreadis et al, 2005; Andreadis and Lettenmaier, 2006) are encouraging.

Furthermore, biases in the modelling of the terrestrial hydrology, through simplifications in the modelling and uncertainties in the parameter data, may result in errors in the representation of the occurrence and trends in drought (Sheffield et al, 2004a). These biases are generally unquantifiable but most importantly are systematic and therefore uniform over time. Therefore it is likely they will not impact the sign or strength of the calculated trends appreciably. We similarly argue that comparable results would be obtained if we used a different model. The comparison of land surface models driven by the same forcings has been carried out regionally (Wood et al, 1998; Mitchell et al, 2004) and globally (Guo and Dirmeyer, 2006), and it has been concluded that, although the models do poorly at reproducing the absolute values of observed soil moisture, they do reasonably well at mimicking the anomalies and interannual variability. They may, therefore, provide useful information on drought occurrence and trends when viewed with respect to their own climatologies.

There are also a number of non-meteorological boundary conditions, such as land cover, that are assumed time invariant in the global simulations (although vegetation parameters such as leaf area index do vary seasonally), but may actually have an appreciable impact on the results. For example, anthropogenic factors, including irrigation, water withdrawals and land-use change, and natural processes such as vegetation dynamics and wildfire, are not modelled explicitly and the impact of these may vary in time also, thus affecting the representation of drought. Estimates of current-day irrigation are that 16.3 per cent of cultivated regions are equipped for irrigation (Siebert et al, 2005), which can have an impact on the water cycle (Haddeland et al, 2007), although historically this may have been offset by changes in land use. Land cover has changed dramatically over the past 300 years (Ramakutty and Foley, 1999) and more so in tropical and developing regions over the 20th century (Goldewijk, 2001). This may have had a substantial impact on the water cycle (Zhang and Schilling, 2006; Scanlon et al, 2007), probably reducing evapotranspiration and increasing run-off with possible implications for the results presented here. Furthermore, elevated levels of CO_2 and increased growing season length may be responsible for recent increases in net primary productivity (NPP) and thus transpiration (Friend et al, 2007; Huntington, 2008). Stomatal closure in response to elevated CO_2 levels may have had the opposite effect (Gedney et al, 2006), although this link is tenuous, given the wealth of contradictory evidence (Huntington, 2008).

Palaeoclimate Data: Reconstructing Drought over Millennia

Scientists use a number of techniques to see how the Earth's climate has changed in the past. These are satellites, instrumental records and proxy data. Models may also be used to merge these data together in a consistent fashion, as discussed above. These various sources of climate data form the basis for reconstructing drought. For the recent past this is based on instrumental and satellite data, and the methods for doing this are well established. However, the satellite record is quite short (less than 30 years). Instrumental records go back to the 19th century, or even earlier in some places. These types of data include thermometers, barometers, rain gauges and historic documents of weather observations. As we go further back in time, the instrumental record dwindles, and we must rely on other sources (or proxies) to infer the changes in climate and therefore drought. These include coral records, sediments, ice cores, pollen records and tree growth rings. Such types of data are known as palaeoclimate data and are used in the discipline of palaeoclimatology to infer and understand the variations in climate over hundreds to millions of years in the past, prior to the widespread availability of instrumental records, although the past few thousand years is probably the best dated and sampled of the palaeoclimatic record.

Palaeoclimatology is central to the science of global climate change because it can help to establish the range of natural climate variability prior to the period of human influence (Anthropocene). As the emergence of widespread instrumental records coincides with the start of the industrial revolution, when the human influence on atmospheric greenhouse gases began in earnest, it is difficult to discern between natural and anthropogenically influenced changes without going back further in time and relying on palaeoclimate data. These data are particularly useful as indicators of major drought events in the past and thus for putting modern droughts and potential future changes into perspective. Some of the most widely used proxy data sources are as follows.

Historical records

These are generally available in the form of written records such as newspapers, shipping logs, farmers' records and censuses. These documents can contain quantitative and qualitative information on climate variability, such as direct information about weather records as gleaned from shipping logs, or indirect information on agricultural yields, which reflect the season's climate. For example, historical records of grape harvest dates and wine quality have been used to reconstruct summer temperatures in the Paris region between 1370 and 1879 (Le Roy Ladurie and Baulant, 1980), a period that extends far beyond the instrumental record. Even everyday documents can also be of use, such as journals and diaries, which can contain direct observations of the weather or give clues to the prevailing climate of the time.

Corals

Many proxy records are based on the composition of chemical isotopes found in proxy material. Coral is a good example of this and is useful for reconstructing sea temperature records at the time at which the coral lived. Corals extract calcium carbonate from seawater to build their skeletons, which contains isotopes of oxygen, whose composition is related to water temperature.

Ocean, lake and river sediments

Layers of sediment that have settled on the beds of lakes and seas contain fossils and chemicals that can be used to infer climate of the time. They may also contain preserved pollen grains from flowering plants. The grains have distinctive shapes for each plant type, and this can be used to infer the prevailing climate when the plant was living, since different plants thrive in different climates. The layers of sediment reveal a history of climate based on the layers of different sediments and pollen types. River sedimentology is an important factor when reconstructing periods of flood and drought. Data from slack water sediment deposition, scour marks from floods, shifts in channel morphology, transport of boulders, or records of exposure of features at low water levels, can all be used to infer past hydrological events, albeit with a large degree of uncertainty because of assumptions that current regimes are the same as those in the past.

Tree rings

In a similar way as climate is inferred from layers of sediment, trees can also reveal variations in climate in their growth rings. Tree growth is affected by variations in climate and thus the characteristics of the tree rings (width and density), and their chemical and isotopic composition will vary depending on the prevailing temperature and rainfall. In areas where there is a distinctive growing season in temperate climate regions, trees will generally produce one growth ring a year, and so variations in climate at annual timescales can be resolved. More importantly, some tree species can live for hundreds to thousands of years and so provide long-term series of inferred climate variation.

Ice cores

Snow falling on the polar ice caps and glaciers accumulates over time, forming a continuous record of ice over many millennia. Cores of ice can be drilled and examined for the air bubbles and layers of dust that they contain. These data can be used to infer not only the local climate but also to reveal information about global conditions. For example, fallout from major volcanic eruptions on the other side of the world can be seen as layers of dust within ice cores, which can then be dated to within a few hundred years from the layering of the ice.

Figure 4.3 shows an example of tree-ring width data taken from the study of Davi et al (2006) for two sites in western Mongolia, which go back to 1638 and 1340. The raw ring width measurements are adjusted to allow for variations in growth, depending on the age of the tree, and are standardized to an index value relative to mean growth rate. The number of samples contributing to an index value may range from one to several tens of samples, with earlier records having fewer samples and therefore more uncertainty. These data and others are discussed further in Chapter 5, which explores the variation in climate and drought over the past 10,000 years.

Quantifying Drought: Characteristics and Indices

For the application of interest, whether it is the condition of a water supply system, the requirements for irrigation or the sustainability of an ecosystem, a quantitative expression for the state of drought is required. This is usually called a drought index and allows a farmer, a manager or policy-maker to objectively analyse a system and make quantitative management and policy decisions. Drought indices are used in both operational drought monitoring and when forecasting drought within a warning system, where an index can provide an objective basis for acting upon the drought. Here the goal is to anticipate the potential losses caused by droughts, for example, by planning and managing water and food resources to mitigate the future potential drought

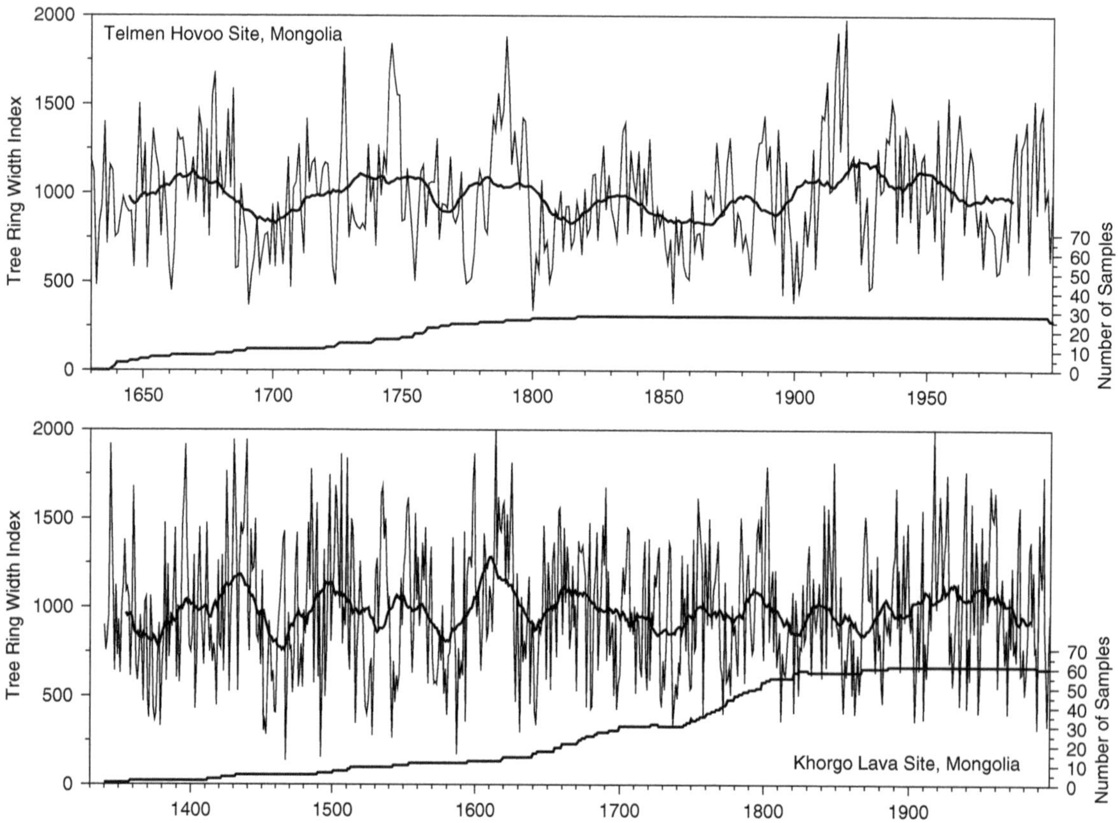

Figure 4.3 *Example of palaeoclimate data from tree-ring measurements at two sites in Mongolia*

Note: The tree-ring widths are standardized to index values that are relative to mean growth value of 1000. The thick black line is the 30-year running mean. The number of samples contributing to each value is also shown.

Source: Data taken from Davi et al (2006), available at www.ncdc.noaa.gov/paleo/paleo.html

impacts (Verdin et al, 2005). Usually a threshold is chosen below which there is drought and above which there is not. This threshold can be fixed but may also vary in time or be fuzzy in the sense that there is a gradation between drought and non-drought, and there may be a number of categories of drought severity, such as 'mild', 'moderate', 'severe' and 'extreme' (e.g. USDM, 2010). Often a drought threshold will be used to declare an official drought in a region and trigger a set of management responses (Steinemann et al, 2005) or implement a drought plan. This could invoke municipal water-use restrictions, a cessation of reservoir releases to maintain water supplies, government aid or insurance payments to farmers, or in vulnerable regions the build-up of food aid to tackle potential future shortages or famine.

Drought characteristics

Describing the state of drought is challenging because it varies in many dimensions. A drought has a spatial extent that describes the area it covers, and a duration that records how long it has persisted. Furthermore, a drought varies in its intensity, which changes in time and space as the drought develops, persists and recedes. These characteristics form the basis of a description of the state of drought, but vary according to the application of interest and the type of drought, for example whether it is agricultural or hydrological. Figure 4.4 shows a time series of streamflow that illustrates some of the characteristics of drought. This could equally be a different type of hydrological quantity at a location, such as precipitation, soil moisture, reservoir storage or well level; or an average value over a region. A drought is defined when the quantity drops below a threshold level, which may be related to plant health or sustainable water supply, for example. The time that the index is below the threshold is the duration of the drought. The level below the threshold at any particular time is the deficit, magnitude or intensity of the drought. The severity describes the combined duration and intensity/magnitude of the drought, calculated as the intensity multiplied by the duration. This is also often referred to as the deficit volume. Droughts can further be characterized by their

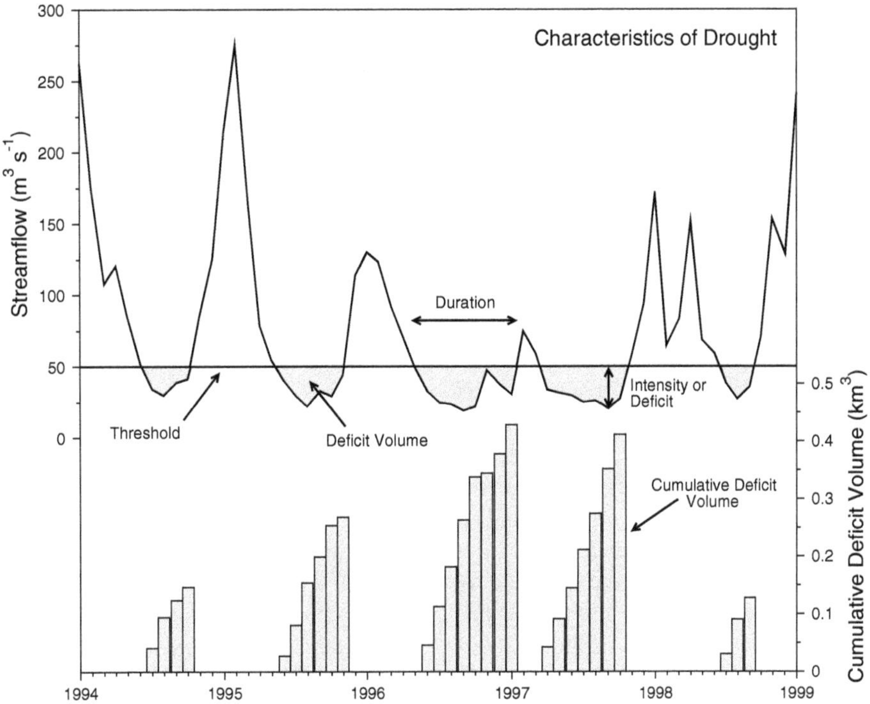

Figure 4.4 *The characteristics of drought*

Source: Adapted from Lehner et al, 2001

spatial extent, for example, by identifying connected regions of low index values, such as modelled soil moisture (Sheffield et al, 2009b), or measured streamflow at various gauging sites (Hisdal et al, 2001).

For hydrological drought in rivers, indices are generally based on low flow quantities, which characterize the drier part of the streamflow regime. Low flows generally occur during the dry season. In cold regions they may occur in the wintertime when precipitation accumulates as snowpack, soils and surface water are frozen, and streamflow is reduced to a minimum. In arid regions the low flows of intermittent and ephemeral streams will be zero or close to zero, and have to be handled differently. Low flows are usually defined in relation to the total streamflow regime, as characterized by the flow duration curve, which shows the percentage of time that a particular flow is exceeded. The curve is calculated by ranking daily flow values and plotting them against the proportion of flows with lower values. A metric of low flow can be selected such as the flow that is exceeded 95 per cent (Q95) of the time. An index of drought can be defined in relation to a low flow metric such as Q95. Examples of indices include the duration of flows continuously below the threshold, the minimum n-day average flow (where n is usually 7 but may vary depending on the application), the number of days per year below the threshold, and the flow deficit volume (the volume of water below the threshold).

Examples of drought indices

Ideally a drought index will capture the physical characteristics and associated impacts of a drought, but also be transferable to other regions and applications. Multiple drought characteristics may have to be taken into account within the index: the start and end of the drought or its duration, its intensity, magnitude and spatial extent, and its impacts. Quantifying these in one index is, however, extremely difficult. Various indices and definitions have been suggested, each of which has advantages and disadvantages, including those by Dracup et al (1980), Shafer and Dezman (1982), Oladipo (1985), McKee et al (1993), Wilhite (2000), Redmond (2002), and Shukla and Wood (2008). Table 4.1 lists some of the popular indices in use today and their advantages and disadvantages. This lack of standardization in the characterization of drought is problematic because it does not allow the consistent identification of droughts across regions and through time, with implications for how losses can be calculated and attributed objectively. The threshold at which a drought is declared is also crucial and thus difficult to set. If a drought is incorrectly called there can be over-compensation for a minor event or economic stress if the drought is worse than predicted, with legal ramifications. In extreme cases, this can lead to under-preparedness and even loss of life.

One of the most widely used measures of drought is the Palmer Drought Severity Index (PDSI; Palmer, 1965), which is popular in the US. It is used by agencies such as the US National Weather Service (NWS) Climate Prediction Center (CPC), as part of the US National Drought Monitor (Svoboda et al, 2002). The PDSI has generally been the tool of choice for observation-based indices of drought, and has been used by Cook et al (1999) and Dai et al (2004), among others, for drought reconstruction. At continental to global scales its simplicity makes it an attractive choice for reconstructing drought records. Other measures of drought, such as the Standardized Precipitation Index (SPI) (McKee et al, 1993), and the satellite-based vegetation and temperature condition index (Kogan, 1997) are also in use operationally. The SPI is a measure of meteorological drought based on the probability of occurrence of precipitation amounts. It indicates how likely the current precipitation (averaged over different time periods from one month (short-term conditions) to 24 months (long-term conditions)) is compared to normal. Two indices that are often used in the context of mapping desertification are the Climatic Moisture Index (CMI) and the Soil Moisture Index (SMI). The CMI is simply the ratio of long-term precipitation to potential evapotranspiration (CMI = P/PET) and represents the long-term status of a region in terms of supply and demand of water. On shorter timescales, the SMI, which is the ratio of actual to potential evapotranspiration (SMI = ET/PET), quantifies the actual conditions on the ground. A description of the drought measures developed over the last century can be found in Heim (2002).

Each of the standard indices in operational use has its advantages and disadvantages (e.g. Keyantash and Dracup, 2004). In the US, the National Drought Monitor is a fairly recent initiative to subjectively blend a number of these different indices into a consistent national indicator. This recognizes the fact that no single index or variable can represent drought for all

Table 4.1 *Popular drought indices, their rationale and advantages/disadvantages*

Index	Description	Advantages	Disadvantages
Meteorological Drought			
Precipitation percentage of normal	Calculated as the actual precipitation divided by the average annual value	Simple and effective for single locations and seasons	The mathematical meaning of normal is different to the general concept of normal weather
Rainfall deciles	Stratifies precipitation into deciles. Used mainly in Australia	Gives an accurate statistical measure of precipitation	Long records of data are required to obtain accurate values
Standardized Precipitation Index (SPI)	Based on the probability of precipitation	Can be computed for different timescales. Adopted as the World Meteorological Organization (WMO) standard meteorological index	Values may change as new data are incorporated. Reflects only precipitation
Agricultural Drought			
Palmer Drought Severity Index (PDSI)	Calculated as the departure of moisture from normal using a simple water balance model. Used by many US agencies as a drought trigger	Takes into account precipitation inputs and outputs from evaporation and run-off	Value may lag behind emerging droughts by several months; does not account for snow; does not handle frequent climatic extremes; complex
Crop Moisture Index (CMI)	Reflects short-term moisture supply for crop regions. Derived from the Palmer indices	Easily computed from precipitation and temperature data	Not applicable to long-term drought. Suffers from the general disadvantages of the Palmer indices
Palmer Moisture Anomaly Index (Z-index)	Same as the moisture anomalies in the PDSI	Quick response time to changing conditions	Same as PDSI; does not consider antecedent conditions
Soil moisture percentiles	Calculated from modelled or observed soil moisture data	Provides a statistically robust measure of soil moisture, which reflects hydrological inputs and outputs	Usually calculated from complex hydrological models; requires detailed input data
Hydrological Drought			
Palmer Hydrological Drought Index (PHDI)	Quantifies the severity of a wet or dry period. The monthly index values are exactly the same as PDSI	Used to monitor long-term drought conditions	Suffers from the same problems as the PDSI
Mean annual minimum *n*-day flow	Calculated from the time series of annual minima of the *n*-day average flow	Can be used to estimate the return period of low flows	Requires long time series and is difficult to apply in intermittent or ephemeral streams
Baseflow index (BFI)	Ratio of baseflow to total flow	Quantifies flow from stored water in the basin	Long records required to separate baseflow from total flow
Surface Water Supply Index (SWSI)	Combination of snowpack, streamflow, precipitation and reservoir storage	Represents water supply conditions, based on multiple variables	The index is unique to each basin, limiting basin comparisons

Table 4.1 *Popular drought indices, their rationale and advantages/disadvantages* (Cont'd)

Ecological Drought			
Normalized Difference Vegetation Index (NDVI)	Difference between maximum absorption of visible and near-infrared radiation	A measure of general vegetative condition. Satellite data provide large areal coverage and high spatial resolution	Requires data from airborne or spaceborne sensors. Difficult to discern other influences on vegetative health
Vegetation Condition Index (VCI)	Normalized version of NDVI	As for NDVI	As for NDVI
Regional Drought			
Regional Drought Area or Deficit Area	Percentage area in drought within a region	Quantifies the spatial extent of drought	Requires spatially continuous or regional data
Regional Drought Severity or Total Deficit Area	The area-weighted intensity over a region	Quantifies the average severity over a region	As above

Source: from NDMC and Keyantash and Dracup, 2004

applications. Of particular concern is the lack of a consistent probabilistic basis for some of these indices, which prevents their application across regions and through time. For example, the probability of an extreme drought as determined by the PDSI is 1 per cent in January in the US Pacific northwest, but more than 10 per cent in July in the Midwest (Guttman, 1998). Similarly, for the SPI, the change in probability between equally spaced index values is not uniform. Thus the probability of a drought changing from moderate to severe is not the same as from severe to extreme. Furthermore, the probability of a severe drought in the SPI is not the same as that in the PDSI, irrespective of other inconsistencies. An approach to overcoming these inconsistencies and providing a sound probabilistic grounding for a drought index is to use percentile values of the drought variable. A percentile value represents the value of the variable with a given probability of occurrence. For example, the 10th percentile is the value of the variable that occurs with a probability of 1 in 10. Time series of data can be converted to percentile values and then compared with percentile series at other points in space as they have exactly the same probability of occurrence (Sheffield and Wood, 2007). The threshold for defining a drought is selected in terms of percentile probability. For example, with a 10th percentile threshold, a drought is defined as having exactly a 10 per cent chance of occurring at each point.

Many of the standard indices are based on climate anomalies for precipitation and temperature. Drought is, however, increasingly being recognized as the compound impact of climate anomalies and human water use. Therefore hydrological indices are gaining in popularity, in part because of the availability of longer time series of measurements of hydrological variables, such as streamflow and soil moisture, and the potential for using increasingly reliable modelled data as a surrogate (Shukla and Wood, 2008). Some of the standard drought indices rely on some form of modelling and take into account hydrological processes to some degree, with varying levels of simplifications and assumptions. This may result in an inadequate representation of drought for some applications and regions, and is especially the case in regions whose hydrology is more complex, such as in cold regions and over varied terrain. For example, the PDSI calculates hydrological water balance terms when estimating cumulative moisture deficits, but does not simulate the impacts of frozen precipitation and snow accumulation and their delaying effects on the hydrological cycle (Shukla and Wood, 2008). The development of drought is a complex process that depends on a variety of factors (Wilhite, 2000). It therefore may be necessary to include as many of the contributing processes as possible when trying to derive an accurate drought index that can be used consistently over a region for a variety of applications. For example, agricultural drought is not only dependent on the temporal availability of precipitation and soil moisture for replacement of evapotranspiration losses, but also depends on the vertical distribution of soil moisture that is available to plants at different phenological stages of development. Soil moisture profiles, although correlated with precipitation at monthly timescales, can show non-linear responses to precipitation at shorter timescales, especially at lower depths and with very dry or very wet antecedent conditions (e.g. Capehart and Carlson, 1997). Additionally, in regions that experience substantial snowfall, the accumulation of the snowpack in the winter and the subsequent melt during the spring may be a major contributing factor to spring/summer drought occurrence (Service, 2004). Therefore, to determine the occurrence and intensity of drought in regions across the world, one has to take into account multiple complex processes.

Heim (2002) details the currently available drought indices for the US but notes that, despite considerable progress being made (e.g. the US Drought Monitor), there are still improvements to be made. These include, incorporating additional indicators such as snowpack data, reservoir and groundwater levels, and the daily variability of precipitation occurrence. Furthermore, Heim suggests the following characteristics are necessary for operational drought monitoring: (1) availability of data on a near real-time basis; (2) national coverage; (3) complete historical datasets to form a basis for comparison; and (4) removal of non-climatic biases. With the emergence of physically based land models over the last decade that simulate the detailed processes of energy and water transfer at the Earth's surface, including detailed soil moisture transport and snow processes, the potential for more accurate and comprehensive drought monitoring is evident (Sheffield et al, 2004a). These can provide an alternative to the standard but simpler indices and give a comprehensive view of the physical hydrological system, rather than an index related to a specific variable or process (Sheffield et al, 2004a; Andreadis et al, 2005). Coupled with the growing availability of remote-sensing products and detailed meteorological data at fine time and space scales with which to force these models, both retrospectively and in real time, there is the potential for analysing drought variability historically and for monitoring at regional and global scales. This is the approach we will take in later chapters to analyse the occurrence and characteristics of drought in the 20th and 21st centuries (see Box 4.1 for details of the datasets and model). Box 4.2 describes the methods for calculating a soil moisture-based drought index from the model output and how the characteristics of drought (intensity, duration, severity, extent) are derived.

Box 4.2 Quantifying drought

In general, drought is described in terms of a hydrological quantity, θ, that is expressed as a drought index, $q(\theta)$. The drought index quantifies some statistical aspect of the hydrological quantity, for example, its percentile or normalized values. In later chapters of this book we will explore the occurrence of drought based on percentiles of soil moisture taken from simulations of a global hydrological model and a suite of climate models. Empirical probability distributions are derived for the modelled soil moisture fields for each month and the current state of drought is characterized by the percentile of the current soil moisture value in relation to the distribution for the 20th century. We are particularly interested in how individual dry months are organized into sequences of consecutive dry months that can then be considered a drought with associated deleterious impacts. We will use the following methods and terminology to describe the characteristics of drought. Formally, a drought is defined as a period of duration D months with a value, $q(\theta)$, less than an arbitrary threshold level, $q_0(\theta)$, preceded and followed by a value above this level. The departure below this level at any particular time is the drought deficit or magnitude:

$$M = q_0(\theta) - q(\theta) \tag{3}$$

The mean magnitude over the drought duration is known as the intensity:

$$I = \frac{1}{D} \sum_{t=t_1}^{t_1+D-1} q_0(\theta) - q(\theta)_t \tag{4}$$

The product of duration and intensity gives the severity, also known as deficit volume or accumulated deficit (see 5 or 6).

$$S = D \times I \tag{5}$$

$$S = \sum_{t=t_1}^{t+D-1} q_0(\theta) - q(\theta)_t \tag{6}$$

Acknowledging that the impacts of drought are a balance between the duration and the intensity of deficits, classes of drought events based on their duration can also be defined as follows, where the subscript to *D* indicates the range of drought duration in months:

D_{1-3}, very short-term: $$1 \le D \le 3, \quad q(\theta) <= q_0(\theta) \tag{7}$$

D_{4-6}, short-term: $$4 \le D \le 6, \quad q(\theta) <= q_0(\theta) \tag{8}$$

D_{7-12}, medium-term: $$7 \le D \le 12, \quad q(\theta) <= q_0(\theta) \tag{9}$$

D_{12+}, long-term: $$D > 12, \quad q(\theta) <= q_0(\theta) \tag{10}$$

Regional drought characteristics include the spatial extent of drought (or deficit area), which is defined as the ratio of the area in deficit to the total area of the region, where *A(i)* is the area of the elementary unit of data values (which may be a model grid cell or a river catchment) and *N* is the total number of units in the region of interest.

$$A = \frac{\sum_{i=1}^{N} A(i) = \{q(\theta) <= q_0(\theta)\}}{\sum_{i=1}^{N} A(i)} \tag{11}$$

We are also interested in the spatial extent of contiguous drought, A_C, calculated using a simple clustering algorithm based on Andreadis et al (2005), which is described below.

Clustering algorithm for drought identification

We identify drought events in time and space using a clustering algorithm (Andreadis et al, 2005) that searches for spatially connected regions (or clusters) of low soil moisture. This assumes that drought happens over a reasonably large spatial area and is driven by the same large-scale climate processes. At each time, individual drought clusters are identified as spatially contiguous areas using a minimum cluster area threshold of 500,000km^2, which filters out minor regional droughts and events that could shrink to a few pixels and traverse continents through tenuous spatial connectivities, thus persisting, incorrectly, for multiple years. The 500,000km^2 threshold ensures that events only persist if there was a reasonable number of adjoining grid cells in drought. Although such long-term (decadal) drought events have been identified in the literature, and are the subject of intensive research, especially for North America (e.g. Cook et al, 2007), these are generally based on deviations in precipitation below the mean. Here we focus on contiguous regions of dry soil moisture identified using a systematic clustering method and a relatively low threshold (20th percentile of soil moisture). The longest duration drought found using this method was about four years, thus precluding such decadal droughts from our analysis. The drought clusters are then tracked through time and can merge with other events or break up into two or more separate events (Figure 4.5). A drought is declared once it has lasted for more than three months.

Severity–area–duration analysis

Each drought is analysed in terms of its coupled characteristics of severity, area and duration as developed by Andreadis et al (2005). The severity–area–duration (SAD) approach is based on the widely used depth–area–duration

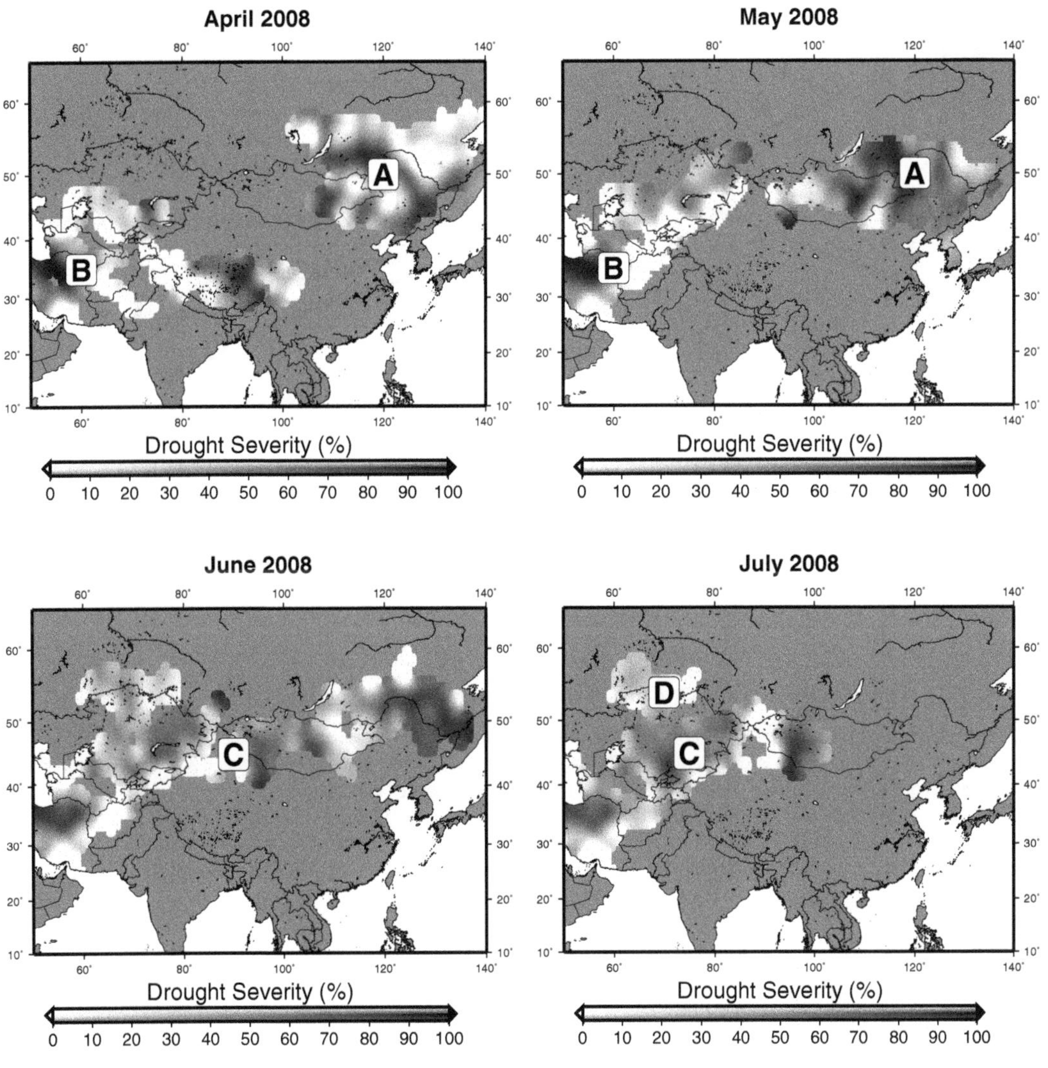

Figure 4.5 *Example of identification and tracking of drought events using a clustering algorithm*

Note: In the first month (April) two events, (A) and (B), are identified by searching for contiguous regions of soil moisture below a threshold. The severity is the percentage deficit below this threshold. During the second month, the two droughts move towards each other and merge in the third month to form a larger drought event (C). In the final month, drought (C) has reduced in size and a smaller drought (D) has broken away to the north.

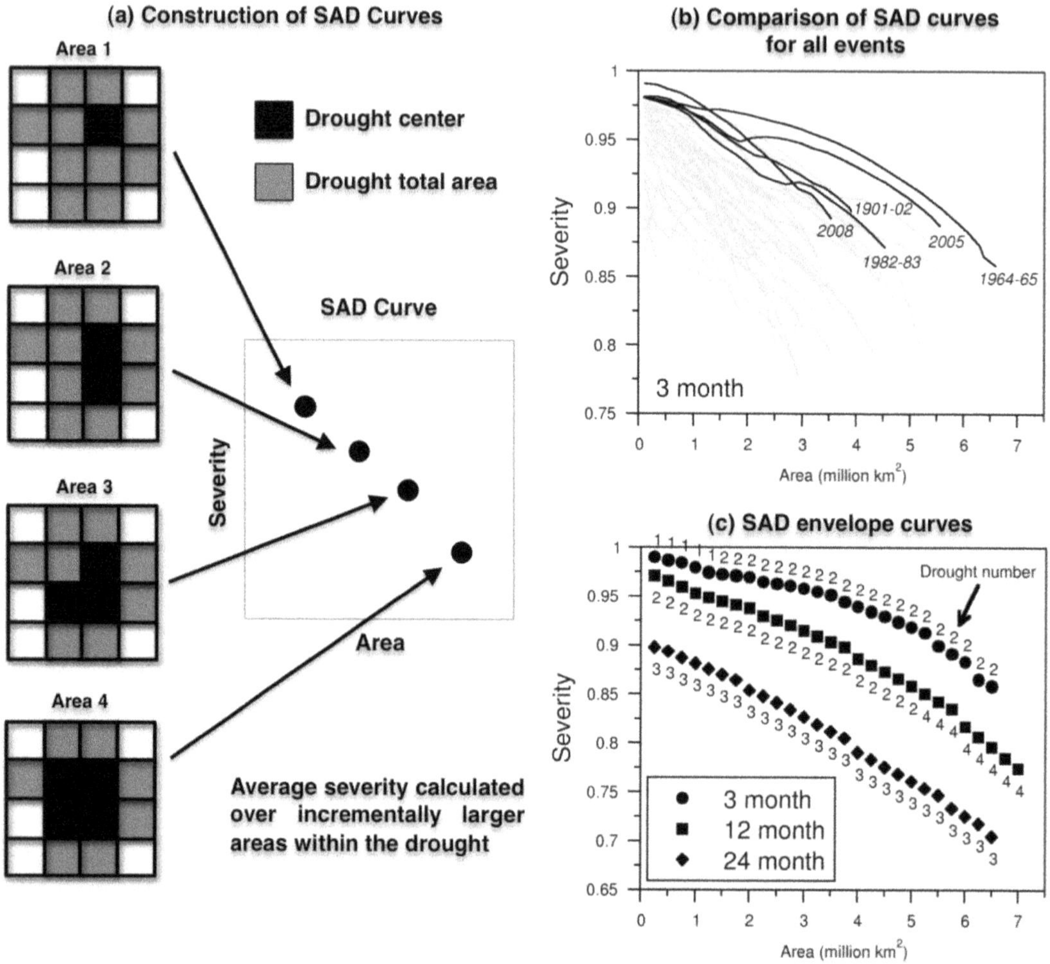

Figure 4.6 *Construction of severity–area–duration (SAD) curves*

Note: (a) The centre of a drought cluster is identified based on the location of maximum severity. For incrementally larger areas, the neighbourhood of the drought centre is identified and the average severity over the neighbourhood is recorded as a severity–area data point on a SAD diagram. This is repeated until the area reaches the maximum extent of the drought. The collection of points then forms the SAD curve, which are calculated separately for different durations (e.g. 3, 6, 12 months) within the total duration of the drought. (b) The SAD curve can be used to compare a drought event with other events. (c) The maximum severity at each area increment over all events forms the SAD envelope curve, which identifies the severest event for each area and duration.

(DAD) analysis (Grebner and Roesch, 1997), to simultaneously evaluate the severity and spatial extent of droughts for different durations.

Average drought severities are calculated for successively larger areas, from a minimum of about 100,000km^2 to the maximum spatial extent of each event (Figure 4.6). The process used to calculate these average severities for each predefined duration (e.g. 3, 6, 12, 18, 24, 36, 48 months) starts with calculating and ranking the severities for each pixel that 'experienced' the drought event in question. The model pixels with the maximum severity are taken as the 'drought centres', corresponding to the storm centres of the DAD analysis (WMO, 1969). The block neighbourhood of the centre is then searched for the next largest severity and that pixel is taken with the centre to form an intermediate drought area. This process is continued until the maximum spatial extent of the drought event is reached, with the severity and extent recorded in area increments of 20 model pixels (~200,000km^2). This is repeated for all overlapping time intervals (equal to a preselected duration) of the particular event. The maximum severity for each area increment is then selected to form the SAD curve for that event (Figure 4.6a), thus providing a way to estimate an absolute drought magnitude without being constrained to an individual basin or area. After calculating the SAD relationships for all identified events (Figure 4.6b), the envelope SAD curve can be formed that represents the maximum bound, or envelope, of severities from the set of all droughts that occurred in the region (Figure 4.6c). These curves allow the identification of the most severe events for each area increment and duration.

Summary

This chapter has summarized the sources of data and the methods used to quantify drought. We have highlighted the wealth of information that is available to explore drought occurrence, its characteristics and related hydrological and climate phenomena. The diversity of information reflects the variety of applications and scientific research that revolves around water and climate. This ranges from palaeoclimate studies on millennial timescales, to reconstructing drought over the past century, to drought monitoring in near real time, forecasting on seasonal timescales, to future climate projections of changes. Together these research areas span the breadth of temporal scales, from millions of years into the past, to hundreds of years into the future, and spatial scales from river basin to global, with regional applications in between. New exciting tools, such as satellite remote sensing, are now providing global pictures of the state of the drought from a variety of perspectives that are enhancing our ability to identify where and when drought is occurring, especially in remote areas where on-the-ground monitoring is non-existent. Furthermore, advances in modelling of the hydrological and climate systems afford the means to merge these different data sources together to obtain a more complete picture of the hydrological cycle and its variation, and help diagnose the mechanisms that force drought.

In terms of drought indices, much progress has been made in recent years. Drought is now being defined objectively through the development of indices that are physically based and offer a robust probabilistic view of drought occurrence. This chapter has presented some of the many indices and metrics that are used to quantify drought. Because drought varies on many scales, and affects different sectors and populations in different ways it is extremely hard to quantify in a simple fashion. Although there is no consensus on a universal drought index that satisfies all stakeholders and encompasses all drought situations, there is increasing agreement on what desirable characteristics a drought index should have. In particular, it is information about physical variables, such as precipitation, soil moisture, streamflow and evapotranspiration that is deemed desirable in most situations (Heim, 2002). With the availability of routine retrievals of hydrological variables and the state of vegetation (natural and agricultural) from remote sensing, there is great potential for monitoring of drought globally. There are now moves to do this in a coordinated fashion through international initiatives such as the intergovernmental Group on Earth Observations (GEO).

In the next chapters we will draw on these datasets and methods to explore how and why drought has varied in the past and may be expected to change in the future. For this we use a conceptual index based on percentile values of soil moisture, supported by evidence from measurements of other hydrological variables. Soil moisture is used because it is a physical variable and gives an aggregate view of the hydrological cycle, reflecting the balance of inputs of precipitation and snowmelt against the outputs of evaporation, run-off and drainage. The soil moisture index is simple and transferable in time and space, and gives a consistent probabilistic basis for drought identification and characterization allowing for the comparison of drought events across regions and through time. It is therefore well suited to evaluation not only of changes in the instrumental period, but also to evaluation of the implications of future climate for drought, which we will discuss in Chapter 8.

References

Alsdorf, D. E. and Lettenmaier, D. P. (2003) 'Tracking fresh water from space', in *Science*, vol 301, pp1491–1494

Alsdorf, D. E., Rodriguez, E. and Lettenmaier, D. P. (2007) 'Measuring surface water from space', in *Reviews of Geophysics*, vol 45, RG2002, p24, doi:10.1029/2006RG000197

Anderson, M. C., Norman, J. M., Diak, G. R., Kustas, W. P. and Mecikalski, J. R. (1997) 'A two-source time-integrated model for estimating surface fluxes using thermal infrared remote sensing', in *Remote Sensing of Environment*, vol 60, pp195–216

Anderson, M., Norman, J., Mecikalski, J., Otkin, J. and Kustas, W. (2007) 'A climatological study of evapotranspiration and moisture stress across the continental United States based on thermal remote sensing: 1. Model formulation', in *Journal of Geophysical Research*, vol 112, D11112, p13

Andreadis, K. M., and Lettenmaier, D. P. (2006) 'Trends in 20th century drought over the continental United States', in *Geophysical Research Letters*, vol 33, L10403, p4

Andreadis, K. M., Clark, E. A., Wood, A. W., Hamlet, A. F. and Lettenmaier, D. P. (2005) '20th century drought in the conterminous United States', in *Journal of Hydrometeorology*, vol 6, no 6, pp985–1001

Bastiaanssen, W., Menenti, R., Feddes, R. and Holtslag, A. (1998) 'A remote sensing surface energy balance algorithm for lands (SEBAL)', in *Journal of Hydrology*, vol 212–213, pp198–212, doi:10.1016/S0022-1694(98)00253-4

Bastiaanssen, W. G. M., Noordman, E. J. M., Pelgrum, H., David, G., Thoreson, B. P. and Allen, R. G. (2005) 'SEBAL model with remotely sensed data to improve water resources management under actual field conditions', in *ASCE: Journal of Irrigation and Drainage Engineering*, vol 131, pp85–93

Berry, P. A. M., Garlick, J. D., Freeman, J. A. et al (2005) 'Global inland water monitoring from multimission altimetry', in *Geophysical Research Letters*, vol 32, no 16, L16401

Birkett, C. M., Reynolds, C., Beckley, B. and Doorn, B. (2009) 'From research to operations: The USDA Global Reservoir and Lake Monitor', in *Coastal Altimetry*, Springer, Berlin, 2011

Bohn, T. J., Lettenmaier, D. P., Sathulur, K., Bowling, L. C., Podest, E., McDonald, K. C. and Friborg, T. (2007) 'Methane emissions from western Siberian wetlands: Heterogeneity and sensitivity to climate change', in *Environmental Research Letters*, vol 2, no 4, p9, doi:10.1088/1748-9326/2/4/045015

Bowling, L. C., Lettenmaier, D. P., Nijssen, B., Graham, L. P. et al (2003) 'Simulation of high latitude hydrological processes in the Torne-Kalix basin: PILPS Phase 2(e) 1: Experiment description and summary intercomparisons', in *Journal of Global and Planetary Change*, vol 38, nos 1–2, pp1–30

Caparrini, F., Castelli, F. and Entekhabi, D. (2004) 'Estimation of surface turbulent fluxes through assimilation of radiometric surface temperature sequences', in *Journal of Hydrometeorology*, vol 5, pp145–159

Capehart, W. J. and Carlson, T. N. (1997) 'Decoupling of surface and near-surface soil water content: A remote sensing perspective', in *Water Resources Research*, vol 33, no 6, pp1383–1395

Cherkauer, K. A. and Lettenmaier, D. P. (1999) 'Hydrological effects of frozen soils in the upper Mississippi River basin', in *Journal of Geophysical Research*, vol 104, D16, pp19,599–19,610

Cherkauer, K. A., Bowling, L. C. and Lettenmaier, D. P. (2002) 'Variable infiltration capacity (VIC) cold land process model updates', in *Global and Planetary Change*, vol 38, nos 1–2, pp151–159

Christensen, N. S., Wood, A. W., Voisin, N., Lettenmaier, D. P. and Palmer, R. N. (2004) 'Effects of climate change on the hydrology and water resources of the Colorado River Basin', in *Climatic Change*, vol 62, nos 1–3, pp337–363

Cook, E. R., Meko, D. M., Stahle, D. W. and Cleaveland, M. K. (1999) 'Drought reconstructions for the continental United States', in *Journal of Climate*, vol 12, pp1145–1162

Cook E. R., Seager, R., Cane, M. A. and Stable, D. W. (2007) 'North American droughts: Reconstructions, causes and consequences', in *Earth-Science Reviews*, vol 81, pp93–134

Dai, A. G., Trenberth, K. E. and Qian, T. T. (2004) 'A global dataset of Palmer Drought Severity Index for 1870–2002: Relationship with soil moisture and effects of surface warming', in *Journal of Hydrometeorology*, vol 5, no 6, pp1117–1130

Davi, N. K, Jacoby, G. C., Curtis, A. E. and Baatarbileg, N. (2006) 'Extension of drought records for Central Asia using tree-rings: West Central Mongolia', in *Journal of Climate*, vol 19, pp288–299

Decharme, B. and Douville, H. (2006) 'Uncertainties in the GSWP-2 precipitation forcing and their impacts on regional and global hydrological simulations', in *Climate Dynamics*, vol 27, pp695–713

Dracup, J. A., Lee, K. S. and Paulson, E. G., Jr. (1980) 'On the definition of droughts', in *Water Resources Research*, vol 16, pp297–302

Entekhabi, D., Rodriguez-Iturbe, I. and Castelli, F. (1996) 'Mutual interaction of soil moisture state and atmospheric processes', in *Journal of Hydrology*, vol 184, pp3–17

Entekhabi, D., Asrar, G. R., Betts, A. K., Beven, K. J., Bras, R. L., Duffy, C. J. et al (1999) 'An agenda for land surface hydrology research and a call for the second international hydrological decade', in *Bulletin of the American Meteorological Society*, vol 80, pp2043–2058

Evensen, G. (2003) 'The ensemble Kalman filter: Theoretical formulation and practical implementation', in *Ocean Dynamics*, vol 53, pp343–367

Fekete, B. M., Vörösmarty, C. J., Roads, J. O. and Willmott, C. J. (2004) 'Uncertainties in precipitation and their impacts on run-off estimates', in *Journal of Climate*, vol 17, no 2, pp294–304

Friend, A. D., Arneth, A., Kiang, N. Y., Lomas, M., Ogee, J., Rodenbeckk, C., Running, S. W., Santaren, J. D., Sitch, S., Viovy, N., Woodward, F. I. and Zaehle, S. (2007) 'FLUXNET and modelling the global carbon cycle', in *Global Change Biology*, vol 13, pp610–633

Gedney, N., Cox, P. M., Betts, R. A., Boucher, O., Huntingford, C. and Stott, P. A. (2006) 'Detection of a direct carbon dioxide effect in continental river run-off records', in *Nature*, vol 439, pp835–838

Giorgi, F. (2002) 'Variability and trends of sub-continental scale surface climate in the twentieth century. Part 1: Observations', in *Climate Dynamics*, vol 18, pp675–691

Goldewijk, K. K. (2001) 'Estimating global land use change over the past 300 years: The HYDE Database', in *Global Biogeochemical Cycles*, vol 15, pp417–433

Grebner, D. and Roesch, T. (1997) 'Regional dependence and application of DAD relationships. Proc. FRIEND '97', in *Regional Hydrology: Concepts and Models for Sustainable Water Resource Management*, International Association of Hydrological Sciences, Postojna, Slovenia, no 246, pp223–230

Guo, Z. C. and Dirmeyer, P. A. (2006) 'Evaluation of the Second Global Soil Wetness Project soil moisture simulations: 1. Intermodel comparison', in *Journal of Geophysical Research-Atmospheres*, vol 111, D22S02, p14

Guo, Z. C., Dirmeyer, P. A., Hu, Z. Z., Gao, X. and Zhao, M. (2006) 'Evaluation of the Second Global Soil Wetness Project Soil Moisture Simulations: 2. Sensitivity to external meteorological forcing', in *Journal of Geophysical Research-Atmospheres*, vol 111, D22S03, p11

Guttman, N. B. (1998) 'Comparing the Palmer Drought Severity Index and the Standardized Precipitation Index', in *Journal of the American Water Resources Association*, vol 34, no 1, pp113–121

Haddeland, I., Skaugen, T. and Lettenmaier, D. P. (2007) 'Hydrological effects of land and water management in North America and Asia: 1700–1992', in *Hydrology and Earth System Science*, vol 11, no 2, pp1035–1045

Hamlet, A. F., Mote, P. W., Clark, M. P. and Lettenmaier, D. P. (2007) '20th Century Trends in Run-off, Evapotranspiration, and Soil Moisture in the Western US', in *Journal of Climate*, vol 20, no 8, pp1468–1486

Hayhoe, K., Wake, C., Huntington, T., Luo, L., Schwartz, M. D., Sheffield, J., Wood, E. F., Anderson, B., Bradbury, J., DeGaetano, A., Troy, T. J. and Wolfe, D. (2007) 'Past and future changes in climate and hydrological indicators in the US Northeast', in *Climate Dynamics*, vol 28, no 4, pp381–407, doi:10.1007/s00382-006-0187-8

Heim, R. R., Jr. (2002) 'A review of twentieth century drought indices used in the United States', *Bulletin of the American Meteorological Society*, vol 83, no 8, pp1149–1165

Hisdal, H., Stahl, K., Tallaksen, L. M. and Demuth, S. (2001) 'Have streamflow droughts in Europe become more severe or frequent?' in *International Journal of Climatology*, vol 21, pp317–333

Hoerling, M. P. and Kumar, A. (2003) 'The perfect ocean for drought', in *Science*, vol 299, pp691–699

Huffman, G. J., Adler, R. F., Bolvin, D. T., Gu, G., Nelkin, E. J., Bowman, K. P., Hong, Y., Stocker, E. F. and Wolff, D. B. (2007) 'The TRMM Multisatellite Precipitation Analysis (TMPA): Quasi-global, multiyear, combined-sensor precipitation estimates at fine scales', in *Journal of Hydrometeorology*, vol 8, pp38–55

Huntington, T. G. (2008) 'CO_2-induced suppression of transpiration cannot explain increasing run-off', in *Hydrological Processes*, vol 22, pp311–314

Joyce, R. J., Janowiak, J. E., Arkin, P. A. and Xie, P. (2004) 'CMORPH: A method that produces global precipitation estimates from passive microwave and infrared data at high spatial and temporal resolution', in *Journal of Hydrometeorology*, vol 5, pp487–503

Kalma, J. D., McVicar, T. R. and McCabe, M. F. (2008) 'Estimating land surface evaporation: A review of methods using remotely sensed surface temperature data', in *Surveys Geophysics*, vol 29, nos 4–5, pp1–49, doi:10.1007/s10712-008-9037-z

Kalnay, E., Kanamitsu, M., Kistler, R., Collins, W., Deaven, D., Gandin, L., Iredell, M., Saha, S., White, G., Woolen, J., Zhu, Y., Leetmaa, A., Reynolds, R., Chelliah, M., Ebisuaki, W., Higgins, W., Janowiak, J., Mo, K. C., Jenne, R. and Joseph, D. (1996) 'The NCEP/NCAR 40-Year Reanalysis Project', in *Bulletin of the American Meteorological Society*, vol 77, pp437–471

Kalnay, E., Hong, L., Miyoshi, T., Yang, S.-C. and Ballabrera-Poy, J. (2007) '4D-Var or ensemble Kalman filter?', in *Tellus*, no 59A, pp758–773

Kerr, R. A. (2005) 'Climate change: Atlantic climate pacemaker for millennia past, decades hence?' in *Science*, vol 309, no 5731, pp41–43

Keyantash, J. A. and Dracup, J. A. (2004) 'An aggregate drought index: Assessing drought severity based on fluctuations in the hydrological cycle and surface water storage', in *Water Resources Research*, vol 40, W09304, p13

Klees, R., Koop, R., Visser, P. and Jessel, I. J. (2000) 'Efficient gravity field recovery from GOCE gravity gradient observations', in *Journal of Geodesy*, vol 74, pp561–571

Knutson, T., Sirutis, J., Garner, S., Held, I. and Tuleya, R. E. (2007) 'Simulation of the recent multidecadal increase of Atlantic hurricane activity using an 18km-grid regional model', in *Bulletin of the American Meteorological Society*, pp1549–1565

Kogan, F. N. (1997) 'Global drought watch from space', in *Bulletin of the American Meteorological Society*, vol 78, pp621–636

Koster, R. D. and Suarez, M. J. (2003) 'Impact of land surface initialization on seasonal precipitation and temperature prediction', in *Journal of Hydrometeorology*, vol 4, pp408–423

Lehner, B., Henrichs, T., Döll, P. and Alcamo, J. (2001) 'EuroWasser – Model-based assessment of European water resources and hydrology in the face of global change', in *Kassel World Water Series 5*, Center for Environmental Systems Research, University of Kassel, Germany

Le Roy Ladurie, E. and Baulant, M. (1980) 'Grape harvests from the fifteenth through the nineteenth centuries', in *Journal of Interdisciplinary History*, vol 10, pp839–849

Liang, X., Lettenmaier, D. P., Wood, E. F. and Burges, S. J. (1994) 'A simple hydrologically based model of land surface water and energy fluxes for general circulation models', in *Journal of Geophysical Research*, vol 99, no 14, pp415–428

Luo, L. and Wood, E. F. (2007) 'Monitoring and predicting the 2007 US drought', in *Geophysical Research Letters*, vol 34, L22702, p6, doi:10.1029/2007GL031673

Maurer, E. P. (2007) 'Uncertainty in hydrological impacts of climate change in the Sierra Nevada, California under two emissions scenarios', in *Climatic Change*, vol 82, nos 3–4, pp309–325, doi:10.1007/s10584-006-9180-9

McKee, T. B., Doesken, N. J. and Kleist, J. (1993) 'The relationship of drought frequency and duration to time scales', in *Preprints of the Eighth Conference on Applied Climatology*, Anaheim, California, 17–22 January, pp179–184

Mesinger, F. et al (2006) 'North American Regional Reanalysis', in *Bulletin of the American Meteorological Society*, vol 87, pp343–360

Mitchell, K. E., Lohmann, D., Houser, P. R., Wood, E. F., Schaake, J. C., Robock, A., Cosgrove, B. A., Sheffield, J., Duan, Q., Luo, L., Higgins, R. W., Pinker, R. T., Tarpley, J. D., Lettenmaier, D. P., Marshall, C. H., Entin, J. K., Pan, M., Shi, W., Koren, V., Meng, J., Ramsay, B. H. and Bailey, A. A. (2004) 'The multi-institution North American Land Data Assimilation System (NLDAS): Utilizing multiple GCIP products and partners in a continental distributed hydrological modelling system', in *Journal of Geophysical Research*, vol 109, D07S90, p32, doi:10.1029/2003JD003823

Mitchell, T. D. and Jones, P. D. (2005) 'An improved method of constructing a database of monthly climate observations and associated high-resolution grids', in *International Journal of Climatology*, vol 25, no 6, pp693–712

Monteith, J. L. (1964) 'Evaporation and environment. The state and movement of water in living organisms', Symposium of the society of experimental biology, Cambridge University Press, Cambridge, vol 19, pp205–234

Mu, A., Heinsch, F., Zhao, M. and Running, S. (2007) 'Development of a global evapotranspiration algorithm based on MODIS and global metereology data', in *Remote Sensing of Environment*, vol 111, pp519–536

Nemani, R. R. and Running, S. W. (1989) 'Estimation of regional surface resistance to evapotranspiration from NDVI and thermal-IR AVHRR data', in *Journal of Applied Meteorology*, vol 28, pp276–84

Nijssen, B., Schnur, R. and Lettenmaier, D. P. (2001) 'Global retrospective estimation of soil moisture using the Variable Infiltration Capacity land surface model, 1980–1993', in *Journal of Climate*, vol 14, pp1790–1808

Nishida, K., Nemani, R. R., Running, S. W. and Glassy, J. M. (2003) 'An operational remote sensing algorithm of land surface evaporation', in *Journal of Geophysical Research*, vol 108, 4270, p14, doi:10.1029/2002JD002062

Norman, J. M., Kustas, W. P. and Humes, K. S. (1995) 'A two-source approach for estimating soil and vegetation energy fluxes from observations of directional radiometric surface temperature', in *Agricultural and Forest Meteorology*, vol 77, pp263–293

Oki, T., Nishimura, T. and Dirmeyer, P. (1999) 'Assessment of annual run-off from land surface models using Total Run-off Integrating Pathways (TRIP)', in *Journal of The Meteorological Society of Japan*, vol 77, pp235–255

Oladipo, E. O. (1985) 'A comparative performance analysis of three meteorological drought indices', in *International Journal of Climatology*, vol 5, pp655–664

Palmer, W. C. (1965) 'Meteorological drought' (Research paper no 45), US Weather Bureau, Washington, DC

Patz, J. A., Hulme, M., Rosenzweig, C., Mitchell, T. D., Goldberg, R. A., Githeko, A. K., Lele, S., McMichael, A. J. and Le Sueur, D. (2002) 'Climate change – Regional warming and malaria resurgence', in *Nature*, vol 420, no 6916, pp627–628

Priestley, C. H. B. and Taylor, R. J. (1972). 'On the assessment of surface heat flux and evaporation using large scale parameters', in *Monthly Weather Review*, vol 100, pp81–92

Ramankutty, N. and Foley, J. A. (1999). 'Estimating historical changes in global land cover: Croplands from 1700 to 1992', in *Global Biogeochemical Cycles*, vol 13, no 4, pp997–1027

Rawls, W. J., Ahuja, L. R., Brakensiek, D. L. and Shirmohammadi, A. (1993) 'Infiltration and soil water movement', in D. R. Maidment (ed) *Handbook of Hydrology*, McGraw-Hill, New York, pp5.1–5.51

Redmond, K. T. (2002) 'The depiction of drought', in *Bulletin of the American Meteorological Society*, vol 83, pp1143–1147

Scanlon, B. R., Jolly, I., Sophocleous, M. and Zhang, L. (2007) 'Global impacts of conversions from natural to agricultural ecosystems on water resources: Quantity versus quality', in *Water Resources Research*, vol 43, W03437, p18

Service, R. F. (2004) 'As the West goes dry', in *Science*, vol 303, pp1124–1127

Shafer, B. A. and Dezman, L. E. (1982) 'Development of a Surface Water Supply Index (SWSI) to assess the severity of drought conditions in snowpack run-off areas', in *Proceedings of the Western Snow Conference*, Colorado State University, Fort Collins, Colorado, pp164–175

Sheffield, J. and Wood, E. F. (2007) 'Characteristics of global and regional drought, 1950–2000: Analysis of soil moisture data from off-line simulation of the terrestrial hydrological cycle', in *Journal of Geophysical Research*, vol 112, D17115, p21, doi:10.1029/2006JD008288

Sheffield, J. and Wood, E. F. (2008) 'Global trends and variability in soil moisture and drought characteristics, 1950–2000, from observation-driven simulations of the terrestrial hydrological cycle', in *Journal of Climate*, vol 21, pp432–458

Sheffield, J., Goteti, G., Wen, F. and Wood, E. F. (2004a) 'A simulated soil moisture based drought analysis for the United States', in *Journal of Geophysical Research*, vol 109, D24108, p19, doi:10.1029/2004JD005182

Sheffield, J., Ziegler, A. D., Wood, E. F. and Chen, Y. (2004b) 'Correction of the high-latitude rain day anomaly in the NCEP/NCAR reanalysis for land surface hydrological modelling', in *Journal of Climate*, vol 17, no 19, pp3814–3828

Sheffield, J., Goteti, G. and Wood, E. F. (2006) 'Development of a 50-year high-resolution global dataset of meteorological forcings for land surface modelling', in *Journal of Climate*, vol 19, pp3088–3111

Sheffield, J., Ferguson, C. R., Troy, T. J., Wood, E. F. and McCabe, M. F. (2009a) 'Closing the terrestrial water budget from satellite remote sensing', in *Geophysical Research Letters*, vol 36, L07403, p5, doi:07410.01029/02009GL037338

Sheffield, J., Andreadis, K. M., Wood, E. F. and Lettenmaier, D. P. (2009b) 'Global and continental drought in the second half of the twentieth century: Severity-area-duration analysis and temporal variability of large-scale events', in *Journal of Climate*, vol 22, no 8, pp1962–1981

Shukla, S. and Wood, A. W. (2008) 'Use of a standardized run-off index for characterizing hydrological drought', in *Geophysical Research Letters*, vol 35, L02405, p7, doi:10.1029/2007GL032487

Siebert, S., Döll, P., Hoogeveen, J., Faures, J.-M., Frenken, K. and Feick, S. (2005) 'Development and validation of the global map of irrigation areas', in *Hydrology and Earth System Science*, vol 9, pp535–547

Steinemann, A. C., Hayes, M. J. and Cavalcanti, L. F. N. (2005) 'Drought indicators and triggers', in D. A. Wilhite (ed) *Drought and Water Crises – Science, Technology, and Management Issues*, CRC Press, Boca Raton, Florida, US, pp71–92

Su, F. and Lettenmaier, D. P. (2009) 'Estimation of surface water budget of La Plata Basin', in *Journal of Hydrometeorology*, vol 10, no 4, pp981–998

Su, H., McCabe, M. F., Wood, E. F., Su, Z., Prueger, J. (2005) 'Modelling evapotranspiration during SMACEX: Comparing two approaches for local and regional scale prediction', in *Journal of Hydrometeorology*, vol 6, no 6, pp910–922

Su, Z. (2002) 'The Surface Energy Balance System (SEBS) for estimation of turbulent heat fluxes', in *Hydrology and Earth System Sciences*, vol 6, pp85–99

Svoboda, M. et al (2002) 'The Drought Monitor', in *Bulletin of the American Meteorological Society*, vol 83, no 8, pp1181–1190

Swenson, S. and Wahr, J. (2002) 'Methods for inferring regional surface-mass anomalies from Gravity Recovery and Climate Experiment (GRACE) measurements of time-variable gravity', in *Journal of Geophysical Research*, vol 107, no 2193, p13, doi:10.1029/2001JB000576

Tapley, B. D., Bettadpur, S., Ries, J. C., Thompson, P. F. and Watkins, M. (2004) 'GRACE measurements of mass variability in the Earth system', in *Science*, vol 305, pp503–505

Uppala, S. M. et al (2005) 'The ERA-40 re-analysis', in *Quarterly Journal of the Royal Meteorological Society*, no 131, pp2961–3012

USDM, 'US Drought Monitor' (2010) 'Explanation of the US Drought Monitor', www.drought.unl.edu/dm/classify.htm

Verdin, J., Funk, C., Senay, G. and Choularton, R. (2005) 'Climate science and famine early warning', in *Philosophical Transactions of the Royal Society B: Biological Sciences*, vol 360, no 1463, pp2155–2168, doi:10.1098/rstb.2005.1754

Wang, A. H., Bohn, T. J., Mahanama, S. P., Koster, R. D. and Lettenmaeir, D. P. (2009) 'Multimodel ensemble reconstruction of drought over the continental United States', in *Journal of Climate*, vol 22, pp2694–2712

Wilhite, D. A. (2000). 'Drought as a natural hazard: Concepts and definitions', in D. A. Wilhite (ed) *Drought: A Global Assessment*, Routledge, London, pp33–40

Wood, E. F., Lettenmaier, D. P., Liang, X., Lohmann, D., Boone, A., Chang, S., Chen, F., Dai, Y. J., Dickinson, R. E., Duan, Q. Y., Ek, M., Gusev, Y. M., Habets, F., Irannejad, P., Koster, R., Mitchell, K. E., Nasonova, O. N., Noilhan, J., Schaake, J., Schlosser, A., Shao, Y. P., Shmakin, A. B., Verseghy, D., Warrach, K., Wetzel, P., Xue, Y. K., Yang, Z. L. and Zeng, Q. C. (1998) 'The Project for Intercomparison of Land-surface Parameterization Schemes (PILPS) phase 2(c) Red-Arkansas River basin experiment: 1. Experiment description and summary intercomparisons', in *Global and Planetary Change*, vol 19, pp115–135

WMO (1969) *Manual for Depth-Area-Duration Analysis of Storm Precipitation*, series no 237.TP.129, World Meteorological Organization, Geneva, Switzerland, p114

Wood, A. W. and Lettenmaier, D. P. (2006) 'A testbed for new seasonal hydrological forecasting approaches in the western US', in *Bulletin of the American Meteorological Society*, vol 87, no 12, doi:10.1175/BAMS-87-12-1699, pp1699–1712

Zhang, Y. K. and Schilling, K. E. (2006) 'Increasing streamflow and baseflow in Mississippi River since the 1940s: Effect of land use change', in *Journal of Hydrology*, vol 324, pp412–422

5

Palaeo-drought: The Occurrence of Drought over Past Millennia

Introduction

Our knowledge of drought over the 20th century is generally good because it is based on the instrumental record (for example, data from weather stations), direct written archives and living memories. Drought during the 20th century is examined in Chapters 6 and 7. To put this analysis into perspective, this chapter gives an overview of drought over the past several millennia, as gleaned from palaeoclimate data. This is important because it provides a benchmark against which current and future projected changes in drought can be assessed. Although many droughts that occurred during the 20th century were severe, they do not represent the full range of severity of historic droughts, and thus what might be expected in future years. Long-term palaeoclimate records from before the industrial revolution provide the opportunity to make better estimates of the range of natural variability of the climate system, and therefore the potential for extreme climate events, such as devastating droughts. Given that drought has occurred over millennia as a natural part of climate variability, and will continue to do so into the future, with or without anthropogenic climate change, the palaeoclimate record can be used to put droughts of the 20th century record and projections for the 21st century into context.

Instrumental data existed generally over the past century, and expanded in coverage and density with the establishment of national meteorological networks, first in Europe and North America, and later in Africa, driven by colonization. For a few individual stations or regions, there are data series that extend much further back, to even the 17th century, including the central England temperature series from 1659, and precipitation series for Kew, England, from 1697, and for Paris from the 1680s. Before this time and elsewhere, climate information must be extracted from non-meteorological sources, including human archives (such as documents) and natural proxy records (such as tree rings). Much of these data relate to the period of pre-anthropogenic influence on climate (generally before the last 150 years), and as such give information on climate variability due to natural effects such as solar variability, volcanoes and ocean–atmosphere interactions. The strength of palaeoclimate data is that they can tell us something about the variation in climate and occurrence of drought over many millennia. Tree rings, ice cores, dune, lake and sea sediments all contain useful information, from which we can infer the climate of the time of tree growth, sedimentation or ice formation (see Chapter 4 for details of how palaeoclimate data are used to reconstruct drought). There now exist, for example, tree-ring records that go back several millennia, with current research looking to go back 10,000 years (Briffa and Matthews, 2002). Sediment and ice cores go even further back, with the longest continuous Antarctic ice core of more than 3000m in depth dating back about 800,000 years (Jouzel et al, 2007). Ice cores are generally limited to providing reconstructions of temperature and atmospheric constituents, and thus do not give a full picture of drought, as this requires information on available water (e.g. precipitation, lake levels and streamflow).

Of great interest are the emerging and related sciences of historical and documentary hydrology and climatology, which are based on pre-instrumental man-made records of past conditions, and are helping to uncover past extreme events and place them in the context of the present (Brázdil et al, 2005; Brázdil and Kundzewicz, 2006). These documentary sources of data may come from historic written documentation (such as an official record, a manuscript, a letter, a newspaper or a book), or from an image or artefact (a piece of art, an inscription, a monument or a water mark). The data may be direct, for example, a narrative of the prevailing weather or climate, or may be indirect, describing the impact of climate anomalies. Examples of indirect or proxy documentary data are descriptions of snow cover and river ice, water levels and dry periods, or the state of the vegetation. They may also describe economic and agricultural factors, such as crop yields and prices, or be records of disasters such as floods, droughts and famines. Indirect data may also reveal subtle representations of climate via social responses, often through religious ceremonies. They can also be very important, in terms of extremes in the context of global warming, because they provide the only evidence of the socio-economic impacts of extreme events such as droughts. Although these records are typically not directly related to climate features, compared to natural proxies, they are often high resolution in time (down to the day of the year) and space (specific towns, cities or waterways) and describe actual socio-economic, hydrological and biological impacts. Given this, documentary evidence has a number of drawbacks and uncertainties that need to be taken into account, most notably the discontinuity of records and the subjectivity of cultural observers and reporters.

As we go back in time, however, there is more uncertainty about how drought has occurred, because the data are more unreliable, cover sparser areas and are not direct measurements of drought. This problem is clearly seen in Plate 5.1, which shows the distribution of palaeoclimate records hosted in the US National Climate Data Center (NCDC) palaeoclimate database. Although there are several hundred records, these are unevenly distributed and virtually non-existent in the tropics and many parts of Asia. Only a small subset of these records was used in the latest report from the Intergovernmental Panel on Climate Change (IPCC, 2007) to assess changes in climate over the past millennium. The further back in time we go, the records become increasingly uncertain, and it is more difficult to reconstruct global and regional climate and hydrological cycles using data from a handful of locations. Drought must also be inferred from surrogate or proxy information. Palaeoclimate data are not a direct measurement of drought or the primary variables of precipitation, temperature and water availability. Instead, these variables of interest must be inferred by translating the raw data into something more tangible. For example, raw tree-ring records may be scaled to account for variations in tree growth with age, and then calibrated to match instrument-based records when they overlap in time, such as the Palmer Drought Severity Index (PDSI; Chapter 4). Other factors that increase uncertainty include external influences, which may distort the record after the event, and the challenges associated with trying to tease out climate variables from a record that accumulates all weather patterns over time. Thus palaeoclimate data cannot provide a snapshot of, say, precipitation in some distant epoch, but gives an overview of the prevailing climate over a period of time, generally from annual to centennial resolution.

For as long as the climate has been varying, drought has occurred in some form, somewhere in the world. Some of these events were long and severe enough to leave a mark on the vegetation, reduce lake levels, or even move sand dunes. Drought has also persistently affected human activity in all parts of the world, and in some cases has hampered development. Because of the central role that water plays in civil society, drought is thought to have been a factor in the downfall of ancient civilizations (e.g. Hodell et al, 1995). Archeological evidence corroborates this fact, and can be explained by climate reconstructions. Although proxy records exist for the last million years and earlier, and cover a number of glacial–interglacial periods, we focus on the Holocene epoch (past 12,000 years), and in particular the last 1000 years. This period covers the main era of human civilization, including the emergence of agriculture, shift to permanent dwellings and the development of arts and scientific knowledge, and thus the greatest impacts of drought. It is during this period that the human influence on the environment and climate, in terms of land-use change and atmospheric emissions, has become noticeable and increasingly important. The palaeoclimate record in the Holocene also provides increasingly significant data, in terms of dating accuracy, temporal and spatial resolution, and overlaps with the instrumental record in the past hundred years. Crucially, it provides the most evidence of variability in a warm climate that is similar to the present day, and thus an expression of what might happen in the future against the background of human-induced changes.

After giving an overview of climate variability over the Holocene, we describe some of the major climate cycles and events, and their likely causes, loosely clustering them according to time period, as dictated by the major climate swings of the era and the availability and type of palaeoclimate data. We also focus on the impacts on human history, as gathered from historical and archaeological evidence. The geophysical evidence for droughts in periods closer to the present (the past several centuries) is bolstered by the increasing availability of written records, which makes for some interesting inferences of climate and drought variability. Importantly, we will see that some of the major drought events of the 20th century (Chapter 7), which caused so much human hardship, are not so unusual in comparison to the past few centuries. The severity of recent droughts pales into insignificance when compared with century-long dry periods during the Holocene geological epoch. However, recent drying in some parts of the world is unprecedented, when placed in the context of the last few thousand years.

Climate of the Holocene

We are currently in an ice age, as defined by the presence of permanent ice sheets, located at the poles and over Greenland. There have been five major ice ages in the history of the Earth, lasting for tens to hundreds of millions of years, which were probably forced by a combination of changes in the Earth's orbit around the sun, solar output, atmospheric composition, ocean circulation, volcanic activity and tectonic activity. In an ice age, there are glacial periods (markedly cooler temperatures and expansions of ice sheets) and shorter interglacial periods (warmer temperatures and ice sheet retreat, but not a disappearance of ice sheets). These have approximate cycles of 26, 41 and 94ky (ky = thousand years), as driven by changes in solar heating caused by variation in the Earth's orbital parameters, know generally as Milankovitch solar cycles, after the Serbian astronomer Milutin Milankovitch. Figure 5.1 shows the temperature record of the past 800,000 years, as reconstructed from ice core data from the Antarctic. This remarkable record shows the fluctuations of the glacial and interglacial periods. The last glacial period (usually referred to as the 'last ice age', but formally called the Weichselian period) peaked at around 21kya (kya = thousand years ago), with a maximum ice sheet extent as low as 40°N in North America, temperatures of about 5–6°C lower and sea levels of up to 150m below those of the present day. This period ended about 10–15kya, as shown by the last 25ky of the Antarctic ice core record in the lower panel of Figure 5.1, and we are now in an interglacial period. Based on calculations of the Earth's orbit, the next glacial period is due to start in 30ky, although this is likely to be delayed if the high levels of current greenhouse gas concentrations persist in future years. As well as temperatures being colder during the last glacial era, the temperature was more variable. Swings of 5–10°C (known as Dansgaard-Oeschger events) occurred over years and decades rather than millennia (Dansgaard et al, 1993), forcing humans to become more adaptive to climate through frequent migration (Fagan, 2008). The current interglacial period is termed the Holocene geological epoch. It is during this time, driven by a more stable climate, that agricultural practices have emerged and civilizations have flourished (Burroughs, 2007).

Despite the relative stability of climate during the Holocene, to qualify it as stable would be a great over-simplification of the temporal and regional variations that have occurred, especially in the first half of the epoch, as driven by orbitally forced changes in solar irradiance. The warming from the last glacial period was not uniform across the world, being punctuated by a sudden cooling around 11.3kya, known as the Younger Dryas cold period, and then tapering off about 5–6kya (Figure 5.2). This period is sometimes referred to as the *climate* or *thermal optimum* (with temperatures about 1–2 °C higher than present), although the evidence indicates that this was not globally simultaneous, with large regional differences, and even cooler temperatures than today in parts of the tropics. Figure 5.2 shows this variation in temperature records from proxies across the world: two ice cores from Antarctica, one from Greenland, and lake and marine sediments in Africa. The conducive

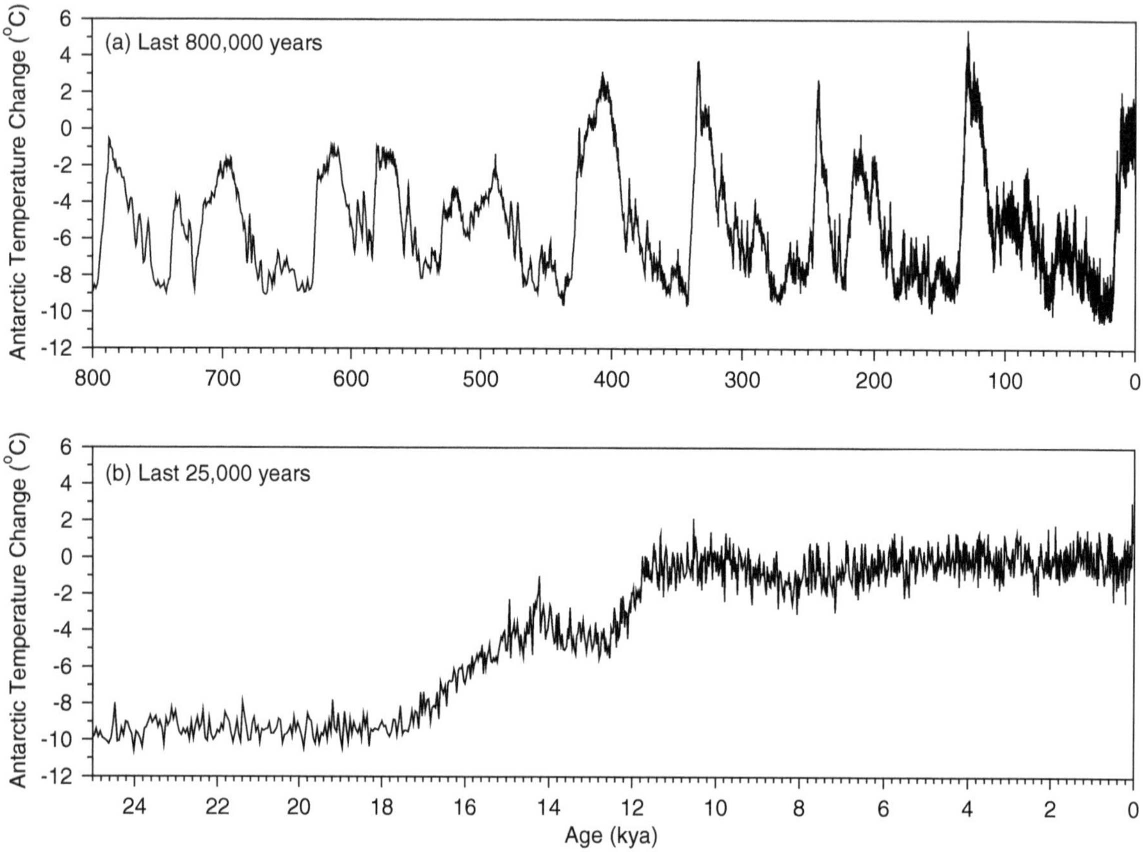

Figure 5.1 *Antarctic temperature estimates for the past (a) 800,000 and (b) 25,000 years from the EPICA Dome C Ice Core*

Source: Jouzel et al, 2007; EPICA Dome C Ice Core 800KYr Deuterium Data and Temperature Estimates. IGBP PAGES/World Data Center for Paleoclimatology Data. Contribution Series # 2007-091 NOAA/NCDC Paleoclimatology Program, Boulder CO, USA. US National Climatic Data Center World Data Center for Paleoclimatology, www.ncdc.noaa.gov/paleo/paleo.html

climate of the mid-Holocene is associated with the northern expansion of temperate forests and the retreat of glaciers (MacDonald et al, 2000; Bigelow et al, 2003), along with the intensification of the world's monsoon systems (Liu et al, 2004; Zhao et al, 2005). It also coincides with the height of ancient civilizations in Africa and Asia. There is also evidence to support a number of long-term and abrupt changes in climate during the Holocene, forced generally by changes in solar orbit and volcanic activity, as well as regional changes in response to the receding ice sheets from the last glacial period. These climate responses include changes not only to global mean temperature, but also to regional temperature and precipitation, the dynamics of monsoon systems and the El Niño Southern Oscillation (ENSO). The Holocene has been punctuated by a series of major abrupt climate deviations. Some of these have been purported to occur on a roughly 1500-year cycle, possibly related to weaker solar activity, which we will refer to as Holocene events (Bond et al, 1997, 2001; Rohling and Palike, 2005). These events are marked by a cooling of the Atlantic, bringing massive ice excursions from the North Atlantic, which left their signature in, among other things, deposited debris in ocean bed sediments. The result was a colder (1–2°C lower) but also drier climate, as the cooling reached as far down as the tropical Atlantic, thereby reducing available moisture from evaporation and instigating drought on land. These

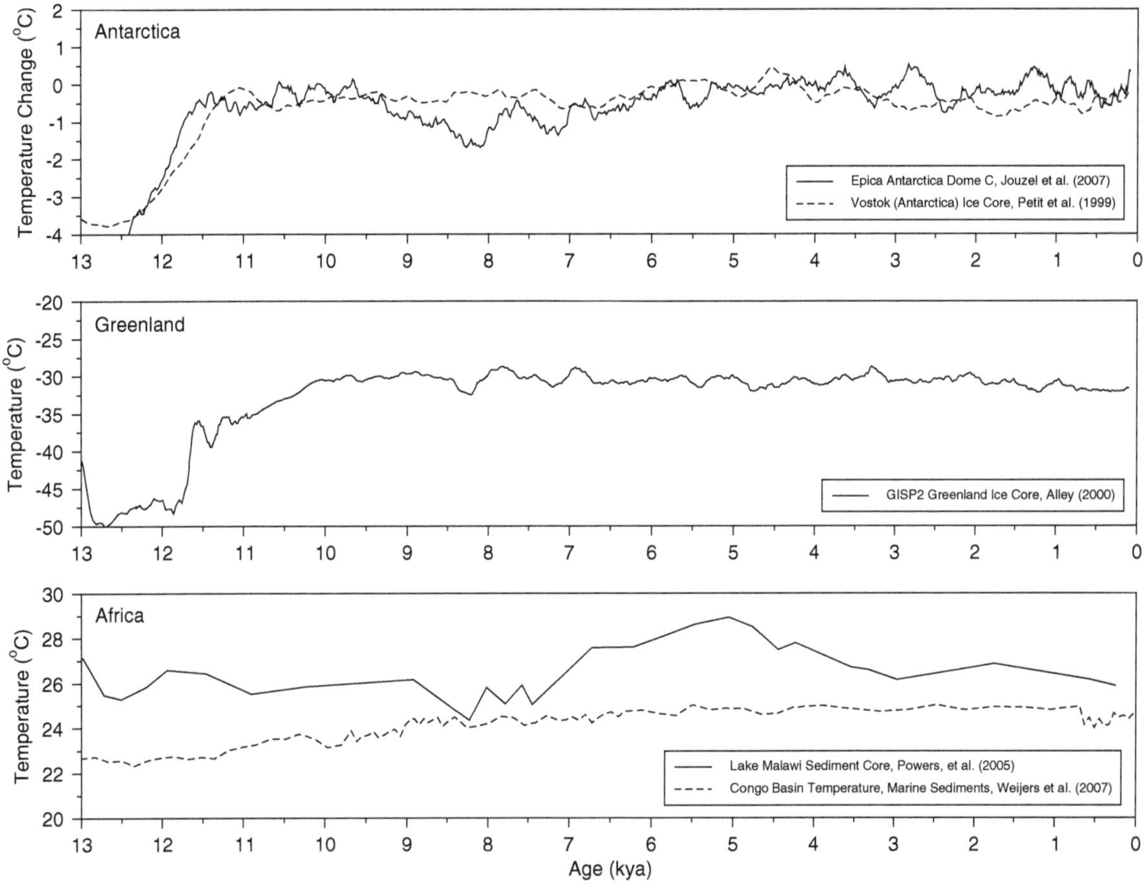

Figure 5.2 *Holocene temperature records from palaeoclimate proxy data from Antarctica, Greenland and Africa*

Source: The data are from the EPICA Dome C ice core record (Jouzel et al, 2007; EPICA Dome C Ice Core 800KYr Deuterium Data and Temperature Estimates.), the Vostok ice core record (Petit et al, 1999; Vostok Ice Core Data for 420,000 Years), the Greenland GISP2 ice core record (Alley, 2000; GISP2 Ice Core Temperature and Accumulation Data), Lake Malawi sediment core record (Powers et al, 2005; Lake Malawi TEX86 Surface Temperature Reconstruction) and the Congo River basin marine sediment core record (Weijers et al, 2007. Congo River Basin 25KYr MBT/CBT Data and Temperature Reconstruction). IGBP PAGES/World Data Center for Paleoclimatology, NOAA/NCDC Paleoclimatology Program, Boulder CO, US, www.ncdc.noaa.gov/paleo/paleo.html.

events are similar to the larger magnitude Dansgaard–Oeschger and Heinrich abrupt climate events that occurred during the last glacial period. However, the validity of the Holocene cycles and the forcing from solar activity is questionable, and it appears that independent climate processes (from oceanic and atmospheric origins) may have actually acted in concert to force these events rather than as a result of a fundamental cyclical process (Schulz et al, 2004; Moros et al, 2006; Debret et al, 2007).

Since then there have been two well-known, but lower-magnitude climate deviations that dominate the record of the last 2000 years: the Medieval Warm Period (MWP), from roughly the 10th to the 14th century, and the Little Ice Age (LIA) from the 13th to the mid-19th century. These events have been associated with variations in the ENSO and the North Atlantic Oscillation (NAO) (Mann et al, 2009). The time series in Figure 5.3 shows a set of reconstructions of northern hemisphere temperature anomalies (relative to 1961–1990) for the last 2000 years (Jones and Mann, 2004). Although there is some disagreement between alternative reconstructions, which are based on different sets of palaeoclimate data and methods to extrapolate from point records to the whole hemisphere, they show warming of the MWP and a gradual descent into the LIA afterwards. As Plate 5.1 shows, the sparsity of evidence means that the weak signature of the MWP in the southern hemisphere and global records is difficult to interpret. The most obvious feature of Figure 5.3, nevertheless, is the rapid rise in temperature over the past 100 years, which is unprecedented within the past 1000-year record and clear for both hemispheres. We will return to this point in Chapter 8, where we discuss changes in drought within the period of anthropogenic warming and look ahead to the future. Data from ice cores and other palaeoclimate records

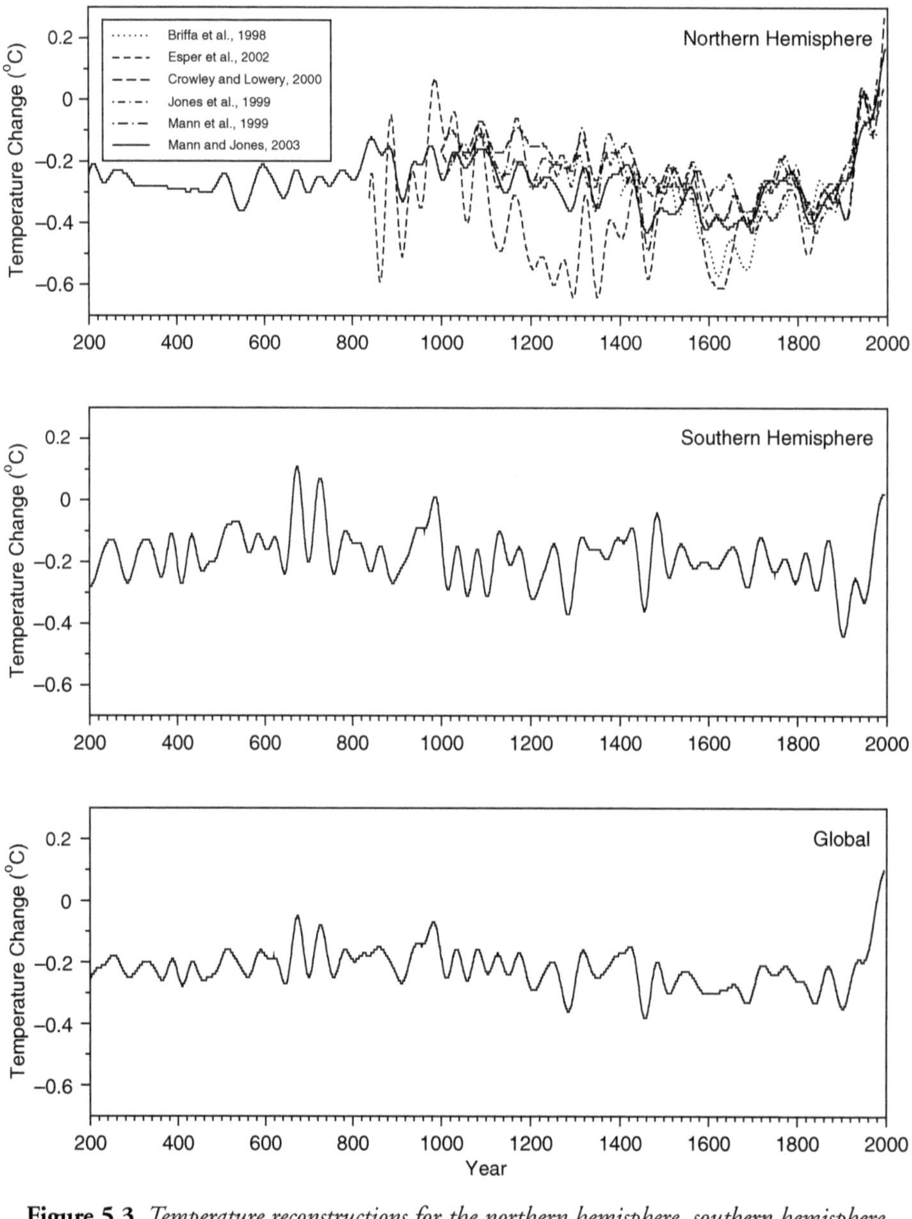

Figure 5.3 *Temperature reconstructions for the northern hemisphere, southern hemisphere and globally for the past 2000 years*

Source: Jones and Mann (2004), downloaded from the US National Climatic Data Center World Data Center for Paleoclimatology at www.ncdc.noaa.gov/paleo/paleo.html

show that the MWP was generally limited to the North Atlantic region and that in fact this period was possibly cooler than today at the global scale. Nevertheless, there is evidence from around the world to indicate that the MWP was pitted by many regional climate anomalies, but the lack of synchronicity of warming makes it difficult to define the MWP as a global event. It is probably more apt to describe the MWP as the Medieval Climate Anomaly (MCA) (Mann et al, 2009). The LIA that followed the MWP is also debatable as a global phenomenon, and its timing is highly regionally specific, but several lines of evidence from around the world, including southern hemisphere ice core and tree-ring data, indicate that it did indeed extend beyond the northern hemisphere.

Drought in the Holocene

Although Holocene temperature records, as derived from a variety of independent sources, are becoming fairly stable and more reliable, especially for the past 1000–2000 years, as more data becomes available, these records are not necessarily useful records of drought. As we have seen in Chapter 3, drought occurs as a result of a lack of precipitation and/or high demand from the atmosphere through evaporation, or from anthropogenic use, such as groundwater abstraction or river diversion. Thus hydroclimate indicators, such as rainfall, evaporation, river and lake levels, and their proxies, are needed to characterize drought in pre-instrumental times. Often proxy records do not yield separate inferences for temperature and precipitation, but a mixture of both, and thus can be difficult to interpret in terms of drought. Furthermore, most records tend to be temperature-orientated. Therefore the palaeoclimate record of drought has to be extracted from a diversity of sources and interpreted in light of the multiple processes that affect its variability.

In general, the early to mid-Holocene was wetter than today with evidence that shows, for example, enhanced summer monsoons across the tropics as a result of orbital changes and increases in land–sea temperature differences (Fleitmann et al, 2003; Wang et al, 2005). Since then there has been a general drying. This is characterized by changes in monsoon activity and atmospheric circulation during the mid- to late Holocene transition, with a northward shift of the ITCZ and intensification of monsoonal activity (Kutzbach and Liu, 1997; Marchant and Hooghiemstra, 2004), leading to a drying of the monsoonal regions of Africa and Asia over the late Holocene (Gasse, 2000; Nicoll, 2004). A number of studies have identified climate anomalies and periods of likely drought that have interspersed the Holocene and extended across continents. Some of the major drought events, of the order of centuries, coincide with the most recent series of Holocene climate events, which occurred at about 8.2, 5.9, 4.2, 2.8 and 1.4kya, and a handful of earlier, but less distinct, events. For example, data from pollen and lake records in a wide swathe, from West Africa through East Africa to India and Tibet corroborate major drought periods during 11–9.5, 8–7 and 4–3kya and a less intense dry period in about 6kya (Gasse and van Campo, 1994). Again, solar variability explains longer-term variations, but short-term fluctuations are linked to variability in North Atlantic sea surface conditions.

The 8.2kya event was an abrupt cooling event of the northern hemisphere that has been detected in multiple independent proxy records, from the Arctic to the tropics (Rohling and Palike, 2005). The event lasted from several decades to centuries, and is thought to have been caused by rapid drainage of the prehistoric Canadian glacial Lake Agassiz into the Hudson Bay, and the forcing of a decrease in the Meridional Overturning Circulation (MOC) in the North Atlantic. The MOC is a major part of the global oceanic circulation pattern and normally brings warm salty water to the North Atlantic and with it milder temperatures. The 8.2kya event is also associated with a series of dry events across the globe, identified by reduced tree-ring widths in central Europe, southerly migration of the ITCZ, reduction in monsoon-fed African lakes, and monsoon records in India and East Asia. Dry conditions were felt as far afield as the southern tip of South America (Mancini et al, 2005). Several of these records indicate anomalies of several centuries, some of which began before the abrupt North Atlantic anomaly, indicating that the 8.2kya event may be a major event embedded in a multiple century scale of climatic deterioration.

The intensity of the event at around 5.9kya was notable because it led to the permanent desertification of the Sahara region. Previously, during 20–12.5kya, there were extreme dry periods when the Sahara migrated southwards into the savannas of the Sahel region. Following that, the region experienced wetting when the Sahara Desert was quite the opposite of its current state. There are suggestions that abundant vegetation and life was present, as gleaned from records showing that lake and river flows were much higher than normal between 9 and 5kya. This includes evidence from Lake Chad in the central Sahel, which was about 35 times greater in area at its maximum around 7.5 to 5kya. Overall, palaeoclimate data across the Sahel, from various stream, dune and lake records, indicate that the Sahel droughts of the 1970s and 1980s may not have been unique in their severity over the past tens of thousands of years (Butzer, 1983). They appear to be of little significance in comparison to the centennial to millennial shifts in climate that have affected this region. More significantly, the record for the last 2500 years indicates variations in climate that are similar to the present, and at least six periods of short-term (less than a decade) drought since 1400 AD, which are comparable to the 20th century Sahel droughts. A recent study on lake sediments in Ghana (Shanahan et al, 2009) corroborates this evidence, showing that severe droughts lasting decades to centuries were the norm over the past 3000 years in West Africa. Sediments from the bottom of Lake Bosumtwi contain deposits of soil and organic matter that, through analysis of oxygen isotopes, reflect variations in climate. The identified droughts, which fluctuate on timescales of 30–40 years, may have been related to the Atlantic Multidecadal Oscillation (AMO; see Chapter 3).

The general drying that occurred between about 6 and 4kya, was characterized by a number of abrupt climate changes globally, including an increase in northern hemisphere sea ice and changes in monsoonal regions. The shift to drier conditions is seen in records over tropical Africa, which indicate that rainfall started to decrease after about 5.8kya (de Menocal et al, 2000), and across many sites between 5 and 3kya in Africa and South America (Bonnefille and Chalié, 2000; Marchant and Hooghiemstra, 2004). The overall drying was characterized by many short-term and severe dry events (Gasse, 2000; Schilman et al, 2001).

The dry period around 4.2kya (Holocene event 3) is particularly interesting because it had profound impacts across the globe (coming towards the end of the Holocene climate optimum) and is linked with the demise and migration of many ancient civilizations (Booth et al, 2005). There is evidence to indicate that changes were felt as far as the mid-latitudes affecting Europe and North America (Booth et al, 2005), whereas high latitudes were wetter than normal. Some of the most profound impacts were felt across North Africa and the Middle East. A multitude of different lines of evidence from various locations around the region are consistent with severe drought events, including cave deposits (speleothems), marine sediments in the Arabian and Red Seas, and even ice cores from Mount Kilimanjaro in Kenya. The Akkadian empire was one of the series of ancient civilizations that flourished in Mesopotamia (present-day Iraq and Syria) between the Tigris and Euphrates rivers. Evidence such as dust deposition within marine sediments and archaeological data (including the lack of archaeological data from this period, suggesting a civilization collapse) indicate that a century-long drought may have caused the abandonment of the agricultural areas in both the rain-fed north and irrigated alluvial plain between the rivers (Weiss et al, 1993; Cullen et al, 2000). Studies estimate that rainfall decreased by 30 per cent during this period. It is suspected that the event was caused by the cooling of the North Atlantic, which is linked to climate variability in Mesopotamia through data connecting the Mediterranean winter rainfall to the headwaters of the Tigris and Euphrates. Other civilizations were severely impacted during the same time period, such as the Old Kingdom of Egypt in the Nile valley (Stanley et al, 2003), which may have dissipated (Salzmann and Waller, 1998; Gasse, 2002; Nicoll, 2004). Changes are also recorded as far afield as Pakistan, India, Southeast Asia and China (Wu and Liu, 2004; Gao et al, 2007). Oxygen isotopic records from caves in southern China, which are proxies for the Asian monsoon over the past 9000 years (Wu and Liu, 2004), reveal a series of eight weak monsoon periods that lasted between one and five centuries and are synchronous with the North Atlantic cooling events, including the 8.2kya event and the 4.2kya event. The 4.2kya event correlates well with the demise of the Neolithic cultures around central China, as indicated by fewer archaeological sites, less sophisticated materials and structures, as well as shifts from dry farming to pastoralism.

Around 3kya across the near east and eastern Mediterranean, the Bronze Age transition into what are termed the ancient Dark Ages may have been accompanied by climate change and drought, including the collapse of the Mycenaean kingdoms, the Hittite empire in Syria and the Egyptian Empire. Although a combination of many factors including migration, societal collapse and outside incursions and raids are likely to have been responsible, there is evidence that drought may have played a part (Weiss, 1982; Fagan, 2003).

Much progress has been made in recent years in reconstructing drought over the western and Midwest US, especially from tree-ring records for the past 2000 years, and this is discussed later. However, palaeo-drought in the eastern US has been studied less. Springer et al (2008) is an example of one of the few studies that have looked at the geological record in the east. Their speleothem record from the Buckeye Creek Cave in West Virginia in the eastern US is arguably one of the most detailed records of climate variation in eastern North America over the last 7000 years. A combination of measurements of strontium (a trace metal found in soils), which are of higher concentration in dry periods, and isotopic carbon ratios, which reflect lower biological activity during dry periods, were used to reconstruct climate records. These indicators are excellent recorders of drought, which show linkages to the 1500-year cycles of Holocene events and especially severe events of decade to century duration between 6.3 to 4.2kya.

The variation in hydrology on millennial timescales over the Holocene can be summarized, in part, by the merged lake status database of Viau and Gajewski (2001) shown in Plate 5.2. These data are an amalgamation of three databases of lake-level proxy data for the world and span the Holocene at 3ky intervals. The spatial coverage of the data is generally lower further back in time, but nevertheless these records highlight the drying of the Sahara, Middle East, Central Asia, east and southern Africa between 6 and 3kya, and the drying in the southwestern US since 9kya. However, the Canadian prairies have flipped from being wetter about 12kya to drier during 9–6kya and then wetter since then.

The Last 2000 Years

As we move into the past two millennia, geophysical evidence is increasingly corroborated by archaeological and written records, and it is often the lack of such human evidence that indicates drought-induced hardship. Some of the best-preserved data comes from the remains of the great civilizations and there are fascinating theories of their demise and linkages to drought. Probably the most famous example is that of the Mayans, which is discussed later, but droughts have been linked to the demise or dissipation of many other cultures and civilizations, which once flourished and provided great advances in agriculture, science, engineering and the arts. Figure 5.4 maps some of the major drought events of the past 2000 years before the instrumental era that are discussed below, and highlights some of the possible consequences.

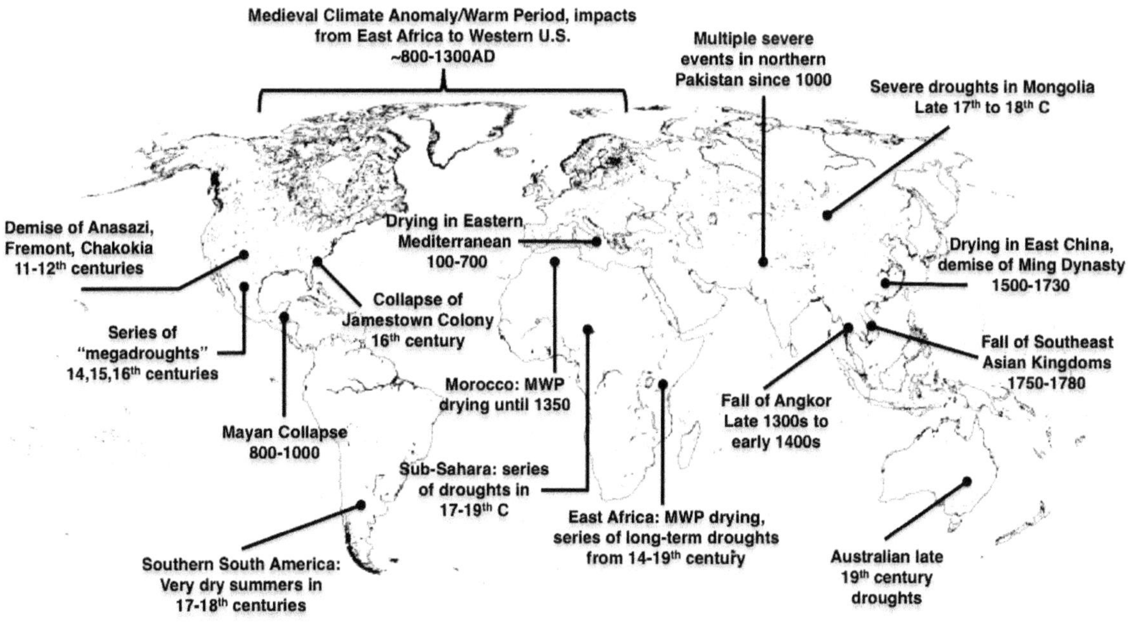

Figure 5.4 *Examples of major drought events in the past 2000 years*

We start with the Roman Empire, which existed in various forms for over two millennia, and is famous for its documented decline and fall. This generally refers to the fall of the Western Roman Empire, which lasted from 27 BC to 476 AD, and was preceded by 500 years of the Roman Republic. Its legacy and that of the subsequent Byzantine Empire or the Eastern Roman Empire is enormous, covering language, law, engineering and architecture. The fall of the Western Empire that stretched from Italy through to northern England is attributed to a variety of causes, including an over-stretching of resources across the vastness of the empire that induced untenable taxation and economic strain. Increasing immigration, incursions and eventually influence by Germanic tribes and others, especially to the east, may also have been important. There is also mounting evidence and interest in the possible influence of environmental factors such as deforestation and drying over several centuries that may have played a role. For example, analysis of the chemical composition of speleothems in caves near Jerusalem indicates a gradual drying over the course of the decline of the Roman and Byzantine Empires. Using an ion microprobe and laser spectroscopy, Orland et al (2009) were able to get exceptionally high-resolution (down to weekly changes) estimates of rainfall in the eastern Mediterranean for 1–2kya. The probe has the capability of distinguishing layers 1/100th mm thick. The record indicates that there was a gradual drying between 100 AD and 700 AD, dropping by some 50 per cent, with dramatic declines at 100 AD and 400 AD.

In Central America, the Mayan civilization flourished from 3kya (2600 BC to 1200 AD), originating in the Yucatan Peninsula and spreading across eastern Mexico, Guatemala and Belize, and reaching a peak between 600 and 800 AD. This civilization developed a sophisticated scientific and social culture based in a series of cities, of which the ruins are still being uncovered now. The Mayans also built reservoirs, canals and dykes to store and divert water for drinking and agriculture, enabling large city populations to be supported. However, in the ninth century, the Mayan civilization suddenly collapsed, bringing an end to the Classic phase. By the time the Spanish had arrived in 1530 AD, the population had dwindled to about 30,000 from a height of perhaps several million (Pearce, 2006), whom by this time had been assimilated into other Mesoamerican civilizations. Although the collapse is an archaeological mystery, palaeoclimate data from tree rings and lake cores suggest a series of devastating droughts during the ninth century may have influenced its demise. Reconstructions of climate from lake sediments across the Yucatan Peninsula have been made using oxygen isotopes in shells of gastropods and ostracods (Hodell et al, 1995; Curtis et al, 1996), which are enriched with heavier oxygen isotopes during periods of intense evaporation (thus giving clues to the balance between precipitation and evaporation) (Figure 5.5). These results were corroborated by the amount of sulphur in the sediment, which precipitates when evaporation intensifies the amount of sulphur in the lake water. The reconstruction showed that the driest period was between 800 and 1000 AD. More recent reconstructions (Haug et al, 2003; Hodell et al, 2005) have been able to identify the most intense droughts in 810, 860 and 910 AD, lending weight to the argument that severe drought may have contributed to the downfall of the Mayan civilization. Although the data suggest that the climate returned to a more normal state around 1000 AD, the Mayans never rebounded to their former glory in the Classic phase.

Great scientific advances have been made in recent years in understanding drought variation over the US West for the past 2000 years (Cook et al, 2007). This has been prompted by a number of long-term tree-ring-based reconstructions and some perceptive climate modelling work that has allowed the palaeoclimate data to be put in the context of the wider climate system. Of particular interest in the West is how this relates to possible future warming and the impacts to water supply, which is an already dwindling resource that is becoming ever more strained. The palaeoclimatic record for North America suggests that droughts of similar severity to that of the 1930s and 1950s have occurred several times over the past few hundred years and that there have been droughts of much greater severity as early as 500 years ago. These records include salinity data from Moon Lake in North Dakota in the Northern Great Plains of the US (Laird et al, 1996). The data are based on the type of diatoms (phytoplankton) found in layered lake sediments, some of which prefer more saline water than others, which in turn reflects drought variability. The record indicates spells of persistent and severe drought during 200–370, 700–850, and 1000–1200 AD, with a relatively abrupt

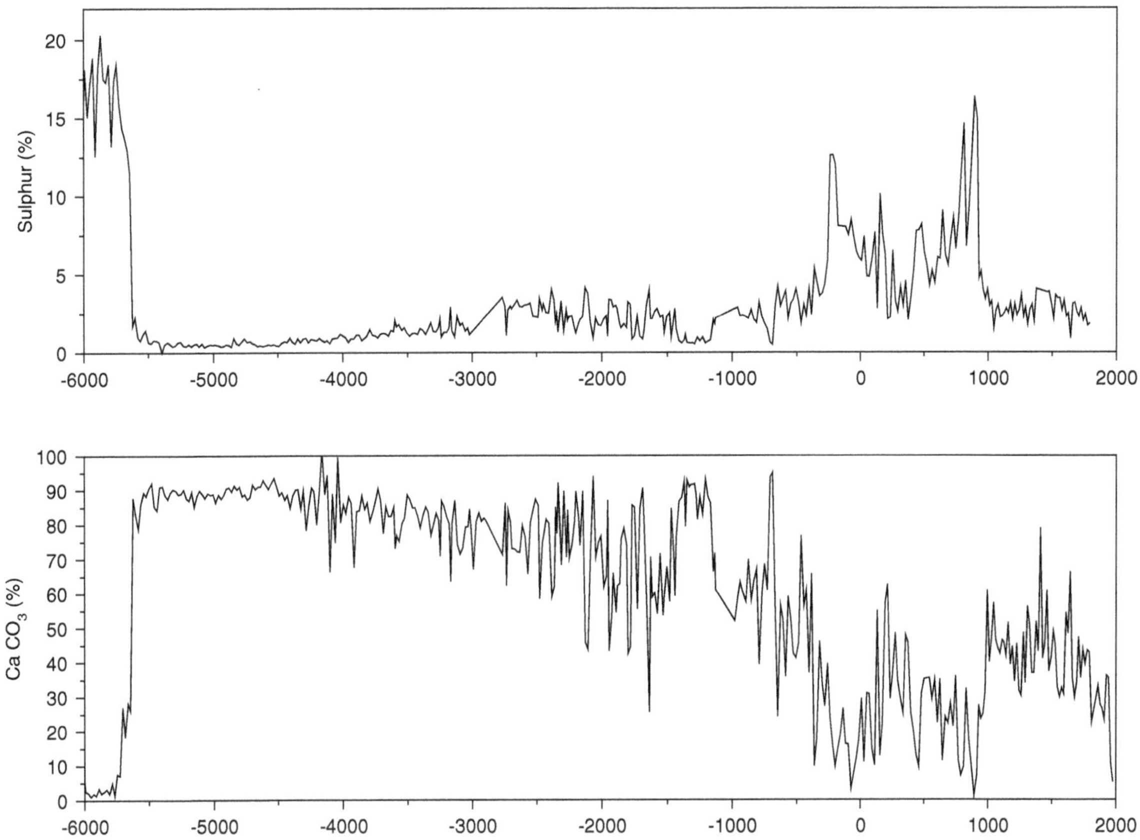

Figure 5.5 *Time series of per cent calcite ($CaCO_3$) and sulphur from sediments of Lake Chichancanab in the Yucatan Peninsula of Mexico as proxies for the climate of the region*

Note: The drop in calcite and corresponding peak in sulphur concentration is indicative of intense evaporation and coincides with the collapse of the Classic Mayan civilization in the ninth century (1100ya).

Source: Hodell et al (1995), downloaded from the NCDC paleoclimate data website

switch to wetter conditions since 1200 AD. In fact, the data suggest that since that time, the Northern Great Plains of the US have experienced a wetter climate, with no drought events coming close to these previous ones. In New Mexico, some of the longest-living trees on Earth have given clues about climate in the region, and have revealed a similar story of persistent droughts that dwarf the 1950s drought, which was probably the most severe in living history. Tree-ring samples from a Douglas fir which is over 1250 years old, in the El Malpais volcanic field, and fossil wood from the same area have provided a 2100-year record going back to 136 BC (Grissino-Mayer, 1996). Again, the most severe drought of the 20th century appears to be fairly typical when compared to the last three centuries, and is minor in comparison to the full record, including a multidecadal drought towards the late 16th century.

The Last 1000 Years

Studies have shown that the MWP was a period of warmth during 800–1300 AD of about 1°C higher than present, mainly centred on the North Atlantic region and impacted from North America, through southern Greenland, Europe and northern Eurasia. Palaeoclimate data indicate that the MWP was driven by La Niña-like cool conditions in the eastern tropical Pacific, coupled with the warm phase of the AMO

(Seager et al, 2007) and positive phase on the NAO/AO (Mann et al, 2009). For some, the MWP brought opportunity. Warmer climates allowed the Vikings to colonize Greenland and establish colonies as far west as Newfoundland, although the subsequent cooling of the LIA drove them back to Europe. For other peoples, the MWP meant significant drought and hardship.

Long periods of warmth associated with the MWP have been noted across the US, and particularly in the West. These compounded into a set of multidecadal droughts, the most severe of which were during the mid-12th and mid-13th centuries (Figure 5.6). A network of tree-ring chronologies for the western US has allowed the exploration of climate variation over multiple millennia (Hughes and Graumlich, 1996), some of which is responsive to winter and summer precipitation, so gives details of seasonal changes as well. Some of the identified events dwarf anything that has happened during the instrumental record, and may have been associated with the dispersion or even collapse of cultures in the Midwest and the lower Mississippi of the US, including the Anasazi, Fremont and Cahokia cultures. The Anasazi, or ancient Puebloans, were cliff dwellers in the Four Corners region of the US; the Fremont were their farmer-forager neighbours from what is now present-day Utah and western Colorado; and the Cahokia farmed the Mississippi River floodplains and valleys (Benson et al, 2006; Kloor, 2007). Of note are the dry conditions that persisted for several decades, including what is termed as a 'megadrought' during 1135–1180 AD, which can be seen in the reconstructed drought time series in Figure 5.6a for the western US, and in Figure 5.6b for Upper Colorado. This is associated with the collapse of agriculture and social order, and may have contributed to cultural demise. The Anasazi abandoned their canyon homes in New Mexico for the Mesa Verde cliffs in Colorado. The Cahokia culture, who were famed for their mound building, also collapsed as the population abandoned its settlements and gradually dispersed. By the end of the 13th century, after another megadrought (1276–1299 AD), the upheaval and violence had grown so much that peoples across the region were hiding in forts, stashing their food in high cliff stores, and the cultures were gone from their ancestral homelands (perhaps compounded by interspersed periods of cold and population booms during years of conducive wetness).

After the 13th century, the US West basked in relative climatic peace, with perhaps only two major events of this severity and duration until the 19th and 20th centuries, when there were several multiyear events culminating in the drought during the late 1990s and early 2000s, which were much shorter than those of the MWP. These will be discussed later and in Chapter 7. In particular, Stahle et al (2007) identified the 16th century megadrought during 1559–1582 (defined by smoothed PDSI data below the mean value) as being more severe and lengthy than anything experienced in the 20th century (identifiable in the reconstructed drought time series for North America in Figure 5.6c). This event was most severe across northern Mexico, but was mild to moderate across the rest of the continent, as shown at its peak in Plate 5.3. The data are more complete for the US West because of the higher density of records further back in time, and show two other megadroughts: the 14th-century (1387–1402) drought across the Colorado Plateau and Great Plains, and the 15th-century (1444–1481) drought across most of the US West, but mainly over the Central Plains (Stahle et al, 2007). The 16th-century continental drought is estimated to have impacted on the US West and northern Mexico during 1571–1586, as discussed below. Plate 5.3 shows maps of these droughts at the height of their spatial extent.

Seager et al (2009) analysed instrumental and tree-ring data over northern Mexico in the context of North American drought. They found that the late 16th-century megadrought was outstanding over Mexico because of its duration (Figure 5.6d). This drought is also related to the massive amount of depopulation that occurred in the region during the aftermath of European conquest, which took the lives of an estimated 7–17 million people in the highlands of Mexico during two epidemics that occurred between 1545 and 1576 (Acuña-Soto, 2002). The losses amounted to about 80 per cent of the population. Although it is generally thought that diseases brought by European invaders, such as smallpox, measles and typhus, were the cause of these depopulations, there is mounting evidence that haemorrhagic fevers derived from indigenous viruses were the cause. The persistent and severe drought conditions at the time probably exasperated the plight of the already-stressed population and made them extremely vulnerable to disease. The 1950s drought that affected much of the rest of North America was likely drier in magnitude, but did not come close to the duration of the 16th-century event.

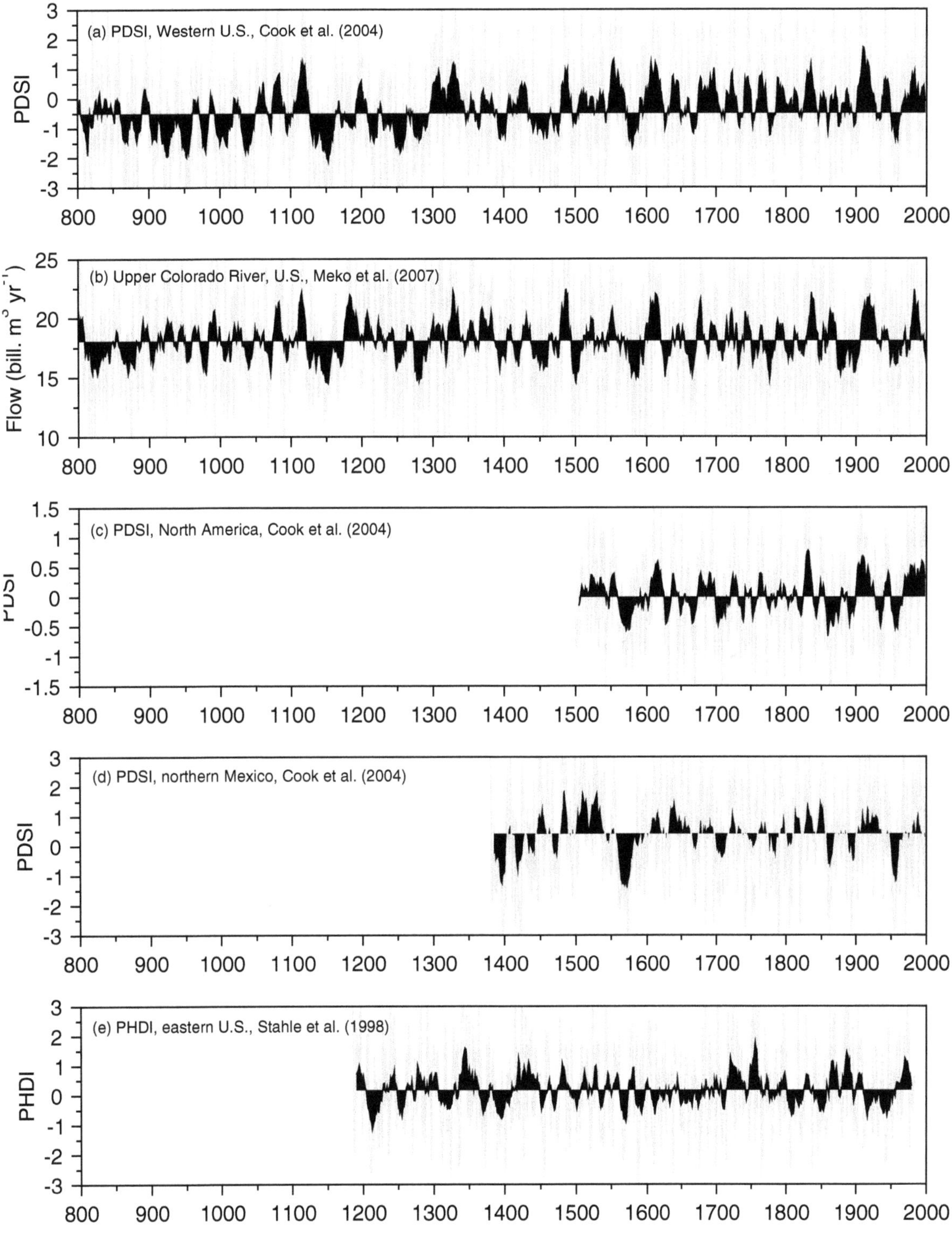

Figure 5.6 *Reconstructions of hydroclimate for North America for the past 1200 years*

In the eastern US, the evidence indicates that drought events of the past millennium were much more severe than anything experienced in the past hundred years, including the 1930s drought. These include an event in the latter half of the 16th century, which may have persisted for up to six years. This appears to be unusual for the eastern part of the US, which has a relatively variable climate, but nevertheless shows the potential for more severe events than have been experienced in living memory. One of the most interesting stories of the human impacts of drought comes from Stahle et al (1998), who describe the possible demise of the colony on Roanoke Island off the coast of North Carolina. This was a precursor of the Jamestown colony established by the English in 1607, and probably the first permanent settlement in what is now the US. Their mysterious disappearance during 1587 may have been forced by severe drought, which appears in a tree-ring-based climate reconstruction back to 1185 (Figure 5.6e and Plate 5.3). It has been established that this drought may have extended from the west to east coasts of the US, and from Mexico in the south to British Columbia in the north. In the southwest US and northwestern Mexico, the drought appears to have lasted several decades (Stahle et al, 2000).

Although it should be stressed that the evidence for the global scope of the MWP is limited, several regions around the world do show warming during this period and signatures of drought (Seager et al, 2007). A study on lake sediments from Lake Naivasha in Kenya shows large swings in climate over the last 1100 years (Verschuren et al, 2000) (Figure 5.7a). The decadal resolution data are based on independent data from sediment stratigraphy and the species compositions of fossil diatoms and midges. The records show the fingerprint of the MWP, which was significantly drier than today, and the LIA, which was relatively wet. The record also reveals drought variations in more recent years, including three prolonged dry periods in 1380–1420, 1560–1620 and 1760–1840. These were deemed by the researchers to be more severe than anything experienced during the 20th century. They also made links to pre-colonial history of the region, as derived from oral histories of agriculture and pastoral society. Evidence of famine, political unrest and migration coincides with the dry periods, whereas periods of high lake levels coincide with eras of prosperity and political stability.

Esper et al (2007) looked at drought in Morocco since 1049 using a reconstruction of the PDSI from Atlas cedar tree-ring data. The time series reveals long-term changes in climate over centennial scales, beginning with drier conditions in the early part of the record about 1350 (Figure 5.7b). This is followed by a transition from 1350–1450 to above-average values during 1450–1980. Recent years are characterized by drier conditions. The severest drought period lasted from 1237 to 1256, which had similar conditions to 1981–2000. Other decadal scale droughts happened in the 1350s, 1520s, 1800s, 1880s and 1980–1990s. In fact, the recent warming is exceptional in the last 500 years, but similar to conditions before 1400. Switching from dry conditions in the MWP to generally wetter conditions in the LIA is consistent with the reconstructions of Verschuren et al (2000), Stine (1994) and Cook et al (2004). These series show drier conditions during the medieval period until the 15th century, and wetter conditions in the LIA, along with drying from anthropogenic warming in recent years. Similar to North American reconstructions, the long-term variation is linked to ENSO variability.

For China, temperature reconstructions for the past two millennia from proxy records obtained from ice cores, tree rings, lake sediments and historical documents indicate that eastern China and the northeast Tibetan Plateau experienced warming during the MWP (Yang et al, 2002). The record also shows that the whole of China experienced cooling during the LIA. Palaeoclimate data records of drought are generally shorter but indicate a series of severe events during the general period of the LIA. Palaeoclimate data from written official records on weather and climate during the main rainy season from county and district levels reveal that eastern China (east of 105°E) has experienced many droughts over the past 500 years, especially during 1500–1730 and from 1900 onwards (Shen et al, 2007; Zhang et al, 2008, Tan et al, 2008). As well as referring to drought directly, some of these historical records refer to plagues of locust or more descriptive accounts, such as 'a thousand miles of barren land'. The records identified severe events in 1586–1589, 1638–1641 and 1965–1966 that covered more than 40 per cent of the region. Precipitation fell by 50 per cent or more. In fact, during 1500–1650, a series of severe droughts occurred, which are associated with economic impacts that may have led to the end of the Ming Dynasty in 1644. For example, the population of China may have decreased by up to 40 per cent during the period 1585–1645 due to various social reasons, including the

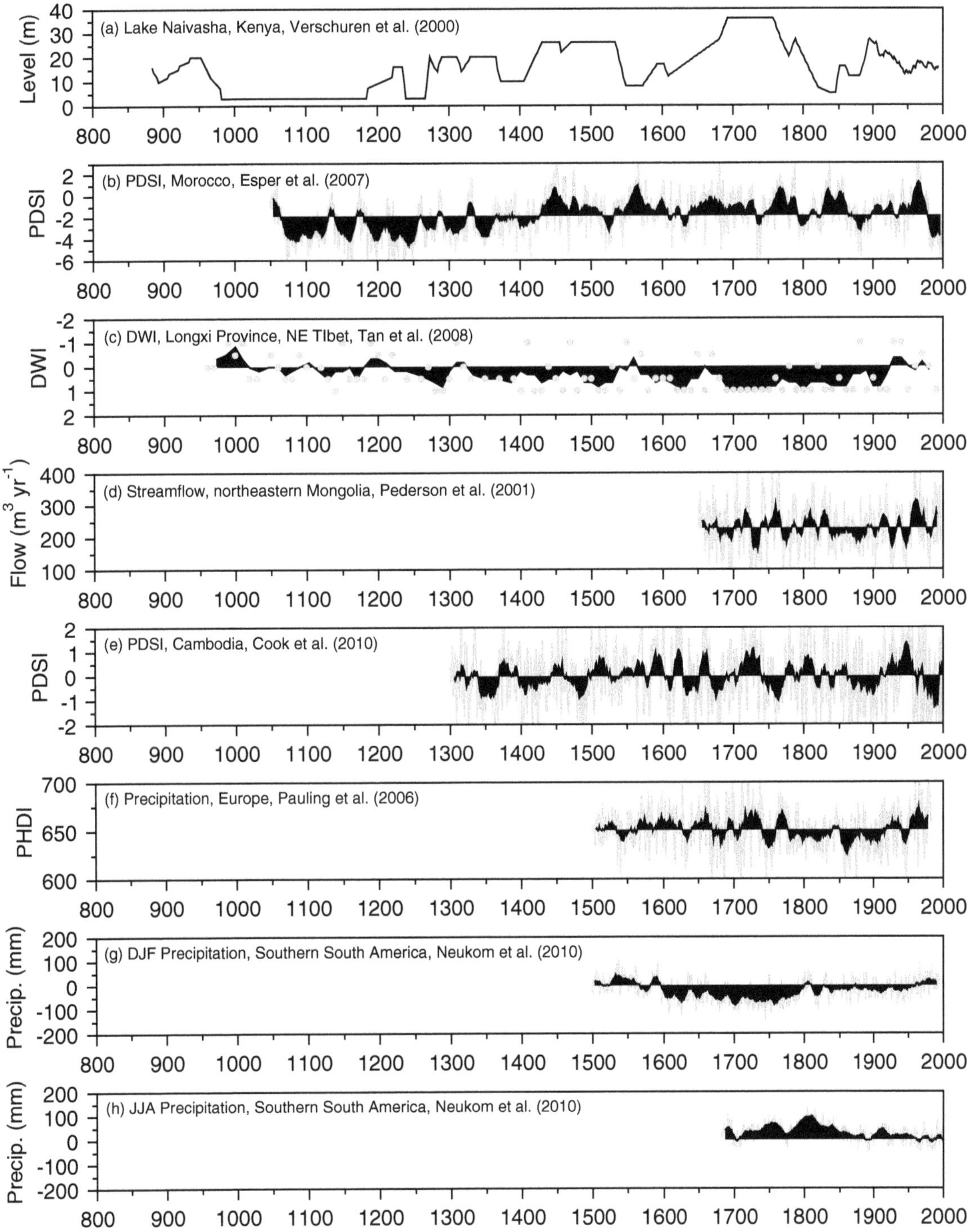

Figure 5.7 *Time series of various hydroloclimate reconstructions across the globe for the past 1200 years: Palmer drought indices, categorical indices, precipitation and streamflow*

Note: The dark shading indicates anomalies relative to either zero or the mean of the dataset, with upper values indicating wet periods and lower values showing dry periods.

impacts of drought. The region is vulnerable to changes in the East Asian monsoon, and a weakened monsoon driven by El Niño conditions of warming in the tropical Pacific may have been responsible for these drought events. There is also evidence to indicate that volcanic eruptions in the preceding years of these droughts played a direct role (by cooling temperatures) or indirect role (by enhancing Eurasian snowpack linked to monsoon variability or inducing the El Niño-like ocean conditions) in forcing drought conditions. Figure 5.7c shows an example of these types of data for Longxi, a small area at the northeast edge of the Tibetan Plateau in China, for which written records extend back over 1000 years (Tan et al, 2008). The historical records have been characterized into a dry–wet intensity (DWI) index that contains five classes, from wet (–1) to dry (1). Sustained drought from about the end of the 17th to the 18th century is evident.

In East Asia (China, Japan, Korea) sets of ancient documents on weather and the occurrence of wet and dry spells (including the DWI data) have been used extensively to analyse climate variation over the past 500 years. For example, Qian et al (2003) combined these with other independent reconstructions for the past 1000 years across East Asia. These include a set of annual dry–wet reconstructions, derived from over 2000 ancient Chinese documents, from which climatologists have extracted records back to 1470. The main sources of records are the official government 'clear and rain records', drought and flood reports from local government, and private diaries. The data have been categorized into an annual summertime (May–September) DWI for 120 locations across China. This is supported by rainfall data from Seoul, Korea, based on one of the longest continuous records in the world, covering 1777 to 1907. For northern China, Qian et al (2003) found no trend in severe droughts over the entire time period, but identified a number of periods with a higher frequency of dryness: 1480–1540s, 1610–1650s, 1720–1760s and 1870s. However, it also emerged that there was an increasing trend for severe wet periods to develop as the time period of 530 years progressed. In the middle and lower Yellow River regions, drought occurred in the 1570–1650s, and from 1910 to the present day, with more droughts in the latter period. In northeast China, drought was most prominent in the 16th century, with an increasing occurrence of drought in the 20th century, which is explored in later chapters. In the middle and lower Yangtze River region, droughts were more prevalent and severe during the 16th and 17th centuries. After the turn of the 19th century, droughts increased in frequency, and severe dry conditions reached a peak of 19 per cent occurrence in the 1970s. This is also true for the south and southwest, where droughts have exceeded a frequency of 10 per cent since the 1970s, although no significant variations are found in records before the 19th century. In many regions that have experienced a rising frequency of droughts since the 19th century, the frequency of wet periods has also increased. Variations in the East Asian monsoon are the main contributors to changes in wet and dry spells. In the 20th century, the weakening of the land–sea temperature contrast is believed to have resulted in a diminished monsoon and a drier northern China.

Further inland in Central Asia, the impact of the Pacific is also felt in concert with a multitude of influences from the westerly jet stream, the Siberian high-pressure cell, monsoonal systems from the south and southwest, and local land–atmosphere feedbacks that influence convective systems in the summer. Reconstructions of regional precipitation and streamflow records for northeastern (Pederson et al, 2001) and west-central Mongolia (Davi et al, 2006) have drawn from Scots pine and Siberian larch tree-ring chronologies across the region. The reconstructed streamflow records for the Kherlen and Selenge rivers (a key source of fresh water for Mongolia and the major input to Lake Baikal), which cover the period of 1650–1995, indicate a higher variability before 1800 (Figure 5.7d), and a series of extreme droughts in both regions during the 1600–1700s (especially in the late 17th century, and during 1725–1740 and 1771–1795). The peak of the late 17th-century drought is shown in Plate 5.4 and highlights the drought in the west-central region. Drier than normal conditions stretched from Mongolia, down to India and Southeast Asia. In the 1800s, there was a considerable reduction in the occurrence of drought in the northeast, while a number of long-term droughts arose in the central region, notably during the early years of the century and the period of 1854–1888. The 20th century was dry in the northeast, whereas it was relatively normal in the central region. The researchers found this pattern of drought corresponded with the Pacific Decadal Oscillation, ENSO and quasi-covariability with solar output.

Drought is also implicated as a factor in the downfall of the well-established Angkor civilization, which existed from the 9th to the 14th centuries in what is now Cambodia. Again, scientists used tree-ring data to infer reconstructions of climate variability (Sano et al, 2009; Buckley et al, 2010). The challenge for tropical regions is that the differences between wet and dry seasons are usually too subtle to be discerned in the annual growth rings. Nevertheless, there are some responsive tree species, such as teak, and other methods like chemical composition of calcium uptake and carbon isotopes do show sensitivity. Cores from the rare tree species *Fokienia hodginsii*, which is the Fujian cypress found in the Vietnam's cloud forests, provided a record that goes back 700 years over Indochina. The record shows multiple decadal droughts of high severity, including two between the late 1300s and early 1400s, which coincide with the collapse of the Angkor culture (Figure 5.7e and Plate 5.4). The Angkor people were probably already in decline because of their fighting with the neighbouring Siamese, and the drought may have weakened the city that relied on irrigation and rice growing. This tree-ring record was corroborated by ancient written and teak tree-ring records from Thailand, and the accounts of foreign merchants that plied their trade across the region. Another significant drought occurred in the mid-18th century, which lasted for over 30 years and coincided with the demise of all the Southeast Asian kingdoms during 1750–1780. The climate in this region is intricately tied to ENSO activity, with annual drought events brought on by the El Niño warm phase. However, the causes of the multidecadal droughts remain unclear.

In the mountains of northern Pakistan, tree-ring data have revealed that the 20th century was probably the wettest period of the last millennium (Treydte et al, 2006). Researchers used oxygen isotopes from juniper tree-ring cellulose at annual resolution to reconstruct precipitation back to 828 AD. At these high elevations, most precipitation occurs in the form of snow falling in the wintertime, but the isotopes provide an integrated record of snowfall, melt water, soil moisture and transpiration from the trees throughout the year. The record identifies a series of drought periods just prior to 1000, and around 1270, 1420, 1600 and 1720. The 1790s and 1890s are also dry periods that were linked with severe famine. Plate 5.4 shows the drought in 1602 was actually part of a much wider event, which stretched from Central Asia to eastern China. The overall long-term trend is towards wet conditions in the 20th century, which are unprecedented in the 1000-year record. The researchers surmised that this trend relates to anthropogenic warming as this is broadly consistent with wetting in long-term precipitation from reconstructions from northeast China, southwest Asia, Germany and the western US.

Multiple reconstructions of precipitation for Europe are available for many regions of Europe, from Scandinavia (Linderholm and Chen, 2005) through to Turkey (D'Arrigo and Cullen, 2001). A recent reconstruction for the whole of Europe, which combines multiple proxy datasets for the past 500 years (Pauling et al, 2006), indicates decadal variations from 1500 to the middle of 19th century, followed by a dry second half of the 19th century (Figure 5.7f). This record was constructed from documentary and natural proxy records that were regressed against the 20th century precipitation record from gauges and then projected onto the proxy data for 1500–1900. The data indicate that the summers of 1666 and 1669 were the driest on record, the former of which coincides with the great fire of London in September of that year. Winter precipitation anomalies were generally out of phase between southwest and central Europe, with one region being wet while the other region was dry. Pauling et al (2006) found that dry periods over southern Spain and Morocco were linked with reconstructed high-pressure data that prevents oceanic moist air reaching the area. For central and eastern Europe, low pressure over the North Sea causes precipitation, while reducing moisture to the Mediterranean and North Africa, and western Scandinavia. In the opposite phase, high pressure over the British Isles causes drying through to central Europe, but western Scandinavia and the Mediterranean are wetter than normal. For a similar time period, Nicault et al (2008) reconstructed the PDSI for the Mediterranean region on the basis of tree-ring data. They noted that the 16th and early 17th centuries were particularly dry and comparable to periods of drought at the end of the 20th century, which in the eastern part of the region appears to be the severest of the record.

In South America, the number of high-resolution palaeoclimate records has increased tremendously in recent years. These include datasets based on tree rings (Boninsegna et al, 2009), ice cores (Vimeux et al, 2009) and lake sediments (e.g. von Gunten et al, 2009). The records tend to be highly sensitive to climate variability and are relatively free of anthropogenic influence, as seen in activities such as logging. Nevertheless, as pointed out by Neukom et al (2009), they tend to be clustered around the Andes Mountains, are generally limited to the Austral summer, and usually diminish in resolution and accuracy further back in history. Even the Andean records are not spatially complete and large geographical gaps persist, although efforts are being made to fill these (Lara et al, 2005). Tree-ring records for the Chilean Andes cover over 2500km of the mountain range and stretch back 1000 years. Reconstructions in the central Andes (32°–39°S) from the long-lived Cordilleran cypress, which are sensitive to wintertime precipitation, reveal two long periods of drought during 1279–1450 and 1600–1650 (Boninsegna et al, 1988; Lara et al, 2005). At higher elevations, a set of reconstructions based on Lenga beech tree rings indicate that the 20th century had the most severe droughts and pluvials since 1837, with 1890–1914 being one of the driest periods (Lara et al, 2001). In northern Patagonia (39°–48°S), records based on the Cordilleran and Patagonian cypress go back over 1000 years (Lara et al, 2000), and remarkably even 5666 years (Wolodarsky-Franke, 2002) and 3622 years (Lara and Villalba, 1993). However, most of these records are associated with temperature reconstructions, but indicate the anomalous warming of the late 20th century (notable in the reduction of Patagonian ice fields) and can identify a prolonged cool period during 1640–1850 that coincides with the LIA (Villalba et al, 2003). A recent reconstruction based on multiple proxy records by Neukom et al (2010) is the first spatially complete estimate of precipitation for southern South America. Figures 5.7g–h show the precipitation anomalies for the summer (DJF) and the winter (JJA), respectively. These indicate dramatically drier summer conditions relative to the 20th century, during the 17th and 18th centuries, and wetter winter conditions in the 18th and early 19th centuries, although the winter record is much shorter.

Drought in Recent Centuries

The palaeoclimate record of the past 1000 years highlights several datasets that make use of historical and documentary data. These types of data have been useful for reconstructions in recent centuries, for which more detailed and spatially extensive records are available, from official sources as well as private documents. Hubert Lamb (e.g. Lamb, 1972, 1995) was probably one of the first modern-day climatologists to garner documentary information to reconstruct climate variations in a meaningful way. Since then, significant progress has been made in using historical information by incorporating new sources of data, improving systems of verification and validation, and using multi-proxy data. Furthermore, these efforts have been extended beyond reconstructing surface conditions to statistically derive synoptic conditions (for example, atmospheric pressure for several centuries in Europe by Luterbacher et al, 2002, 2004). In addition, it has become possible to discern the occurrence of extreme events in general, along with their socio-economic impacts, and thus the ability of humans to adapt. This approach has also helped to define the temporal and spatial characteristics of the MWP and LIA.

Multiple research studies exist that indicate the occurrence of severe and prolonged drought events across Europe over the past few centuries. In general, reconstructing these events from geophysical evidence has been aided by the relative abundance of written evidence. An example of the use of documentary data is by Pfister et al (2006), who reconstructed wintertime droughts in the Upper Rhine basin around the city of Basel in Switzerland. They used a combination of water stage measurements since 1808 and written evidence of the appearance of rocks during low water flows back to 1540. The record shows that almost 30 droughts have occurred since that time and been related to lower than normal precipitation for the prior four months, probably forced by persistent anticyclones centred on western Europe. Interestingly, this type of drought has been rare during the 20th century because of the increase in wintertime precipitation and temperature.

In some cases, written documents are the only source of evidence. For example, the study of Do O' and Roxo (2008) used documented records (newspapers, archives and even memories) to decipher the occurrence and impacts of drought in the southern Portuguese region of Algarve, from the 12th century to the end of the 19th century. The painstaking reconstruction identified a number of dry spells, although there are always difficulties in interpreting subjective material. Another interesting study comes from neighbouring Spain, where researchers have reconstructed drought occurrence from 1506 to 1900, based on Cathedral records of *rogativas* (ceremonies related to agricultural rites), which therefore give an indication of the prevailing climate (Domínguez-Castro et al, 2008). For example, *rogavitas* were recorded that said prayers for better weather or gave thanks for the end of a bad weather spell. Severe droughts were identified mainly between the end of the 16th century and the 18th century. Interestingly, from the 1560s to the 1620s much of Europe was gripped by witchcraft persecutions that have been linked to a climate-induced decline in the living conditions (Brázdil et al, 2005).

The natural palaeoclimatic record is sparse for South America and mainly focused around the Andes. However, the plethora of documentary evidence offers great opportunity to supplement natural records and improve the temporal and spatial completeness of the reconstructions, although this process is still in its infancy (Prieto and Herrera, 2009), in comparison to research in Europe (Brázdil et al, 2005). The records have benefited from the legacy of Spanish colonialism, starting in the 16th century in Central America and expanding southwards across the continent over the next decades. This left behind a wealth of written documentation, which can be related to periods of drought. For example, historians have been able to reconstruct drought over the Yucatan Peninsula that covers parts of Mexico and Belize using historical documentation, including diaries, archives, chronicles and newspapers, as well as iconographical and bibliographical data from a catalogue of agricultural disasters for Mexico, and other documentation on epidemics and famine in colonial Yucatan. The study of Mendoza et al (2007) identified 25 droughts, ranging from one- to ten-year events, which were severe enough to have been recorded in some way. The longest events were in 1535–1541, 1650–1653, 1725–1727, 1765–1774 and 1800–1805. The study surmised that Atlantic surface temperatures, as characterized by the AMO, plus solar variation were the leading forcings of drought in the region. Prieto and Herrera (2009) give a detailed summary of studies of hydroclimate variability for South America, based on documentary evidence. For example, there was increased drought occurrence in the 18th century in Buenos Aries, and particularly during 1710–1760, in comparison to the 17th century, also seen in the reconstructions of droughts for the La Plata River.

Li et al (2007) and Fang et al (2009) reconstructed the PDSI for northwestern China using tree-ring data from spruce and conifer species, and found regional differences between the western and eastern parts of the region. The records go back to 1675 and show a series of wet and dry epochs, interspersed with short-term drought events. For the eastern part of the region, Fang et al (2009) showed low-magnitude decadal variability up until the early 20th century, when pronounced drying during 1922–1934 occurred with the driest year in 1929. This was followed by a dry period again in the 1990s and early 21st century. The western part of the region, however, behaved quite differently, with increased wetting during the 20th century. This region was fairly normal in the 17th century, apart from an extremely dry event in 1690. However, then wet conditions persisted during the 18th century with no drought events, drier conditions in the 19th century and finally increased wetting during the 20th century. The 19th century was particularly dry during the 1820s and appears to have been spatially coherent with dry spells in western Mongolia, discussed in the previous section (Pederson et al, 2001; Davi et al, 2006). There were extremely dry years in 1823, 1843, 1885 and 1895, which were also recorded in historical archives of the region. Despite the general wetting during the 20th century, some extreme drought events occurred in 1943–1945 (also in Pederson et al, 2001) 1900, 1910 and 1916–1919 (Fang et al, 2009).

Australia is the driest inhabited continent on Earth, but is also prone to extremes of climate, including excessive precipitation and flooding. Instrumental records since the end of the 19th century indicate large decadal swings in precipitation but little is known before that time. Only now are palaeoclimate records being developed to infer climate variability over the past millennium. Reconstructions based on tree-ring data have also been absent in tropical and subtropical regions of Australia. Again this is because of the lack of species that produce annually resolvable rings. Heinrich et al (2009) used rings from the Australian red cedar, which is one of the few tropical species that does show distinct annual rings, and reconstructed climate back to 1854. The series shows signals that are associated with ENSO and the Interdecadal Pacific Oscillation (IPO). The reconstructed record of precipitation is consistent with the instrumental record over the 20th-century, identifying the 'Federation' drought during 1895–1902, the World War II drought in the 1940s and the 1991–1994 drought (see Chapter 7 for an analysis of 20th-century droughts). Dry conditions were also identified in the 1880s, with an exceptionally dry year in 1876. D'Arrigo et al (2008) compared a set of independent proxy records of tree-ring widths for northern Australia, tree-ring and coral records for Java, and precipitation reconstructions for northeastern Australia from Great Barrier Reef coral data. They used this to develop an annual PDSI reconstruction for 1787–2002, for the September–January Australasian monsoon in this region. The initial results indicate the influence of ENSO on episodes of drought, as in 1877–1878. Droughts in Australia during the 18th and 19th centuries are documented in part because of the failure of white settlement. For example, lack of water partly caused the failure of the first attempted settlement in Victoria in 1803. Later on, drought was officially denied to entice more European settlers.

As well as official records from colonial rule, there also exist related observations and records from the missionaries that accompanied colonial expansion. In fact, the theories and understanding of the geophysical environment of Africa was often developed by the missionaries. This was initially practised by Christian missionaries, and subsequently by missionary scientists, the most famous of whom was David Livingstone, who played major roles in opening up the continent's interior and shaped thinking on environmental conservation. Several studies have recognized the worth of documentary records in reconstructing pre-instrumental data in Africa (e.g. Nicholson, 1979; Lindesay and Vogel, 1990), drawing from official colonial era records, as well as diaries, travel logs, geographic descriptions and historical works. Nicholson (1979) gives an excellent overview of climate variation across sub-Saharan Africa, from the 8th century to the 20th century, with particular attention to the historic literature (some of which goes back to the 16th and 17th centuries). This reveals that the region was probably wetter than today between the 8th and the 13th centuries, and again during the 16th to 18th centuries, when historical information became more plentiful. A series of severe droughts occurred in the 1680s, 1738–1756, 1770s and 1790s to 1830s, leading to the present-day arid conditions, as confirmed by reports of many famines during this later time and the fall of Lake Chad levels. A return to wetter conditions occurred at the end of the 19th century, followed by drier conditions that culminated in a particularly severe event in 1913.

A good example of how these records can be used to reconstruct drought is the work of Endfield and Nash (2002), who explored the relationship between drought and environmental thinking in 19th-century southern Africa from missionary correspondence. These records, which consist of letters, personal papers and journals of missionaries sent to the London Missionary Society during 1815–1900, reveal six drought periods affecting large parts of the Kalahari region. The most widespread were those during 1844–1851, 1857–1865, 1884–1886 and 1894–1899. In the context of the ongoing discussion at the time of the desiccation of Africa, it appears that the debate was misguidedly fuelled by observations derived from these short-term drought periods. This can be juxtaposed against the 'myth of the garden' in the US West, which was used to encourage expansion and open up the Great Plains. Overall, Endfield and Nash conclude that the climate variations of the 19th century were similar to those of today: generally dry conditions punctuated by short-term wet periods.

In central and northern Mexico, there are numerous sources of historic documentation that have been tied to the natural palaeoclimate record (O'Hara and Metcalfe, 1995; Endfield and O'Hara, 1997). O'Hara and Metcalfe (1995) used a variety of historical sources to reconstruct climate over the past 600 years across central Mexico, covering the rise and height of the Aztec Empire before colonial conquest. The Aztec period and the immediate colonial period were relatively wet up to about the 1640s, from whence there was a return to drier conditions until the early 1900s with a series of severe and widespread droughts. The most severe of these happened during the mid- to late 1700s and the late 1800s and had the largest impacts on the indigenous population (Endfield and O'Hara, 1997) as aggravated by the monopolization of resources by the church among others. Plate 5.5 shows that at its peak this event extended northwards into the Great Plains.

The 19th century brought a series of major droughts that had significant consequences on the social and environmental fabric of the West and Midwest of the US. Newly expanded tree-ring reconstructions of drought across the US (Cook et al, 2004) indicate a severe event in the mid-1850s to mid-1860s, known as the Civil War drought. The summer of 1855 is estimated to be one of the driest of the past 500 years across the southern Great Plains (Stahle et al, 2007) as shown in Plate 5.5. One of the major consequences of this drought was a contribution to the demise of the vast bison population that roamed the grassy plains at the time. Expansion of European settlers westwards and the increase in the Indian population before them was amassing pressure on resources across the plains. In times of drought the bison herds were forced to compete for fewer resources near to rivers and lakes that were increasingly being settled by the human population. The drought has been shown through climate model simulations to be forced by La Niña-like conditions.

Another La Niña was probably responsible for the widespread drought of the 1870s that stretched from the southeast US to the Canadian prairies. The drought was responsible for the unimaginable massive swarms of Rocky Mountain locusts, which Lockwood (2004) recounts from newspaper reports: 'The sunlight dimmed, and the air took on the thick veil that he associated with the smoke of a prairie fire. The roaring crackle of a million wings sounded like a horrific blaze.' During drought times locusts are attracted by the concentrated nutrients of crops and the combined heat and grouping that sparse vegetation brings, which aids maturation and migration in swarms. It is estimated that the swarm of 1875 numbered 3.5 trillion insects and covered about 700,000 km^2 of the West, an area larger than France. Estimates of damages from the series of outbreaks between 1874 and 1877 are about $200 million to the West. This is equivalent to $166 billion in today's money, or just under half of the total value of current US agriculture (Lockwood, 2004).

Towards the end of the century, a further drought was partly responsible for changing the way the West was eventually settled and the intervention of the federal government as a pivotal role in this. A wetter period following the 1870s drought had fuelled widespread migration that was also promoted by the railroad companies. This was the misbelief that 'rain follows the plough'. However the drought of the 1890s showed that dryland farming was insufficient to support isolated frontier farm families and eventually led to the Federal Reclamation Act of 1902 that brought irrigated agriculture and eventually full development of the West. Climate model simulations again implicate La Niña-like conditions as mainly responsible. One legacy of this drought and the preceding period of settlement was the removal of natural grass vegetation and the increase in dust storms that was felt through to the great dust storms of the 1930s drought.

Global Mechanisms and Connections

There have been many changes in climate globally during the Holocene, manifesting in dry periods on different timescales, including abrupt shifts from wet to dry conditions (over decades), rapid descents into arid conditions (such as over North Africa) and so-called megadroughts that last for decades. At millennial timescales, these events were driven by orbital changes in solar irradiance, which redistributed solar irradiance latitudinally and seasonally. The impacts of these changes are complex, however, because of the feedbacks with the climate system that may alter the local sensitivity of the land surface, or alter atmospheric circulation and the redistribution of energy. The role of vegetation as a feedback to the climate system is increasingly being recognized as a modulator and instigator of climate change that may drive abrupt changes in the climate system and a switch to arid conditions. Changes in solar activity have added to orbitally forced variability, which has possibly varied by 0.25 per cent since the 'Maunder Minimum' around the late 17th century, and is estimated to have translated into a mean northern hemisphere temperature change of 0.2–0.4°C. Inferred estimates from cosmogenic isotopes show that solar irradiance may have varied by +/–0.4 per cent over the Holocene and was high from 1100 to 1250 AD and at a minimum from the 15th century to the end of the 17th century, and has increased again over the 20th century (Bard et al, 2000). This has potentially influenced the major climate anomalies of the Holocene, including many of the dry periods described in this chapter and the approximate 1450-year to 1500-year periodicity identified by many studies. At shorter timescales, the influence of volcanic eruptions has been important, not only as a cooling influence at global and hemispheric scales, but also influencing atmospheric circulation and forcing warming in some regions and cooling in others.

The ocean plays an important role on annual to decadal timescales due to its thermal inertia and redistribution of heat throughout the globe via oceanic circulation patterns. ENSO is especially important and has guided interannual climate variability since well before the mid-Holocene, although it is becoming increasingly clear that such modes of climate variability vary greatly in their intensity and periodicity over time. The best palaeoclimatic records for ENSO come from the Galapagos Islands, which experience strong drought during El Niños (Bush, 2003), and show that 435 events occurred during the past 7100 years (Riedinger et al, 2002). Multiple independent palaeoclimate records show that ENSO activity was relatively weak during the early to mid-Holocene. For example, lake records from Ecuador (Sanweiss et al, 2001) indicate that El Niños have increased in intensity from about 7kya, reaching their modern time quasi-periodicity around 5.5kya, but waxed and waned since then, becoming more active during 7.1 to 5.9, 4.6 to 3.1 and 1.9 to 0.9kya.

The state of the North Atlantic, in terms of the NAO, is a recurring modulator of climate across the northern hemisphere. Dry conditions during the MWP are linked with the warming of the North Atlantic and the intensification of the MOC (Scourse et al, 2006), which ultimately were driven by high solar irradiance and low volcanic activity (Emile-Geay et al, 2007). The subsequent wet period of the LIA is associated with much colder SSTs and reduced MOC (Hebbeln et al, 2006). These may be connected to La Niña-like conditions in the tropical Pacific and the poleward shift of storm tracks, which brings drought to North America (Seager et al, 2007). This may similarly have impacted on northern Africa (Esper et al, 2007) and be part of a hemispheric symmetric pattern that equally impacts on mid-latitudes across the northern hemisphere, as well as in the southern hemisphere (Herweijer and Seager, 2008).

Summary

The Holocene is considered climatically stable, in comparison to changes over longer timescales. The extremes of the ice ages are testament to the large variations in climate that occur over very long (10^4–10^5 years) time periods that have the power to extinguish species and shape terrain. Nevertheless there has been surprisingly large climate variability and change in the Holocene, which has brought about large swings in global temperatures, shifts in ENSO and monsoon systems, and transitions between wet and dry climates. Although there are significant levels of uncertainty over palaeoclimate records, especially when going further back in time, huge progress has been made in recent years in combining these disparate records, in order to improve their resolution in time and space, and provide more robust estimates of past variability at regional to global scales. What is certain is that temperatures have increased since the last glacial period (or ice age), reaching a climate optimum around 4000–6000 years ago, which provided the impetus for human migration, the development of widespread agriculture and water management, and the flourishing of civilizations.

Embedded in this longer-term variability have been several transitions from wet to dry conditions and many periods of drought, lasting for decades to centuries, as well as presumably shorter events, masked by the generally low temporal resolution of palaeoclimate data. These droughts have had widespread environmental impacts and have been associated with the demise and dispersal of civilizations. Overall, the early to mid-Holocene was generally wetter than today, with enhanced summer monsoons across the tropics driven by orbital changes and land–sea interactions. In particular, reconstructions of lake levels and vegetation change point to long-term wet conditions existing over North Africa.

At higher latitudes, warming temperatures led to the expansion of northern hemisphere forests further northwards, enhanced by snow and vegetation albedo feedbacks. Africa and parts of Asia have gradually dried out since that time. Many of the dry periods were the result of abrupt (of the order of decades to centuries) climate change that brought drier conditions to a region or even permanent aridity, in the case of the Sahara, for example. Changes in climate on a 1500-year cycle appear to have played a role in instigating dry periods globally and have been labelled as Holocene 'events'. These have been identified at approximately 1.4, 2.8, 4.2, 5.9, 6.2 and 8.2kya, although there is much debate as to the cause and timing of these events. The 8.2kya event brought about the cooling of the northern hemisphere, with drying in central Europe, a more southerly ITCZ and drying of African lakes, Indian and East Asian monsoons. The 5.9kya event resulted in the desertification of the Sahara. Between 5 and 4kya, multiple abrupt changes have been identified, including the 4.2kya event that affected much of Europe, North America, North Africa and the Middle East, and is associated with the demise of several civilizations.

In the last few millennia, the climate has been relatively stable, yet severe and lengthy droughts have occurred. The palaeoclimate evidence is much more robust during this period, both in relation to the diversity of independent records, but also in terms of their temporal resolution and spatial coverage, and is increasingly supported by documentary evidence and instrumental data over the past few hundred years. The

MWP and LIA appear to be the two most obvious large-scale climate anomalies of the period, although there is still much uncertainty as to their spatial coverage, timing and magnitude. It is increasingly clear, however, that late 20th-century temperatures and their rate of increase are larger than anything in the past 1000 years. The MWP is associated with long-term drought conditions in western North America, North and East Africa, and evidence of drought recurs for other regions of the world. Many other droughts of multiyear to multidecadal duration have been identified around the world, of which numerous droughts are longer and more severe than anything experienced in the 20th century. Exciting reconstructions of spatially continuous fields of drought proxies are becoming available, which draw on the expanding databases of proxy data and indicate the regional signatures of drought. Synthesis of these reconstructions with archeological data are opening up possible links between long-term drought and the demise of civilizations, such as the collapse of the Angkor culture in Southeast Asia during the 14th and 15th centuries.

References

Acuña-Soto R., Stable, D. W., Cleveland, M. K. and Therrell, M. D. (2002) 'Megadrought and megadeath in 16th century México', in *Emerging Infectious Diseases*, vol 8, pp60–362

Alley, R. B. (2000) 'The Younger Dryas cold interval as viewed from central Greenland', in *Quaternary Science Reviews*, vol 19, pp213–226

Bard, E., Raisbeck, G., Yiou, F. and Jouzel, J. (2000) 'Solar irradiance during the last 1200 years based on cosmogenic nuclides', in *Tellus*, 52B, pp985–992

Benson, L. V. et al (2006) 'Possible impacts of early-11th-, middle-12th-, and late-13th-century droughts on western Native Americans and the Mississippian Cahokians', in *Quaternary Science Reviews*, vol 26, no 3–4, pp336–350, doi:10.1016/j.quascirev.2006.08.001

Bigelow, N. et al (2003) 'Climate change and Arctic ecosystems: 1. Vegetation changes north of 55 degrees N between the last glacial maximum, mid-Holocene, and present', *Journal of Geophysical Research*, vol 108, 8170, p25, doi:10.1029/2002JD002558

Bond, G. et al (1997) 'A pervasive millennial-scale cycle in North Atlantic Holocene and glacial climates', in *Science*, vol 278, no 5341, pp1257–1266, doi:10.1126/science.278.5341.1257

Bond, G., Kromer, B., Beer, J., Muscheler, R., Evans, M.N., Showers, W., Hoffmann, S., Lotti-Bond, R., Hajdas, I. and Bonani, G. (2001) 'Persistent solar influence on North Atlantic climate during the Holocene', in *Science*, vol 294, no 5549, pp2130–2136, doi:10.1126/science.1065680

Boninsegna, J. A. (1988) 'Santiago de Chile winter rainfall since 1220 as being reconstructed by tree rings', in *Quaternary of South America and Antarctic Peninsula*, vol 6, pp67–87

Boninsegna, J. A., Argollo, J., Aravena, J. C., Barichivich, J., Christie, D., Ferrero, M. E., Lara, A., Le Quesne, C., Luckman, B. H., Masiokas, M., Morales, M., Oliveira, J. M., Roig, F., Srur, A. and Villalba, R. (2009) 'Dendroclimatological reconstructions in South America: A review', in *Palaeogeography, Palaeoclimatology, Palaeoecology*, vol 281, pp210–228

Bonnefille, R. and Chalié, F. (2000) 'Pollen-inferred precipitation time-series from equatorial mountains, Africa, the last 40 kyr BP', in *Global and Planetary Change*, vol 26, pp25–50

Booth, R. K. et al (2005) 'A severe centennial-scale drought in midcontinental North America 4,200 years ago and apparent global linkages', in *The Holocene*, vol 15, no 3, pp321–328, doi:10.1191/0959683605hl825ft

Brázdil, R. and Kundzewicz, W. (2006) 'Historical hydrology', in *Hydrological Sciences–Journal–des Sciences Hydrologiques*, vol 51, no 5, pp733–738

Brázdil, R., Pfister, C., Wanner, H., von Storch, H. and Luterbacher, J. (2005) 'Historical climatology in Europe: The state of the art', in *Climate Change*, vol 70, pp363–430

Briffa, K. R. and Matthews, J. A. (2002) 'ADVANCE-10K: A European contribution towards a hemispheric dendroclimatology for the Holocene', in *The Holocene*, vol 12, pp639–642

Buckley, B. M., Anchukaitis, K. J., Penny, D., Fletcher, R., Cook, E. R., Sano, M., Nam, L. C., Wichienkeeo, A., Minh, T. T. and Hong, T. M. (2010) 'Climate as a contributing factor in the demise of Angkor, Cambodia', in *Proceedings of the National Academy of Sciences*, vol 107, no 15, pp6748–6752

Burroughs, W. J. (2008) 'Climate change in prehistory: The end of the reign of chaos', Cambridge University Press, Cambridge, UK, p372.

Bush, M. (2003) 'Holocene climates of the lowland tropical forests', in A. Mackay, R. Battarbee, J. Birks and F. Oldfield (eds) *Global Change in the Holocene*, Hodder Arnold, London, pp293–306

Butzer, K. W. (1983) 'Paleo-Environmental Perspectives on the Sahel Drought of 1968–73', in *GeoJournal*, vol 7.4, pp369–374

Cook, E. R. and Krusic, P. J. (2004) *The North American Drought Atlas*, Lamont-Doherty Earth Observatory and the National Science Foundation

Cook E. R., Woodhouse, C., Eakin, C. M., Meko, D. M. and Stable, D. W. (2004) 'Long term aridity changes in the western United States', in *Science*, vol 306, pp1015–1018

Cook, E. R., Seager, R., Cane, M. A. and Stable, D. W. (2007) 'North American droughts: Reconstructions, causes and consequences', in *Earth-Science Reviews*, vol 81, pp93–134

Cook, E. R., Anchukaitis, K. J., Buckley, B. M., D'Arrigo, R. D., Jacoby, G. C. and Wright, W. E. (2010) 'Asian monsoon failure and megadrought during the last millennium', in *Science*, vol 328, no 5977, pp486–489, doi:10.1126/science.1185188

Cullen, H. M., de Menocal, P. B., Hemming, S., Hemming, G., Brown, F. H., Guiderson, T. and Siroko, F. (2000) 'Climate change and the collapse of the Akkadian empire: Evidence from deep sea', in *Geology*, vol 28, pp379–382

Curtis, J. H., Hodell, D. A. and Brenner, M. (1996) 'Climate variability on the Yucatan Peninsula (Mexico) during the past 3500 years, and implications for Maya cultural evolution', in *Quaternary Research*, vol 46, pp37–47

Dansgaard, W., Johnsen, S. J., Clausen, H. B. et al (1993) 'Evidence of general instability of past climate from a 250-year ice-core record', in *Nature*, vol 364, pp218–220

D'Arrigo R. D. and Cullen H. M. (2001) 'A 350-year (AD 1628–1980) reconstruction of Turkish precipitation', in *Dendrochronologia*, vol 19, pp167–177

D'Arrigo, R., Baker, P., Palmer, J., Anchukaitis, K. and Cook, G. (2008) 'Experimental reconstruction of monsoon drought variability for Australasia using tree rings and corals', in *Geophysical Research Letters*, vol 35, L12709, p6, doi:10.1029/2008GL034393

Davi, N. K., Jacoby, G. C., Curtis, A. E. and Bachin, B. (2006) 'Extension of drought records for central Asia using tree rings: West central Mongolia', in *Journal of Climate*, vol 19, pp288–299

Debret, M., Bout-Roumazeilles, V., Grousset, F., Desmet, M., McManus, J. F., Massei, N., Sebag, D., Petit, J. R., Copard, Y. and Trentesaux, A. (2007) 'The origin of the 1500-year climate cycles in Holocene North-Atlantic records', in *Climate of the Past*, vol 3, pp569–575

de Menocal, P., Ortiz, J., Guilderson, T., Adkins, J., Sarnthein, M., Baker, L. and Yarusinsky, M. (2000) 'Abrupt onset and termination of the African Humid Period: Rapid climate responses to gradual insolation forcing', in *Quaternary Science Reviews*, vol 19, pp347–361

Do O', A. and Roxo, M. J. (2008) 'Drought events in Southern Portugal from the 12th to the 19th centuries: Integrated research from descriptive sources', in *Natural Hazards*, vol 47, pp55–63, doi:10.1007/s11069-007-9196-0

Domínguez-Castro, F., Santisteban, J. I., Barriendos, M. and Mediavilla, R. (2008) 'Reconstruction of drought episodes for central Spain from rogation ceremonies recorded at the Toledo Cathedral from 1506 to 1900: A methodological approach', in *Global and Planetary Change*, vol 63, nos 2–3, pp230–242

Emile-Geay, J., Cane, M., Seager, R., Kaplan, A. and Almasi, P. (2007) 'El Niño as a mediator of the solar influence on climate', in *Paleoceanography*, vol 22, PA3210, p16, doi:10.1029/2006PA001304

Endfield G. H. and Nash, D. J. (2002) 'Drought, desiccation and discourse: Missionary correspondence and nineteenth century climate change in central southern Africa', in *Geographical Journal*, vol 161, no 1, pp33–47

Endfield, G. H. and O'Hara, S. L. (1997) 'Conflicts over water in "The Little Drought Age" in central Mexico', in *Environment and History*, vol 3, pp255–72

Esper, J., Frank, D. C., Büntgen, U., Verstege, A., Luterbacher, J. and Xoplaki, E. (2007) 'Long-term drought severity variations in Morocco', in *Geophysical Research Letters*, vol 34, L17702, p5, doi:10.1029/2007GL030844

Fagan, B. (2003) *The Long Summer: How Climate Changed Civilization*, Basic Books, New York

Fagan, B. (2008) *The Great Warming: Climate Change and the Rise and Fall of Civilizations*, Bloomsbury Press, London

Fang, K., Gou, X., Chen, F., Yang, M., Li, J., He, M., Zhang, Y., Tian, Q. and Peng, J. (2009) 'Drought variations in the eastern part of Northwest China over the past two centuries: Evidence from tree rings', *Climate Research*, vol 38, no 2, pp129–135

Fleitmann, D. et al (2003) 'Holocene forcing of the Indian monsoon recorded in a stalagmite from southern Oman', in *Science*, vol 300, pp1737–1740

Gasse, F. (2000) 'Hydrological changes in the African tropics since the Last Glacial Maximum', in *Quaternary Science Reviews*, vol 19, pp189–211

Gasse, F. (2002) 'Diatom-inferred salinity and carbonate oxygen isotopes in Holocene waterbodies in Western Sahara and Sahel (Africa)', in *Quaternary Science Reviews*, vol 21, pp737–767

Gasse, F. and van Campo, E. (1994) 'Abrupt postglacial climate events in West Asia and North-Africa monsoon domains', in *Earth and Planetary Science Letters*, vol 126, pp435–456

Gao, H., Zhu, C. and Xu, W. (2007) 'Environmental change and cultural response around 4200 cal. yr BP in the Yishu River Basin, Shandong', in *Journal of Geographical Sciences*, vol 17, no 3, pp285–292, doi:10.1007/s11442-007-0285-5

Grissino-Mayer, H. (1996) 'A 2129-year reconstruction of precipitation for northwestern New Mexico, U.S.A.', in J. S. Dean, D. M. Meko and T. W. Swetnam (eds) *Tree Rings, Environment and Humanity*, Radiocarbon, Tucson, Arizona, pp191–204

Haug, G., Gunther, D., Peterson, L., Sigman, D., Hughen, K. and Aeschlimann, B. (2003) 'Climate and the collapse of Maya civilization', in *Science*, vol 299, pp1731–35

Hebbeln, D. et al (2006) 'Late Holocene coastal hydrographic and climate changes in the eastern North Sea', in *Holocene*, vol 16, pp987–1001

Heinrich, I., Weidner, K., Helle, G., Vos, H., Lindesay, J. and Banks, J. C. G. (2009) 'Interdecadal modulation of the relationship between ENSO, IPO and precipitation: Insights from tree rings in Australia', in *Climate Dynamics*, vol 33, no 1, p63, doi:10.1007/s00382-009-0544-5

Herweijer, C. and Seager, R. (2008) 'The global footprint of persistent extra-tropical drought in the instrumental era', in *International Journal of Climatology*, vol 28, pp1761–1774, doi:10.1002/joc.1590

Hodell, D. A., Curtis, J. H. and Brenner, M. (1995) 'Possible role of climate in the collapse of the Classic Maya Civilization', in *Nature*, vol 375, pp391–394

Hodell, D. A., Brenner, M. and Curtis, J. H. (2005) 'Terminal classic drought in the northern Maya lowlands inferred from multiple sediment cores in Lake Chichancanab (Mexico)', in *Quarterly Science Reviews*, vol 24, nos 12–13, pp1413–1427

Hughes, M. K. and Graumlich, L. J. (1996) 'Climatic variations and forcing mechanisms of the last 2,000 years', in *Multi-millenial Dendroclimatic Studies from the Western United States*, NATO ASI Series, vol 141, pp109–124

IPCC (2007) S. Solomon, D. Qin, M. Manning, Z. Chen, M. Marquis, K. B. Averyt, M. Tignor and H. L. Miller (eds) *Climate Change 2007: The Physical Science Basis. Contribution of Working Group I to the Fourth Assessment Report of the Intergovernmental Panel on Climate Change*, Cambridge University Press, Cambridge and New York

Jones, P. D. and Mann, M. E. (2004) 'Climate over past millennia', in *Reviews of Geophysics*, vol 42, RG2002, p42, doi:10.1029/2003RG000143

Jouzel, J. et al (2007) 'Orbital and millennial Antarctic climate variability over the past 800,000 years', in *Science*, vol 317, no 5839, pp793–796, doi:10.1126/science.1141038

Kloor, K. (2007) 'Archaeology: The vanishing Fremont', in *Science*, vol 318, no 5856, pp1540–1543, doi:10.1126/science.318.5856.1540

Kutzbach, J. E. and Liu, Z. (1997) 'Response of the African monsoon to orbital forcing and ocean feedbacks in the Middle Holocene', in *Science*, vol 278, pp440–443

Laird, K. R., Fritz, S. C., Maasch, K. A. and Cumming, B. F. (1996) 'Greater drought intensity and frequency before AD 1200 in the Northern Great Plains, USA', in *Nature*, vol 384, pp552–554

Lamb, H. H. (1972) *Climate Present Past and Future*, Methuen, London

Lamb, H. H. (1995) *Climate, History and the Modern World*, Routledge, New York, p464

Lara, A. and Villalba, R. (1993) 'A 3620-year temperature record from *Fitzroya cupressoides* tree rings in southern South America', in *Science*, vol 260, pp1104–1106

Lara, A., Villalba, R., Aravena, J. C., Wolodarsky, A. and Neira, E. (2000) 'Desarrollo de una red de cronologías de *Fitzroya cupressoides* (Alerce) para Chile y Argentina', in F. A. Roig (ed) *Dendrocronología en América Latina*, EDIUNC, Mendoza, Argentina, pp217–244

Lara, A., Aravena, J. C., Villalba, R., Wolodarsky-Franke, A., Luckman, B. and Wilson, R. (2001) 'Dendroclimatology of high-elevation *Nothofagus pumilio* forests at their northern distribution limit in the central Andes of Chile', in *Canadian Journal of Forest Research*, vol 31, no 6, pp925–936

Lara, A., Wolodarsky-Franke, A., Aravena, J. C., Villalba, R., Solari, M. E., Rivera, A. and Le Quesne, C. (2005) 'Climate fluctuations derived from tree-rings and other proxy records in the Chilean Andes: State of the art and future prospects', in U. M. Huber, H. K. M. Bugmann and M. A. Reasoner (eds) *Global Change and Mountain Regions: An Overview of Current Knowledge*, Advances in Global Change Research Series, Kluwer Academic Press, New York, vol 23, pp145–156

Li, J., Chen, F., Cook, E. R., Gou, X. and Zhang, Y. (2007) 'Drought reconstruction for north central China from tree rings: The value of the Palmer Drought Severity Index', in *International Journal of Climatology*, vol 27, pp903–909

Linderholm, H. W. and Chen D. (2005) 'Central Scandinavian winter precipitation variability during the last five centuries reconstructed from *Pinus sylvestris* tree rings', in *Boreas*, vol 34, pp43–52

Lindesay, J. A. and Vogel, C. H. (1990) 'Historical evidence for Southern Oscillation-Southern African rainfall relationships', in *International Journal of Climatology*, vol 10, no 7, pp679–689

Liu, Z., Harrison, S. P., Kutzbach, J. E. and Otto-Bleisner, B. (2004) 'Global monsoons in the mid-Holocene and oceanic feedback', in *Climate Dynamics*, vol 22, pp157–182

Lockwood, J. A. (2004) *Locust: The Devastating Rise and Mysterious Disappearance of the Insect that Shaped the American Frontier*, Basic Books, New York

Luterbacher, J., Xoplaki, E., Dietrich, D., Rickli, R., Jacobeit, J., Beck, C., Gyalistras, D., Schmutz, C. and Wanner, H. (2002) 'Reconstruction of sea level pressure fields over the eastern North Atlantic and Europe back to 1500', in *Climate Dynamics*, vol 18, pp545–561

Luterbacher, J., Dietrich, D., Xoplaki, E., Grosjean, M. and Wanner, H. (2004) 'European seasonal and annual temperature variability, trends and extremes since 1500', in *Science*, vol 303, pp1499–1503

MacDonald, G. M. et al (2000) 'Holocene treeline history and climate change across northern Eurasia', in *Quaternary Research*, vol 53, pp302–311

Mancini, M. V., Paez, M. M., Prieto, A. R., Stutz, S., Tonello, M. and Vilanova, I. (2005) 'Mid-Holocene climatic variability reconstruction from pollen records (32°–52°S, Argentina)', in *Quaternary International*, vol 132, nos 47–59, pp617–24

Mann, M. E., Zhang, Z., Rutherford, S., Bradley, R. S., Hughes, M. K., Shindell, D., Ammann, C., Faluvegi, G. et al (2009) 'Global signatures and dynamical origins of the Little Ice Age and Medieval Climate Anomaly', in *Science*, vol 326, no 5957, pp1256–1260, doi:10.1126/science.1177303

Marchant, R. and Hooghiemstra, H. (2004) 'Rapid environmental change in African and South American tropics around 4000 years before present: A review', in *Earth-Science Reviews*, vol 66, pp217–260

Meko, D. M., Woodhouse, C. A., Baisan, C. A., Knight, T., Lukas, J. J., Hughes, M. K. and Salzer, M. W. (2007) 'Medieval drought in the upper Colorado River Basin', in *Geophysical Research Letters*, vol 34, L10705, doi:10.1029/2007GL029988

Mendoza, B., Garcia-Acosta, V., Velasco, V., Jauregui, E. and Diaz-Sandoval, R. (2007) 'Frequency and duration of historical droughts from the 16th to the 19th centuries in the Mexican Maya lands, Yucatan Peninsula', in *Climatic Change*, vol 83, nos 1–2, pp151–168

Moros, M., Andrews, J. T., Eberl, D. E. and Jansen, E. (2006) 'Holocene history of drift ice in the northern North Atlantic: Evidence for different spatial and temporal modes', in *Paleoceanography*, vol 21, PA2017, doi:10.1029/2005PA001214

Neukom, R., del Rosario Prieto, M., Moyano, R., Luterbacher, J., Pfister, C., Villalba, R., Jones, P. D. and Wanner, H. (2009) 'An extended network of documentary data from South America and its potential for quantitative precipitation reconstructions back to the 16th century', in *Geophysical Research Letters*, vol 36, L12703, p5, doi:10.1029/2009GL038351

Neukom, R., Luterbacher, J., Villalba, R., Küttel, M., Frank, D., Jones, P. D., Grosjean, M., Esper, J., Lopez, L. and Wanner, H. (2010) 'Multi-centennial summer and winter precipitation variability in southern South America', in *Geophysical Research Letters*, vol 37, L14708, p6, doi:10.1029/2010GL043680

Nicault, A., Alleaume, S., Brewer, S., Carrer, M., Nola, P. and Guiot, J. (2008) 'Mediterranean drought fluctuation during the last 500 years based on tree-ring data', in *Climate Dynamics*, vol 31, pp227–245

Nicoll, K. (2004) 'Recent environmental change and prehistoric human activity in Egypt and Northern Sudan', in *Quaternary Science Reviews*, vol 23, pp561–580

Nicholson, S. E. (1979) 'The methodology of historical climatic reconstruction and its application to Africa', in *Journal of African History*, vol 20, pp31–49

O'Hara, S. L. and Metcalfe, S. E. (1995) 'Reconstructing the climate of Mexico from historical records', in *The Holocene*, no 5, pp485–90

Orland, I. J., Bar-Matthews, M., Kita, N. T., Ayalon, A., Matthews, A. and Valley, J. W. (2009) 'Climate deterioration in the eastern Mediterranean from 200 BC to 1100 AD as revealed by ion microprobe analysis of speleothems from Soreq Cave, Israel', *Quaternary Research*, vol 71, pp27–35

Pauling, A., Luterbacher, J., Casty, C. and Wanner, H. (2006) 'Five hundred years of gridded high-resolution precipitation reconstructions over Europe and the connection to large-scale circulation', in *Climate Dynamics*, vol 26, pp387–405

Pearce, F. (2006) *When the Rivers Run Dry: Water–The Defining Crisis of the Twenty-First Century*, Beacon Press, East Sussex

Pederson, N., Jacoby, G. C., D'Arrigo, R. D., Cook, E. R., Buckley, B. M., Dugarjav, C. and Mijiddorj, R. (2001) 'Hydrometeorological reconstructions for Northeastern Mongolia derived from tree rings: AD 1651–1995', in *Journal of Climate*, vol 14, pp872–881

Petit, J. R., Jouzel, J., Raynaud, D., Barkov, N. I., Barnola, J. M., Basile, I., Bender, M., Chappellaz, J., Davis, J., Delaygue, G., Delmotte, M., Kotlyakov, V. M., Legrand, M., Lipenkov, V., Lorius, C., Pépin, L., Ritz, C., Saltzman, E. and Stievenard, M. (1999) 'Climate and atmospheric history of the past 420,000 years from the Vostok ice core, Antarctica', in *Nature*, vol 399, pp429–436

Pfister, P., Weingartner, R. and Luterbacher, J. (2006) 'Hydrological winter droughts over the last 450 years in the Upper Rhine basin: A methodological approach', in *Hydrological Sciences–Journal–des Sciences Hydrologiques*, vol 51, no 5, pp966–985

Powers, L. A., Johnson, T. C., Werne, J. P., Castañeda, I. S., Hopmans, E. C., Sinninghe Damsté, J. S. and Schouten, S. (2005) 'Large temperature variability in the southern African tropics since the Last Glacial Maximum', in *Geophysical Research Letters*, vol 32, L08706, p4, doi:10.1029/2004GL022014

Prieto, M. R. and Herrera, R. G. (2009) 'Documentary sources from South America: Potential for climate reconstruction', in *Palaeogeogr. Palaeoclimatol. Palaeoecol.* vol 281, pp196–209

Qian, W., Hu, Q., Zhu, Y. and Lee, D.-K. (2003) 'Centennial-scale dry-wet variations in East-Asia', in *Climate Dynamics*, vol 21, pp77–89

Riedinger, M. A., Steinitz-Kannan, M., Last, W. M. and Brenner, M. (2002) 'A ~6100 14C yr record of El Niño activity from the Galápagos Islands', in *Journal of Paleolimnology*, vol 27, pp1–7

Rohling, E. J. and Palike, H. (2005) 'Centennial-scale climate cooling with a sudden cold event around 8,200 years ago', in *Nature*, vol 434, pp975–979

Salzmann, U. and Waller, M. (1998) 'Holocene vegetation history of the Nigerian Sahel, based on multiple pollen profiles', in *Review of Palaeobotany and Palinology*, vol 100, pp39–72

Sanweiss, D. H., Maasch, K. A. and Anderson, D. G. (2001) 'Variations in Holocene El Nino frequencies: Climate records and cultural consequences in Ancient Peru', in *Geology*, vol 7, pp603–606

Sano, M., Buckley, B. M. and Sweda, T. (2009) 'Tree-ring based hydroclimate reconstruction over northern Vietnam from *Fokienia hodginsii*: Eighteenth century mega-drought and tropical Pacific influence', in *Climate Dynamics*, vol 33, nos 2–3, pp331–340

Schilman, B., Bar-Matthews, M., Almogi-Labin, A. and Luz, B. (2001) 'Global climate instability reflected by Eastern Mediterranean marine records during the late Holocene', in *Palaeogeography, Palaeoclimatology, Palaeoecology*, vol 176, pp157–176

Schulz, M., Paul, A. and Timmermann, A. (2004) 'Glacial-interglacial contrast in climate variability at centennial-to-millennial time scales: Observations and conceptual model', in *Quarterly Science Review*, vol 23, pp2219–2230

Scourse, J., Sejrup, H. P. and Jones, P. D. (2006) 'Late Holocene oceanographic and climate change from the western European margin: The results of the Holsmeer project', in *Holocene*, vol 16, pp931–935

Seager, R., Graham, N., Herweijer, C., Gordon, A. L., Kushnir, Y. and Cook, E. (2007) 'Blueprints for medieval hydroclimate', in *Quarterly Science Review*, vol 26, pp2322–2336

Seager, R., Ting, M., Davis, M., Cane, M., Naik, N., Nakamura, J., Li, C., Cook, E. and Stahle, D. W. (2009) 'Mexican drought: An observational modelling and tree ring study of variability and climate change', in *Atmosfera*, vol 22, no 1, pp1–31

Shanahan, T. M., Overpeck, J. T., Anchukaitis, K. J., Beck, J. W., Cole, J. E., Dettman, D. L., Peck, J. A., Scholz, C. A. and King, J. W. (2009) 'Atlantic forcing of persistent drought in West Africa', in *Science*, vol 324, pp377–380

Shen, C., Wang, W. C., Hao, Z. and Gong, W. (2007) 'Exceptional drought events over eastern China during the last five centuries', in *Climatic Change*, vol 85, pp453–471

Springer, G. S., Rowe, H. D., Hardt, B., Edwards, R. L. and Cheng, H. (2008) 'Solar forcing of Holocene droughts in a stalagmite record from West Virginia in east-central North America', in *Geophysical Research Letters*, vol 35, L17703, p5, doi:10.1029/2008GL034971

Stahle, D. W., Cleaveland, M. K., Blanton, D. B., Therrell, M. D. and Gay, D. A. (1998) 'The lost colony and Jamestown droughts', in *Science*, vol 280, pp564–567

Stahle D. W., Cook, E. R., Cleaveland, M. K., Therrel, M. D., Meko, D., Grissino-Mayer, H., Watson, E. and Luckman, B. H. (2000) 'Tree ring data document 16th century megadrought over North America', in *Eos, Transactions*, American Geophysical Union, vol 81, no 12, p121

Stahle, D. W., Fye, F. K., Cook, E. R. and Griffin, R. D. (2007) 'Tree-ring reconstructed megadroughts over North America since A.D. 1300', in *Climatic Change*, vol 83, pp133–149, doi:10.1007/s10584-006-9171-x

Stanley, J.-D. et al (2003) 'Nile flow failure at the end of the Old Kingdom, Egypt: Strontium isotopic and petrologic evidence', in *Geoarchaeology*, vol 18, no 3, pp395–402, doi:10.1002/gea.10065

Stine, S. (1994) 'Extreme and persistent drought in California and Patagonia during mediaeval time', in *Nature*, vol 369, no 6481, pp546–549

Tan, L. C., An, Z. S., Cai, Y. J. and Long, H. (2008) 'Hydrological representation of the 4.2 ka BP event in China and its global linkages', in *Geological Review*, vol 54, no 1, pp94–104

Treydte, K., Schleser, G. H., Helle, G., Frank, D. C., Winiger, M., Haug, G. H. and Esper, J. (2006) 'The twentieth century was the wettest period in northern Pakistan over the past millennium', in *Nature*, vol 440, pp1179–1182

Verschuren, D., Laird, K. R. and Cumming, B. F. (2000) 'Rainfall and drought in equatorial east Africa during the past 1,100 years', in *Nature*, vol 403, nos 410–414, doi:10.1038/35000179

Viau, A. E. and Gajewski, K. (2001) 'Holocene variations in the global hydrological cycle quantified by objective gridding of lake level databases', in *Journal of Geophysical Research-Atmosphere*, vol 106, D23, pp31,703–31,716

Villalba, R., Lara, A., Boninsegna, J. A., Masiokas, M., Delgado, S., Aravena, J. C., Roig, F. A., Schmelter, A., Wolodarsky, A. and Ripalta, A. (2003) 'Large-scale temperature changes across the Southern Andes: 20th century variations in the context of the past 400 years', in *Climatic Change*, vol 59, no 1–2, pp177–232, doi:10.1023/A:1024452701153

Vimeux, F., Ginot, P., Schwikowski, M., Vuille, M., Hoffmann, G., Thompson, L. G. and Schotterer, U. (2009) 'Climate variability during the last 1,000 years inferred from Andean ice cores: A review of methodology and recent results', in *Palaeogeography, Palaeoclimatology, Palaeoecology*, vol 281, pp229–241

von Gunten, L., Grosjean, M., Rein, B., Urrutia, R. and Appleby, P. (2009) 'A quantitative high-resolution summer temperature reconstruction based on sedimentary pigments from Laguna Aculeo, Central Chile, back to AD 850', in *The Holocene*, vol 19, no 6, pp873–881

Wang, Y. J. et al (2005) 'Holocene Asian monsoon: Links to solar changes and North Atlantic climate', in *Science*, vol 308, pp854–857

Weijers, J. W. H., Schefuss, E., Schouten, S. and Sinninghe Damsté, J. S. (2007) 'Coupled thermal and hydrological evolution of tropical Africa over the last deglaciation', in *Science*, vol. 315, pp1701–1704, doi:10.1126/science.1138131

Weiss, B. (1982) 'The decline of the Late Bronze Age civilization as a possible response to climate change', in *Climatic Change*, vol 4, pp173–198

Weiss, H. et al (1993) 'The genesis and collapse of third millennium north Mesopotamian civilization', in *Science*, vol 261, no 5124, pp995–1004, doi:10.1126/science.261.5124.995

Wolodarsky-Franke, A. (2002). 'Fluctuaciones ambientales de los últimos 1000 años a partir de anillos de crecimiento de Fitzroya cupressoides (Molina) Johnston) en el área del Volcán Apagado, Décima Región, Chile' (MSc thesis), Facultad de Ciencias, Universidad Austral de Chile

Wu, W. and Liu, T. (2004) 'Possible role of the "Holocene Event 3" on the collapse of Neolithic cultures around the central plain of China', in *Quaternary International*, vol 117, no 1, pp153–166, doi:10.1016/S1040-6182(03)00125-3

Yang, B., Braeuning, A., Johnson, K. R. and Yafeng, S. (2002) 'General characteristics of temperature variation in China during the last two millennia', in *Geophysical Research Letters*, vol 29, no 9, p1324, doi:10.1029/2001GL014485

Zhang, Q., Gemmer, M. and Chen, J. Q. (2008) 'Climate changes and flood/drought risk in the Yangtze Delta, China, during the past millennium', in *Quaternary International*, vol 176, pp62–69

Zhao, Y. et al (2005) 'A multi-model analysis of the role of the ocean on the African and Indian monsoon during the mid-Holocene', in *Climate Dynamics*, vol 25, nos 7–8, pp777–800

6

Drought in the 20th Century

Introduction

This chapter describes the occurrence of drought across the globe during the 20th century. There have been many hundreds of severe droughts over the past 100 years, which have had various impacts, ranging from loss of life to huge economic losses. Some of these have been recorded in terms of their impacts (numbers killed, numbers affected, economic costs), whereas others have little information associated with them. For example, the Sahel drought in the 1970s and 1980s claimed hundreds of thousands of lives and the 1930s Dustbowl drought in the US is estimated to have cost the government an equivalent of US$13 billion in assistance alone, and prolonged the economic depression of the time.

As highlighted in Chapter 4, identifying droughts is challenging because of the lack of data and definitive methods for their identification across time and space scales, and especially their direct and indirect impacts. Our knowledge of drought over the 20th century is generally good, as compared to the palaeoclimate record, because it is based on the instrumental data record (e.g. from weather stations). Nevertheless, huge progress has been made in unearthing the occurrence of drought over the past millennia, on regional to global scales through palaeoclimate reconstructions (see Chapter 5). Of particular importance is the growing indication from palaeoclimate data that in many regions drought in the 20th century pales into insignificance, as compared to some of the drought events over the past 10,000 years. Some of these palaeo-droughts have dried up rivers and lakes, caused vegetation to die off and deserts to expand, as well as contributing to the demise of great civilizations. Therefore, an analysis of 20th-century drought only offers a glimpse of the possible severity that natural climate variability may induce. This is worrisome because the 20th century has witnessed the devastating impacts of droughts like those in the Sahel, despite the relative mildness of these events. However, many of these drought events are still in living memory. The occurrence and severity of 20th-century drought must form the basis of an assessment of how we can cope with large-scale drought events and thus our capacity for coping in the future. The descriptions of 20th-century drought variability in this chapter and of individual large-scale events in Chapter 7 set the stage for quantifying and understanding how drought has been changing in recent years, and what the future might hold under global warming (Chapter 8).

As we move into the instrumental period (from the late 19th century onwards), direct evidence has increased tremendously, and researchers have been able to reconstruct the hydrology and atmospheric drivers of individual drought events in more detail. The presence of instrumental data allows for the exploration of the details of drought occurrence through direct analysis of the data and via computer modelling. Palaeoclimate reconstruction has significantly progressed in recent years, through the modelling of global climate conditions to extrapolate information from the sparse data that can be deduced from corals and lake sediments. These data provide a glimpse of sea surface temperatures and volcanic activity at the time, but at present can only offer suggestions of what might have happened at large scales and over many decades to centuries.

Climate modelling builds up a global picture from these sparse data. The density of modern observations enables us to evaluate and extrapolate data in space (through modelling) at a much higher resolution, down to tens of kilometres and monthly to sub-daily timescales, and to do this with much greater confidence. These types of data provide better quantitative information on climate and drought, and therefore enable us to improve the analysis of temporal and spatial patterns of variation, including any cyclical behaviour and trends, and their driving mechanisms.

A useful approach to quantify drought is to employ a conceptual definition or a drought index (see Chapter 4) based on physical quantities, such as precipitation and soil moisture, which can be used to identify individual drought events and to analyse their space–time characteristics. This is what is done in palaeoclimate reconstructions, whereby a physical quantity (such as an isotope concentration or a tree-ring width) is calibrated to known climate variations, and then specific events are compared to long-term records, to reveal the severity of these events. However, for palaeoclimate studies, the uncertainty and coarse resolution of the proxy data can only give an indication of wetter or drier than normal conditions, rather than specific values of severity, and how drought varies over scales as fine as kilometres and days. In this chapter, we describe the occurrence of drought over the 20th century, on the basis of observations of precipitation, temperature and other meteorological data, and hydrological quantities such as streamflow and soil moisture, as derived from measurements and hydrological modelling (see Box 4.1, Chapter 4). This approach combines observations of data with modelling to offer a continuous description of hydrology across the world over the past 100 years. We begin by describing the characteristics of climate and drought over the 20th century, on global and regional levels, and then focus on the associations between climate and drought.

Climate of the 20th Century

Observations from around the world show that near-surface air temperature has been increasing over the past century (Figure 6.1). This is partly in response to rises in CO_2 and other greenhouse gases (IPCC, 2007). This trend levelled off during the 1940s and 1950s, probably as a result of the cooling effect of anthropogenic aerosols. However, during the past 30 years, the warming trend has accelerated, with 1998 and 2005 being the warmest two years on record, and 2000–2009 being the warmest decade on record. The latest IPCC report sums up these trends in the following statement: 'Warming of the climate system is unequivocal, as is now evident from observations of increases in global average air and ocean temperatures, widespread melting of snow and ice and rising global average sea level.' Long-term changes in temperature are associated with

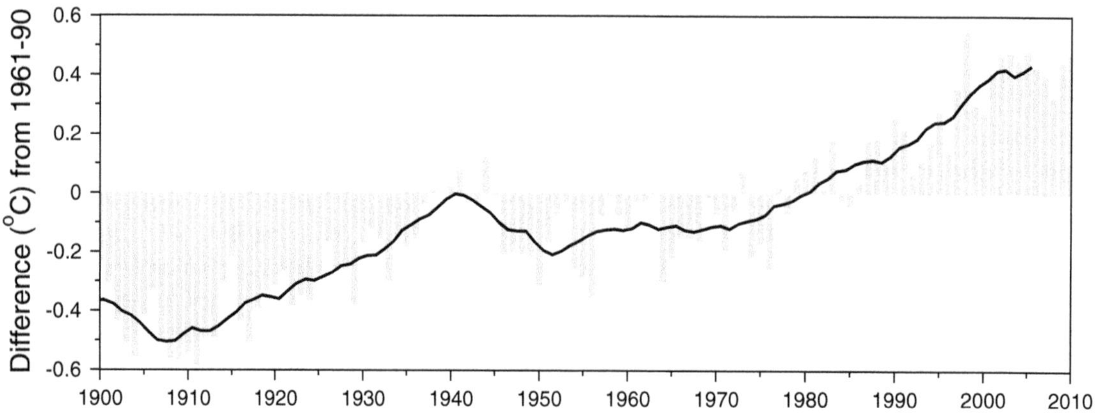

Figure 6.1 *Annual anomalies of global land surface air temperature (°C), 1900–2009, relative to the mean of 1961–1990 for the CRUTEM3 dataset of Brohan et al (2006)*

Note: The black line is the data smoothed using a 10-year moving average.

changes in other components of the climate system, and especially the hydrology of the land surface on which we depend. Global warming also brings changes to atmospheric circulation and moisture, the characteristics of precipitation and the energy balance at the Earth's surface, which influences levels of evaporation. Precipitation can be expected to increase in intensity, but with fewer events and more dry periods in between storms, although this is complicated by the uncertainty surrounding the effect of aerosols. These changes impact on the hydrological cycle and their combined effect is generally to increase the severity of extreme weather events, such as storms, floods and droughts.

For drought, rising temperatures are generally expected to increase levels of drying because of the associated increased potential for evaporation, which we will return to in Chapter 8, where we discuss recent trends in drought and future climate projections. Precipitation is the main driver of drought, however, and has generally increased across global land areas over the last century (Figure 6.2). There is some uncertainty in the estimates of precipitation, especially in the earlier part of the 20th century, in remote regions of the Amazon, central Africa, and northern and central Asia, when the number of rain gauges and their spatial coverage was low. This is reflected in differences with other estimates of global precipitation because of the large spatial variability in precipitation (especially compared to temperature). Nevertheless, there is general agreement on decadal changes among different datasets.

Maps of trends over the second half of the 20th century from observations and modelling of the main components of the hydrological cycle are shown in Plate 6.1. In addition to the global warming trend, precipitation has decreased in many regions, including the Sahel and southern Africa, eastern Australia, Southeast Asia, the Mediterranean, and parts of Central and South America. Increases have been seen in eastern North America, southern South America, northern Europe and northwestern Australia. These changes are reflected in commensurate decreases and increases in run-off and snow (in high latitudes), and similar shifts in evapotranspiration and soil moisture. However, increases in temperature and changes in wind speed, radiation and humidity have tempered or enhanced these patterns regionally.

Despite clear trends for some regions, great variability at interannual to decadal timescales is buried within them, as shown in the global time series of precipitation in Figure 6.2. These variations are linked to a range of internal and external drivers, as discussed in Chapter 3, including solar activity, internal

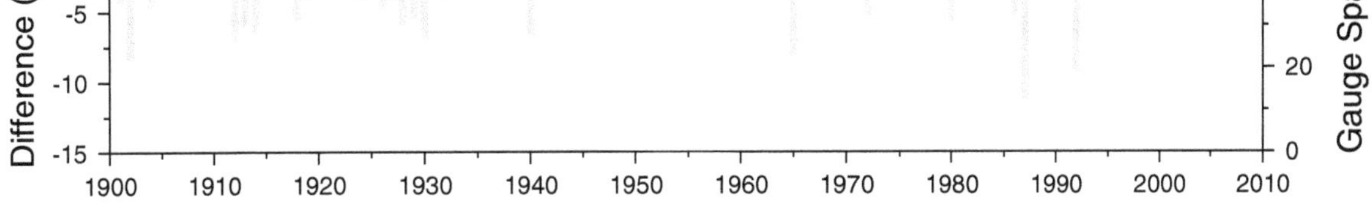

Figure 6.2 *Annual anomalies of global land precipitation (mm), 1900–2008, relative to the mean of 1961–1990*

Note: The black line is the data smoothed using a 10-year moving average. The upper black line is the spatial coverage of the precipitation gauges that contribute to the dataset as the percentage of total land area, assuming a correlative decay distance of 450km.

Source: From the updated dataset of Sheffield et al (2006), which combines data from the CRU TS3.0 (Mitchell and Jones, 2005) and Willmott and Matsuura (2009) datasets ocean–atmosphere variations (such as the ENSO), and the chemical composition of the atmosphere, as driven by volcanic and anthropogenic emissions. It is these types of variations, on decadal and smaller timescales, as well as overall drying trends, that lead to occurrences of drought. In particular, marginal regions at the transition between wet and dry climates are most prone to drought because of the high variability of precipitation in relation to the average precipitation (see Chapter 3). This can be quantified by the coefficient of variation, which is calculated as the interannual variability divided by, and thus normalized by, the long-term mean precipitation. The top panel of Plate 6.2 shows the average annual pattern of precipitation over the second half of the 20th century. The highest values are in the tropics (Amazon, central Africa and Southeast Asia). There are also regions with relatively humid climates, such as in North America (eastern and Pacific northwest), northern Europe and northeast Asia (southeast China, Korea and Japan). At the other extreme, there are deserts that are permanently dry (at least in the latter half of the Holocene), such as the Sahara and Gobi, and the Atacama in South America. At the transition between wet and dry regions are the marginal climates, such as in the Sahel, eastern and southern Africa, southwest North America and southeast Australia, as well as India, northern China and northeastern Brazil. These regions lie at the limits of monsoonal systems that contribute the majority of total annual precipitation for the region, and therefore suffer significant impacts if less precipitation is delivered, or the monsoon circulation is weaker than normal. As we shall see, drought has played a major role in the 20th century history of these regions.

The Global Extent of 20th-Century Drought

The climate has varied greatly over the past 50–100 years, with large decadal changes in precipitation according to region, and an underlying trend towards warmer temperatures globally. We now explore how these variations manifest in the occurrence and characteristics of drought. We focus on soil moisture, since this reflects changes in the supply of water to the land surface (precipitation and snowmelt) and demand/loss (evaporation and run-off), and so provides an overarching picture of drought. We also relate variations in soil moisture to changes in the other environmental variables. The soil moisture data are taken from global hydrological model simulations that provide a spatially and temporally consistent view of hydrological variation for the past 100 years (Sheffield and Wood, 2007). This is a physically based approach that considers the dynamics of the large-scale water cycle and identifies well-known, as well as less well-documented, drought events across the globe in a reliable fashion (see Box 4.1 for details of the hydrological modelling and input data).

Drought is analysed in two ways: first at the local scale, in terms of the time series at each point across the globe (on the data grid of 1° latitude and 1° longitude, about 100km at the equator). From this we can derive the long-term variation at local scales and average this to regional, continental and global scales, to develop a broader picture of drought. Second, we look at individual drought events that are characterized by spatially contiguous areas in drought that persist for several months to years (see Box 4.2). This approach recognizes the fact that drought is the result of regional-scale climate processes that induce drought conditions over large areas. These large-scale drought events are analysed in this chapter as a whole, in terms of their long-term variation, whereas in Chapter 7 they are analysed individually in greater detail. Many studies have considered large-scale drought, either as an individual event, or from a regional, climatological perspective, and the impacts of drought are well documented for more recent events. In the remainder of this chapter and the next we qualitatively compare results from these studies to gain insight into the robustness and consistency of depictions of variation in drought over the 20th century. In particular, we measure these results against the Palmer Drought Severity Index (PDSI), an often-used drought index (see Chapter 4), which is popular in the US for drought monitoring and palaeoclimate reconstructions, as shown in Chapter 5.

Figure 6.3 gives an overview of global drought over the 20th century. Drought is defined when soil moisture drops below a threshold (in this case the 20th percentile, which is the level of soil moisture that is reached, on average, once every five years). As discussed in Chapter 4, the probabilistic nature of the index normalizes the data to the local climatology, allowing for the comparison of events across regions and through time. Drought is

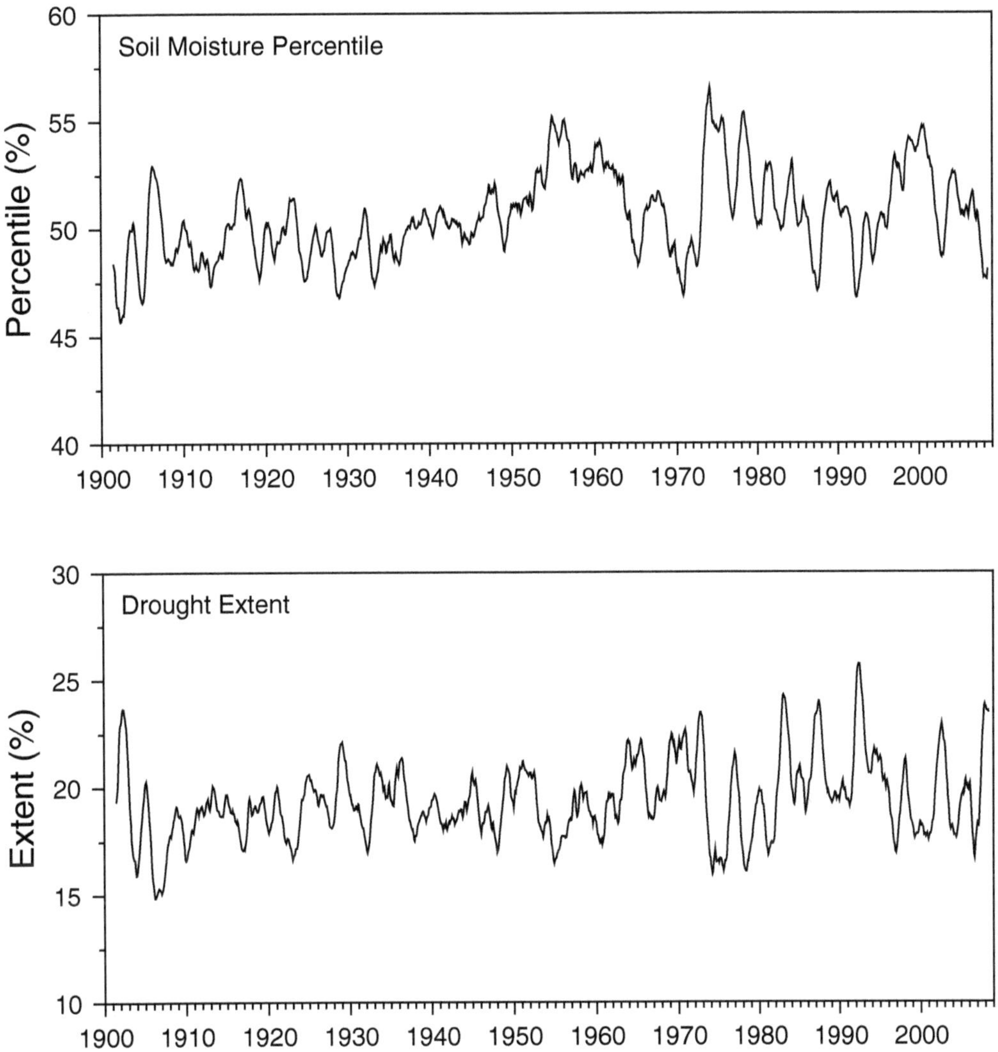

Figure 6.3 *Global monthly time series of area-averaged soil moisture percentile, and percentage area in drought*

Note: The data are smoothed in time to aid visualization. A drought is defined as soil moisture below the 20th percentile.

Source: Data are taken from an updated analysis of Sheffield et al (2009)

given in terms of the area-averaged soil moisture percentile and area in drought. By definition, the mean soil moisture percentile over the full period is 50 per cent (the median value) and the mean area in drought is 20 per cent. Globally, the variation in the extent of drought is reduced due to the spatial averaging, but generally follows the decadal pattern of precipitation (Figure 6.2). There is a general wetting until the mid-1950s, after which there has been a more pronounced decadal variation forced by long-term changes in precipitation. This may also be partly due to the uncertainty in the data in the earlier part of the century, when precipitation gauges were sparse. Most notably, there is a slight trend towards dry conditions in recent years. We will return to this in Chapter 8 when we discuss recent events and future predictions.

Figure 6.4 shows the same data, but for the continents as well as the world. The variability of the time series for the world and Asia is damped because it presents an average over large spatial scales. In comparison, the smaller continents (Europe and Oceania) show much higher variability, but this is also driven by the climate variability in those regions. A number of periods stand out as experiencing prolonged and widespread drought. For example, the droughts of the early 1970s and mid-1980s are clearly visible in the African time series, with more than 40 per cent of the continent experiencing drought conditions at these

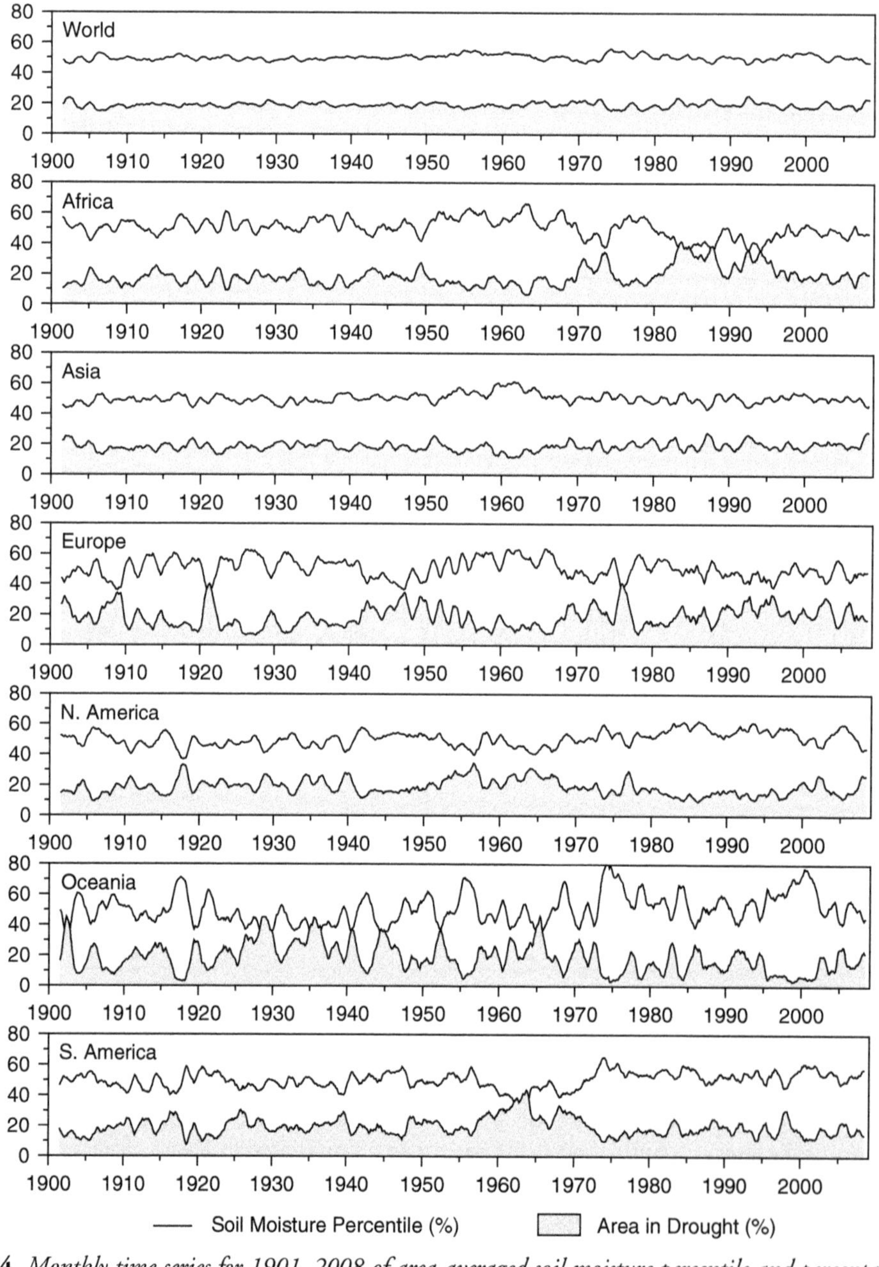

Figure 6.4 *Monthly time series for 1901–2008 of area-averaged soil moisture percentile and percentage area in drought for the world and the continents*

Note: A drought is defined by soil moisture below the 20th percentile. The data have been smoothed to aid visualization.

Source: Adapted from Sheffield et al (2009)

times. In Europe, high variability in the 1950s is associated with multiple periods of large drought extent, which are not repeated until the mid-1970s. In North America, the dry spells in the 1930s and 1950s show up clearly as prolonged sequences of large drought extent, reaching 36 per cent of the total continental area. Other notable dry periods occur around 1976–1977, the late 1990s and in more recent years. For Oceania, there are several events with a high percentage extent in drought (e.g. 82 per cent in 1965) and also a near-complete lack of drought in the mid-1970s. Drought extent is also very low (< 20%) during the late 1990s, in contrast with the extended drought conditions since the turn of the 21st century in the eastern part of Australia (BOM, 2010). South America is characterized by extended dry periods in the early 1950s, 1960s and 1990s, with the mid-1970s and early 2000s being less drought-prone.

Comparison with the PDSI

The PDSI has been extensively analysed and applied in research studies (e.g. Dai et al, 2004, van der Schrier et al, 2006) and is arguably the most prevalent drought index in operational use. It is a proxy for soil moisture that correlates well with soil moisture and streamflow observations (Dai et al, 2004), but has been criticized for its hydrological simplicity and lack of spatial consistency (Alley, 1984). In this section, we compare the VIC model soil moisture index with a PDSI dataset driven by the same precipitation and temperature data. To calculate the PDSI we use the self-calibrating algorithm of Wells et al (2004), which removes some of the spatial inconsistencies of the original index.

At global scales, the soil moisture index correlates with the PDSI dataset (Figure 6.5), and both indices appear to have decreased since the mid-1970s. However,

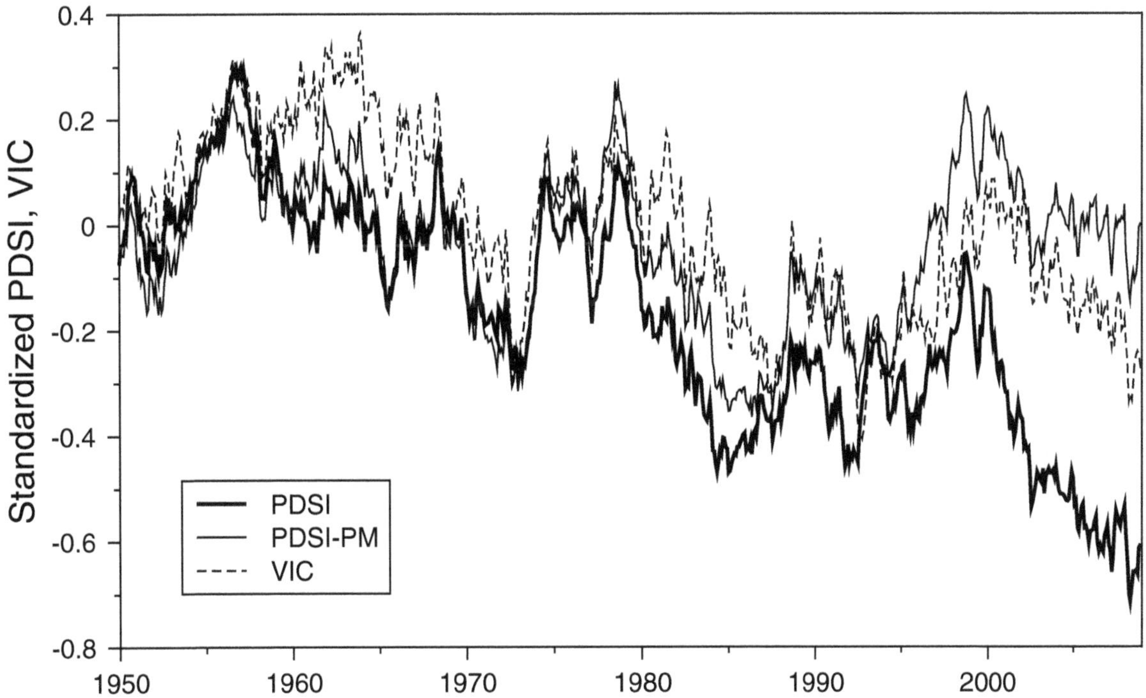

Figure 6.5 *Time series of globally averaged VIC soil moisture index and two PDSI datasets for 1950–2008*

Note: The data are averaged over global land masses, excluding Greenland and Antarctica. All datasets have been standardized (subtract the mean and divide by the standard deviation).

Source: The PDSI dataset is calculated using the self-calibrating algorithm of Wells et al (2004), forced by the same precipitation and temperature data as the VIC soil moisture index. The $PDSI_{PM}$ dataset uses the more physically realistic Penman-Monteith algorithm (Monteith, 1965) for calculating potential evaporation.

they diverge in their magnitude of decline, especially in the last decade, when the PDSI falls dramatically and indicates a rise in drought that has been attributed to global warming (e.g. Dai et al, 2004; Briffa et al, 2009). The reason for this divergence is partly due to the simplified representation of potential evaporation in the PDSI model, which is based on temperature only, and so responds to recent global warming (Sheffield and Wood, 2007). However, evaporation is a complex process that is sensitive to changes in radiation, humidity and wind speed. Figure 6.5 shows PDSI values using the more comprehensive Penman-Monteith potential evaporation model (Monteith, 1965) indicated as $PDSI_{PM}$. This shows the dependency of the decline on the temperature-based formulation, and thus the overestimation of the rise in drought during the late 20th and early 21st centuries by the PDSI. This has ramifications for how the impacts of global warming are monitored and placed into a historical context, which of necessity may be based on simplified methods. At the grid scale, the VIC and PDSI datasets are well correlated but tend to diverge in cooler seasons and high latitudes, and substantially differ in dry regions (Plate 6.3). The soil moisture index also shows greater month-to-month variability in comparison to the PDSI datasets (not shown). This has been attributed, in part, to the PDSI ignoring the daily variation of precipitation, the effects of snow and the use of time-invariant vegetative cover, which will tend to dampen the index over seasonal scales (Sheffield et al, 2004). In Chapter 8, we discuss the impact of changing snowpack on drought in the context of global warming.

Characteristics of Global Drought

Chapter 4 describes some of the main characteristics of drought, including the duration, magnitude and severity. To recap, the severity or magnitude is the deficit of a hydrological quantity such as soil moisture or streamflow below a threshold and the intensity is the mean severity over the duration of the drought. The geographic variation of drought characteristics can be represented by a summary statistic, such as the mean value calculated over all drought events. Plate 6.4 shows the total number of droughts for each 1° grid cell of any duration (Plate 6.4a) and the mean duration of these droughts (Plate 6.4b). The data are calculated for the period of 1950–2000, when the information is at its most reliable. The total time in drought is equivalent to the total number of droughts multiplied by the mean drought length, with an expected value of 20 per cent, as defined by the threshold value. The total number of droughts is minimal in arid regions, such as the Sahara and the Taklimakan desert in western China, and these regions have been masked out. The highest values (> 25 droughts over 1950–2000) occur in more humid regions such as eastern North America, the Amazon and Paraná river basins, tropical Africa, Europe and Southeast Asia. Perhaps counter-intuitively, this definition of drought shows that semi-arid regions have the least number of droughts and humid regions have the most. However, this is because the definition is based only on the number of droughts and does not take into consideration the duration of the droughts. For example, the Sahel, parts of central Asia and high northern latitudes stand out as having high average drought durations (Plate 6.4b), a result of the higher frequency of long-term drought conditions. Regions of higher average drought durations are collocated with regions of higher precipitation variability, as shown by the coefficient of variation (CV) values in the lower panel of Plate 6.2.

The global distribution of frequencies of droughts of different duration for 1950–2000 is shown in Plate 6.4c–f. Very short-term droughts (one to three months) are most frequent in the eastern US, parts of the western Amazon, Argentina, tropical Africa and Southeast Asia, with over 30 droughts per 50 years. The highest frequencies of short-term droughts (four to six months) show much greater spatial variability and are generally located in mid-latitudes and parts of the tropics. Medium-term droughts (7 to 12 months) occur most frequently in high latitudes, especially in northern Canada and in eastern Siberia. These regions are characterized by freezing temperatures and thus are prone to persistent soil moisture anomalies over the cold season. Long-term droughts (longer than 12 months) are restricted to a few areas, including higher northern latitudes, Central Asia and the Sahel, but are absent in the tropics. For the tropics, precipitation variability is highest on shorter timescales, thus preventing droughts from persisting longer than a few months. Overall, shorter droughts are more common in humid regions and longer droughts are more common in arid and semi-arid regions (see Chapter 3).

Finally, Plate 6.4g–h shows the global distribution of mean drought intensity and severity. The data were calculated for a percentile threshold of 50 per cent to better capture the statistics of all drought events, irrespective of their magnitude. However, the results using other threshold values show similar patterns but different amplitudes. Mean drought intensity is 15–25 per cent (departure below the threshold) for humid regions and 5–15 per cent for drier regions, as a result of the greater frequencies of short-term droughts in more humid regions and of long-term droughts in drier regions. The eastern US, central and eastern Europe, Southeast Asia, China and tropical Africa exhibit the largest intensity values. For mean drought severity, the variation is more distinct, and several regions stand out as having experienced relatively severe conditions, driven more by long durations than by high intensities. Accumulated deficit/severity is greater than 200 per cent over the Sahel, northern Canada, northern Siberia and the Taklimakan desert north of the Himalayas, with severities exceeding 500 per cent for parts of the Sahel.

Regional Drought over the 20th Century

Further insight into the temporal variation of drought can be found at regional scales, where we see the impact of different climate regimes on drought. The regionally averaged time series of soil moisture percentile and spatial extent of drought are illustrated in Figure 6.6. The regions are broad rectangular areas shown in Figure 6.7 and defined in Table 6.1 in terms of area (Giorgi and Francisco, 2000). Note that the time series in Figure 6.6 are smoothed to aid visualization, which will tend to diminish the extreme values. For some regions, such as the Amazon, the density of rain gauges was minimal in the early part of the 20th century, and so there is a large degree of uncertainty in the precipitation datasets for this period. Therefore the results are only shown for when a reasonable spatial coverage of gauge data is available. The maximum drought extent values from the original unsmoothed monthly time series are given in Table 6.1.

Globally, the variation is much smaller in comparison to the regional scale because of the spatial averaging (Figure 6.6), which tends to cancel out the regional variations. The regional time series can be summarized as follows. Northern Europe shows higher variability than the Mediterranean region, and dry conditions in northern Europe are most spatially extensive in the early 1920s, 1940s, 1950s, mid-1970s and mid-1990s, and reach a sharp peak in 2003. The Mediterranean shows an increase in drought in recent years. West Africa is dominated by the long-term drought period in the Sahel from the late 1960s to the mid-1980s, which is also reflected but to a lesser extent in eastern Africa. In southern Africa, there are a number of peaks that cover up to 40 per cent of the region and a particularly extensive wet period in the 1970s. The large size of the northern Asian region tends to dampen the variation in soil moisture and spatial extent, although an underlying decadal variability is apparent. Spatially extensive drought conditions in the late 1950s to early 1960s over the Tibetan Plateau are followed by increasingly wet conditions until the 1990s, from when there is a sharp increase in drought extent. East Asia is characterized by decadal variability in drought extent, with a trend to drier conditions in the last 10 to 20 years. Southeast Asia has experienced several periods of large drought extent in the early 1970s and 1980s. The Australian data are dominated by an extensive dry period in the 1940s and 1960s, followed by an upward jump in soil moisture around 1973, also seen in the precipitation data. The extent of drought in western North America shows long-term variation with long-term dry conditions in the 1930s and 1950s, and a relatively wet period in the 1980s. There is much greater variability in central North America and a number of events that surpass 40 per cent coverage. The data for the Amazon region are damped because of the averaging over such a large area, but show long-term drought in the 1960s, and peak extents in the 1980s and 1990s.

Figure 6.6 gives a general overview of the monthly spatial extent of soil moisture deficits on a regional basis and hints at time periods that had more or less drought activity. However, this analysis does not take into account the severity or duration, which is relevant for evaluating the impacts. We will explore this in more detail in Chapter 7, in which we consider individual large-scale events, but first we shall highlight periods of higher and lower drought activity. Figure 6.8 shows regional time series of the spatial extent of drought in a similar manner to Figure 6.6. However, the results are filtered to reveal drought periods that were of high intensity. The filtering method is applied using a moving window of a three-month duration and a 10 per cent threshold, and leaves only those periods where the deficits were severe and spatially extensive.

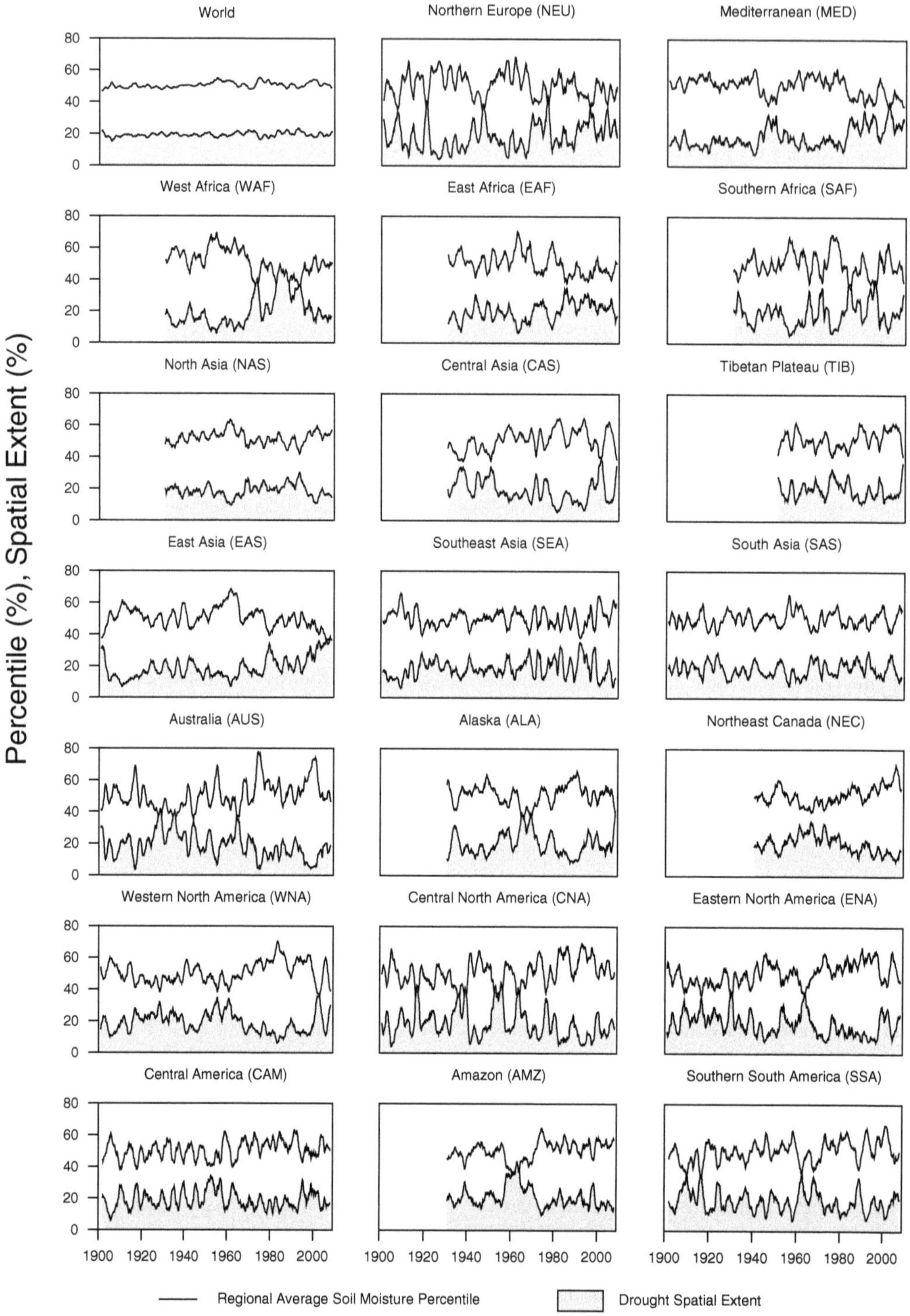

Figure 6.6 *Time series of regionally averaged soil moisture percentile and extent of drought ($q(\theta)$ < 20 per cent) for 1901–2008*

Note: The data are smoothed from the original monthly data with a 25-month moving average.

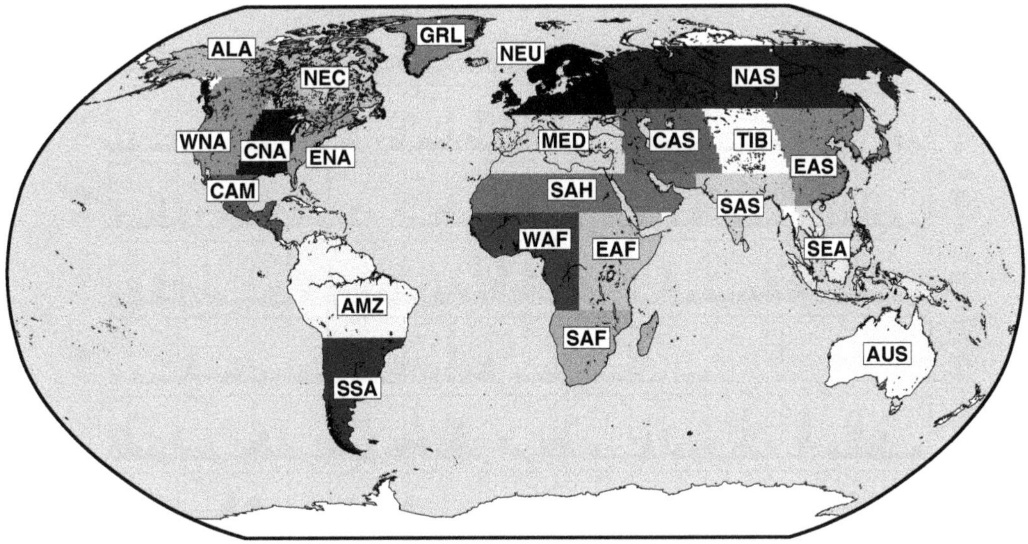

Figure 6.7 *Map of regions used in the analysis*

Source: Based on Giorgi and Francisco (2000)

Table 6.1 *List of regions used in this study, their total area and the maximum spatial extent of drought, using the formula A ($q_0(\theta)$ = 20 per cent), and when it occurred*

Region	Total Area (km² x 10^{12})	A_{max} (%)
World	124.9	29.2 (Jun 1992)
Northern Europe (NEU)	5.1	79.6 (Dec 1920)
Mediterranean (MED)	6.3	60.5 (Jun 1945)
Western Africa (WAF)	8.2	65.2 (Oct 1983)
Eastern Africa (EAF)	8.6	53.7 (Aug 1984)
Southern Africa (SAF)	6.2	77.7 (May 1970)
Northern Asia (NAS)	14.7	51.6 (May 1969)
Central Asia (CAS)	6.3	63.1 (Aug 2008)
Tibetan Plateau (TIB)	4.7	63.3 (Apr 1905)
East Asia (EAS)	8.5	58.7 (Dec 2007)
Southeast Asia (SEA)	6.5	66.8 (Oct 1997)
Southern Asia (SAS)	5.4	60.4 (Jul 1992)
Australia (AUS)	8.6	76 (Feb 1965)
Alaska (ALA)	3.5	59 (May 1964)
Northeastern Canada (NEC)	4.5	58.7 (July 1964)
Western North America (WNA)	5.6	61.2 (May 2002)
Central North America (CNA)	3.4	86.2 (Oct 1952)
Eastern North America (ENA)	2.7	70.1 (Jan 1931)
Central America (CAM)	3	62.5 (Jun 1998)
Amazon (AMZ)	12.9	73.4 (Oct 1963)
Southern South America (SSA)	5.9	63.2 (Nov 1910)

Note: The regions are taken from Giorgi and Francisco (2000) but exclude the Greenland region, which has been replaced by the northeastern Canada (NEC) region. The World region excludes Greenland and Antarctica.

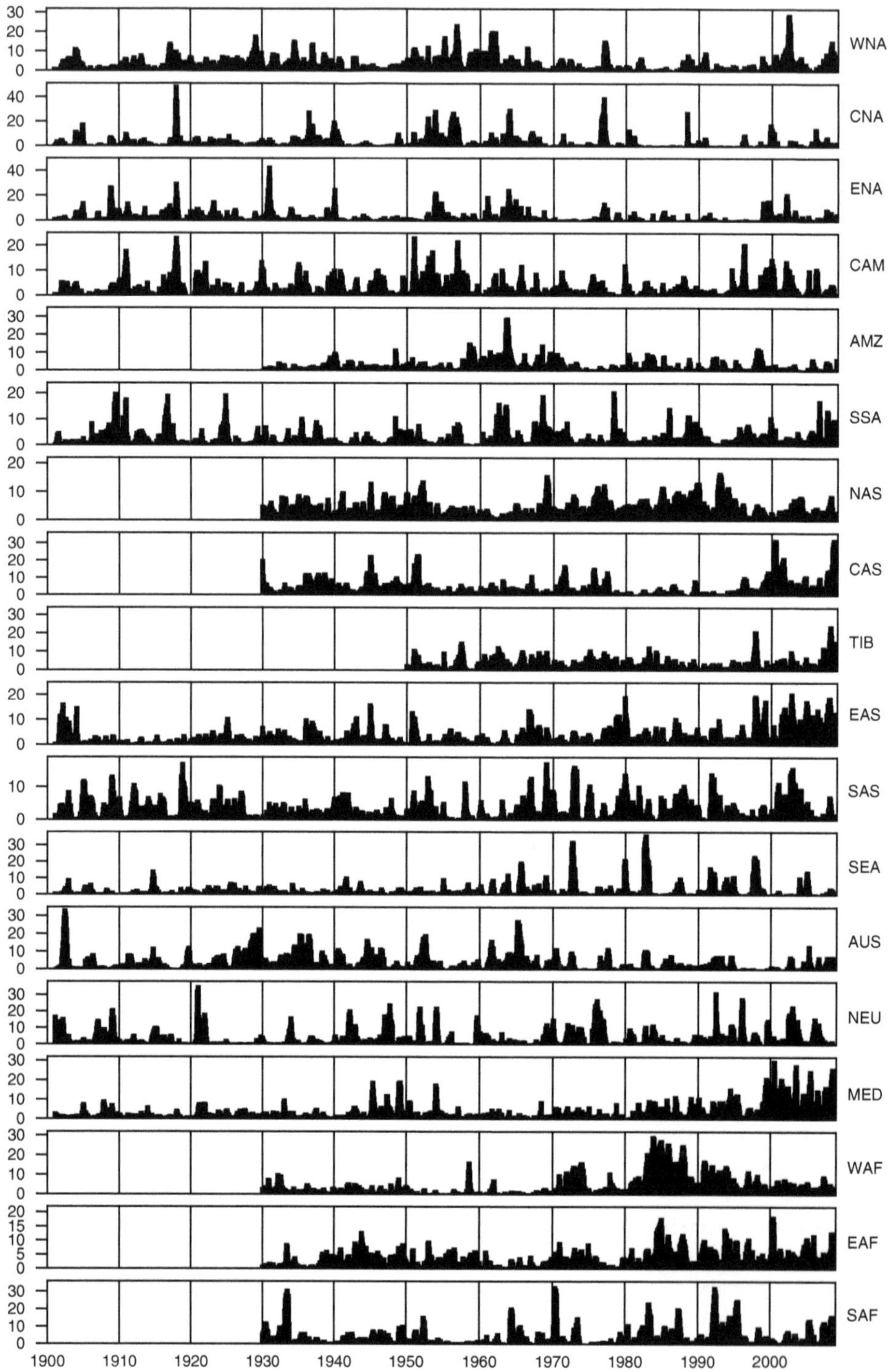

Figure 6.8 *Time series of regional spatial extent of short but intense drought periods*

Note: This is defined as the soil moisture percentile q(θ) < 10% over a three-month moving window {$q_0(\theta)$ = 10, D = 3}. Note that the y-axis scales differ between regions.

Source: reproduced with permission from Sheffield and Wood (2007)

Over North America, the 1950s was the decade of most spatially extensive and prolonged deficits in the west and central regions. This is expected and is only matched over the US by the 1930s drought during this century (Andreadis and Lettenmaier, 2006). In terms of individual years, the winter drought of 1976–1977 was considerably more extensive than the drought of 1988, which was purported to be the worst natural disaster in US history (Trenberth and Branstator, 1992). Recent drought in the west shows up as a very large spike in 2002. The most extensive drought period in Central America occurred in the 1950s (and was linked to the North American droughts of this time), which is consistent with reported conditions (Liverman, 1999) and with gauge-based precipitation records that show an increasing trend since the early 1960s (Aguilar et al, 2005). Drought resurged in the late 1990s and early 2000s. In the Amazon, the series is dominated by dry conditions during the mid-1960s, and other peak extents coincide with strong El Niño events, occurring in 1957–1958, 1965, 1972–1973, 1992 and 1997. The climate of this region is well known for its connectivity with ENSO (see Chapter 3), bringing dry and warm conditions, especially in the northern part of the region (Foley et al, 2002). In southern South America, a multitude of spatially extensive drought years stand out, including events in 1909, 1916, 1924, 1962–1963, 1968, 1978, 1985–1986, 1988 and 2006–2007. Some of these years coincide with La Niña events, which are known to cause dry conditions in the east of this region (Boulanger et al, 2005), although other La Niña events are not reflected in the drought record, which may partially be a result of averaging over the whole region.

The series for northern Asia is interesting for two reasons. First, the extent of severe drought appears to vary over multidecadal timescales, which is possibly a result, in part, of linkages to long-term variability in the NAO (see Chapter 3) (Visbeck et al, 2001). Dry periods are prevalent in the 1940s to early 1950s, and the late 1960s to early 1990s. Second, closer inspection of data shows that the peaks tend to occur at regular intervals in the autumn and winter. As the soil column freezes, the soil moisture at any location tends to remain fairly constant over the winter period, which will tend to prolong any preceding drought conditions. Central Asia shows major dry periods during the mid-1940s and early 1950s, and in the latter years of the record, analysed by Barlow et al (2002). However, there is a spate of less extensive events in the 1970s. In terms of the Tibetan Plateau, peaks in 1997 and 2007 dominate the series. In East Asia, which covers most of eastern China and Japan, spatially extensive severe drought is limited mostly to a prolonged dry period from the late 1990s onward. Southern Asia shows a few major dry periods, which tend to occur in the latter half of the record. There are distinct major events in 1972, 1982–1983 and 1997, which coincide with major El Niño episodes. For Australia, although 1902 stands out with the highest spatial extent, drought periods are more prevalent in the middle part of the record, most notably in the late 1930s, mid-1940s, 1952 and 1965. This may be due to changes in the influence of ENSO on the Australian climate since the 1970s (Nicholls et al, 1996), as well as the influences from outside of the Pacific (Verdon-Kidd and Kiem, 2009).

For Europe, the northern region has far more spatially extensive droughts than the Mediterranean region. Spatially extensive short-term dry periods in northern Europe occur in 1921, the 1940s–1950s, 1976 and the last two decades. The late 1940s and 1954 are the most extensive periods in the Mediterranean region in the early part of the record, but drought appears to be much more extensive in the late 1990s and 2000s. This is generally consistent with Lloyd-Hughes and Saunders (2002) and van der Schrier et al (2006), who found that the 1940–1950s and 1990s (their analyses ended in the early 2000s) were the most drought-prone periods across the whole of Europe in terms of PDSI and 3- and 12-month SPI. Western Africa is dominated by prolonged drought in the 1970s, 1980s and early 1990s (Hulme, 1992; L'Hôte et al, 2002), which reached peak extent in 1984 and 1972. This is reflected in the East African region, but the spatial coverage is lower. Dry conditions persist from the 1980s to the 2000s. In southern Africa, peak events occur in 1933, 1970 and 1992, which corresponds with the SPI analysis of Rouault and Richard (2005), who also determined that the majority of major drought events coincide with El Niño episodes.

Global Variability in Soil Moisture and Links with Large-Scale Climate Variability

We have seen that drought varies considerably at different timescales and space scales. At regional scales, drought varies on decadal timescales but also on shorter timescales, which, in some regions, appear to be in tune with the ENSO. In other regions, there are variations that appear to be linked with other factors. Much research has been carried out on discerning the covariability of climate with terrestrial hydrology and its extremes, including drought. Rajagopalan et al (2000) found relatively robust relationships in the southwestern US between the PDSI and several climate indices, although these varied in time. Modelling studies have confirmed the existence of connections with the ENSO, seen especially in the correlation between the US West and La Niña (Schubert et al, 2004; Seager et al, 2005), and medieval droughts over longer timescales in the US (Herweijer et al, 2007; Seager et al, 2008). McCabe et al (2004) looked at the association between the PDO and AMO with decadal variability in the North American PDSI. Globally, Dai et al (2004) found the PDSI to be coupled to ENSO variability, and this was confirmed by studies of Herweijer and Seager (2008), and Sheffield and Wood (2008). Jiang et al (2006) analysed proxy indicators for the Yangtze River during 1868–2003 and found these to be related to the ENSO. McCabe and Palecki (2006) explored covariability between the PDSI and global SSTs on decadal to multidecadal timescales, and revealed strong relationships and a dominance of the PDO and AMO, in explaining most of the variability. Sheffield and Wood (2008) explored the variability of the soil moisture dataset used here and showed that it was mainly related to the ENSO, with the NAO and the AMO possibly playing secondary roles. We summarize and extend this analysis and look specifically at the occurrence of major drought events (identified by the cluster analysis) and large-scale climatic variability.

Principal component analysis (PCA) can be applied to the monthly soil moisture data to unravel its major modes of variability. Essentially, PCA decomposes the data into new variables that are uncorrelated (independent) but can be recomposed to the original data when linearly combined. The first few variables or components in PCA will generally explain most of the variability in the whole dataset, and can thus be used to identify connections with major modes of climate variability, and where these are strongest globally. Plate 6.5 shows the spatial loadings of the first three principal components of monthly soil moisture percentiles and represents the major modes of variability in global drought and wet spells. These maps identify where the variance of each principal component has the largest imprint, although this will manifest in different directions (drying or wetting) in different locations. Figure 6.9 shows smoothed time series of these principal components. Although the variance explained by the first three components is small, the value decreases rapidly for higher-order components, and so only the first three components are of interest. The first principal component (PC1) explains 8.1 per cent of the global variability in soil moisture and is negative in the Amazon, the Sahel, southern Africa, northeastern Siberia, Southeast Asia and Australia. PC1 is positive in southern South America, central and southern US, northern Canada and Central and East Asia. The spatial distribution of loadings for PC1 reflect known regions of AMO impacts (Figure 3.6, Chapter 3). The time series in Figure 6.9 correlates with Nino 3.4 SST variability (r = 0.50), indicating that tropical Pacific temperatures are the primary driver of global variability in soil moisture, a result also shown by Dai et al (2004) for the PDSI data. PC2 explains 6.4 per cent of the variance and shows strong positive loadings over northern Canada, the Amazon, southern Africa and central Australia. Negative loadings are highest in Alaska, northern Europe and far eastern Asia. The time series of PC2 shows a low-frequency multidecadal oscillation that co-varies well with the AMO index (r = 0.67) and is consistent with the PDSI analysis of McCabe and Palecki (2006). The third component, PC3, explains 6 per cent of the variance and shows strong positive loadings over central Europe, through Russia, Australia, South America and Alaska, and strong negative loadings over northern Siberia, East Africa and northeast Canada. The decadal variability follows the NAO (Figure 6.9), but the correlation is insignificant (r = 0.21) and especially weak during the mid-1980s to mid-1990s. It is likely that any real physical association is masked somewhat in this analysis of annual soil moisture values, because the NAO is thought to impact mostly in the wintertime.

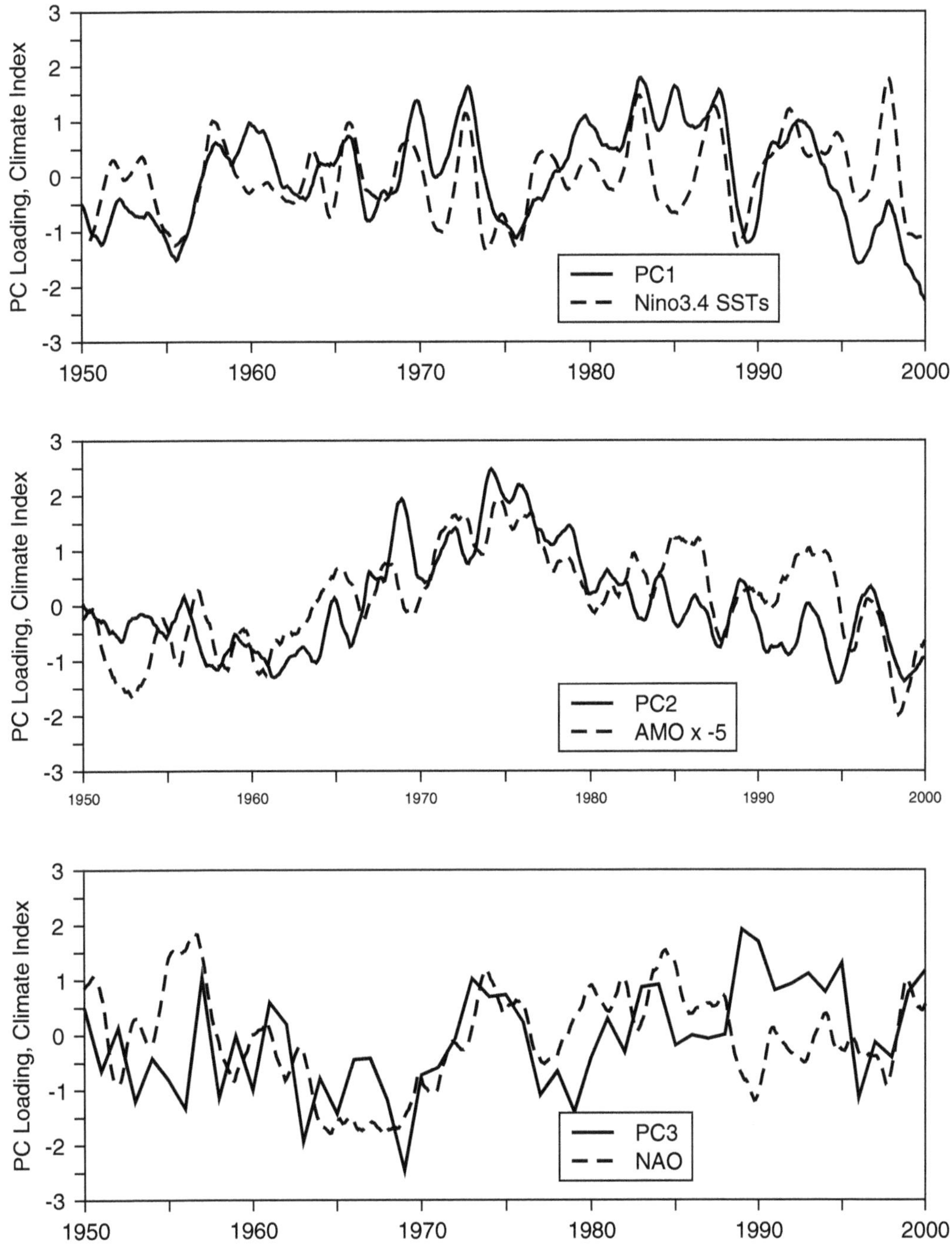

Figure 6.9 *Smoothed time series of the first three principal components of monthly soil moisture percentile for 1950–2000 in comparison to climate indices*

Note: The Nino3.4 SSTs are an index of ENSO and are defined as more than 5°S–5°N, 120°W–170°W. The AMO is defined as the detrended area-weighted average SSTs over the North Atlantic, at latitudes of 0° to 70°N. The NAO is the wintertime difference of normalized sea level pressure between stations in the Azores islands near Portugal and Iceland.

Source: Data taken from Sheffield and Wood (2008)

Global Variability in Drought and Links with Large-Scale Climate Variability

Soil moisture co-varies with large-scale modes of climate variability, and especially on shorter timescales with the ENSO. A question is how does this translate into the occurrence of drought at large scales? We answer this by counting the number of large-scale drought events globally over the 20th century (Table 6.2) and compare these with the same indices of climate variability. As mentioned before, the droughts are identified from the soil moisture data using a cluster analysis that tracks contiguous areas of soil moisture deficits greater than 500,000km^2 and persisting for three months or more (Box 4.2). Based on these criteria, there have been 563 droughts greater than 500,000km^2 and longer than three months, on a global scale during 1901–2008. The uncertainty in precipitation in the earlier part of the record means that this is likely to be a conservative estimate, especially in South America, Africa, Central Asia and in the pan-Arctic. More than half of these events lasted for six months or less, and half were smaller than 3 million km^2. More than 10 per cent of the events lasted for 12 months or more, and 25 droughts were greater than 7 million km^2. Table 6.2 summarizes the occurrence of drought globally and for each continent, including the longest and most spatially extensive events. The drought that lasted the longest was 42 months (4 years) in Asia during 1986–1990, closely followed by the 1954–1957 North American drought (41 months). The most spatially extensive was the South American drought of 1963–1964, which reached its peak extent of 10.7 million km^2.

Compare this to Table 6.3, which summarizes the number of reported droughts from the perspective of the impacts. These data are based on reports from UN agencies, national governments, non-governmental organizations (NGOs), reinsurance companies and the media. The table reports not only the number of events in each region but also the number of deaths, the affected population and the economic losses. The mismatch between Tables 6.2 and 6.3 in terms of the number of droughts is due to the differences between how the events are identified and the underlying data. Here, we use physically based modelling that is subject to biases in the model and the records of meteorological data. The impacts of drought in historical reports are subjectively recorded, and there are likely to be fewer records in earlier periods. The data in Table 6.3 are focused on the impacts to human populations, rather than climatological phenomena, and so display a higher proportion of events in Africa and Asia where the populations are most vulnerable.

Figure 6.10 shows the monthly time series of the number of large-scale drought events, as defined by the cluster analysis. The number of droughts is counted separately in each month, so a single drought event may contribute to several months in the time series (the variability of droughts of different durations is explored later in this section). Also shown is the number of droughts filtered by their maximum spatial extent, where the data for each line is calculated for incremental thresholds of spatial extent, to see how the temporal variation changes for larger drought events. Data for before 1930 are not shown because of the uncertainty in the precipitation. Even the data for the 1940s and 1950s should be viewed with a degree of caution. The mean number of global droughts greater than 500,000km^2 in any month is about 4.5 (or 55 per year), with a standard deviation of 1.6. When coupled with estimates of the rate of flooding globally (about

Table 6.2 *Summary of large-scale drought occurrence for the six continents during 1901–2008*

Region	Number of droughts	Number of droughts <= 6 months	Number of droughts >= 12 months	Longest duration (months)	Maximum spatial extent (million km^2)
Africa	74 (13%)	46	8	21 (1992–1993)	9.9 (33%, Apr 1983)
Asia	183 (33%)	121	19	42 (1986–1990)	9.8 (22%, Jun 1987)
Europe	78 (14%)	45	9	25 (1975–1977)	5.1 (51%, Dec 1920)
North America	102 (18%)	65	15	41 (1954–1957)	7.5 (31%, Oct 1956)
Oceania	44 (8%)	27	7	24 (1928–1930)	5.9 (77%, Feb 1965)
South America	82 (15%)	58	10	19 (1982–1983)	10.8 (61%, Oct 1963)
World	563	362	68	42 (1986–1990)	9.9 (33%, Apr 1983)

Note: For the last column, the extent as a percentage of total area and the date when the maximum spatial extent was attained are given in brackets. Oceania is defined as Australia, New Zealand, Papua New Guinea and the Pacific Islands.

Table 6.3 *Summary of global drought events for 1900–2004 and impacts*

Region	Number of events	Deaths	Affected Population	Economic Losses (US$ 1000)
Africa	139 (35.5%)	1,119,007	243,267,579	5,271,332
Asia	113 (28.8%)	9,663,043	1,541,783,427	24,026,695
Europe	33 (8.4%)	1,200,002	19,865,575	17,888,600
Americas	90 (23%)	73	61,003,351	18.377,739
Oceania	17 (4.3%)	660	8,027,635	13,303,000
Total	392	11,992,785	1,873,947,567	78,867,366

Source: Adapted from Below et al, 2007

one per day), this highlights the extent of extreme weather events. The time series of drought numbers is variable and indicates several periods of increased global drought activity: the 1960s, early 1980s, early 1990s and the 2000s for droughts covering more than 500,000km^2. The mid-1970s to early 1980s are characterized by the lowest occurrence of droughts. Furthermore, 1975 stands out as the year with the fewest droughts of all extents (26 drought months). The year with the most drought months is 1992 (86). Table 6.4 shows the top five ranked years in terms of the number of months with a drought extent of more than 500,000 km^2. Interestingly, all but one of these years (1970) is associated with intense El Niño events. This is explored in more detail later.

When filtered by spatial extent, the number of droughts between 500,000km^2 and 1,000,000km^2 shows similar magnitude and variability (Figure 6.10). For droughts of a greater extent than 5,000,000km^2, there is a large reduction in the number of events and a change in the temporal variation. In fact, for these very large droughts, the series shows a decadal variation, with five to ten years of activity interspersed with a few years of no drought events. The drop in the number of droughts greater than 5,000,000km^2 is an interesting result that probably reflects the geographical or climatic

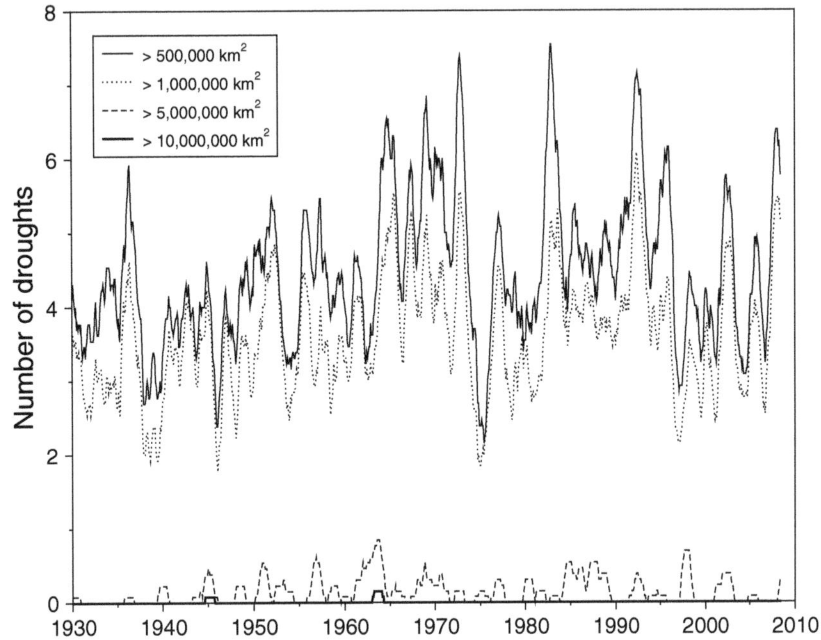

Figure 6.10 *Monthly smoothed time series of the number of global droughts for different areal extent thresholds*

Note: An individual drought is defined as a contiguous region of soil moisture deficit below the 20th percentile, which will probably persist for multiple months (its extent and severity will change through time), and so it will contribute to several months within the time series.

Table 6.4 *Top ranked global drought years based on the number of months with a drought extent of more than 500,000 km²*

Year	Number of droughts
1992	86
1965, 1972, 1983	77
1964	75
1982	74
1970	73

factors that control the size of drought events. The number of very large droughts (more than 5,000,000km²) is significantly lower than the total number of droughts, which is partly a result of the size and shape of continental land masses, which limit the maximum size of a drought, but which may also be related to the length scales of climate anomalies, as driven by planetary (Rossby) waves (Herweijer and Seager, 2008). These large-scale atmospheric disturbances emerge in the meanders of the jet stream and account for much of the variability in high- and low-pressure systems, and therefore mid-latitude weather.

At interannual timescales, the main influence on global climate is the ENSO, and this has impacts on soil moisture. Presumably, this has a trickle-down effect on the number of large-scale drought events across the world. Figure 6.11 compares the number of droughts with the time series of SSTs in the Nino3.4 region of the tropical Pacific (5°N–5°S, 120–170°W), which is an indicator of ENSO variability. As ENSO episodes (warm El Niño or cool La Niña) and their impacts tend to persist for several months to two years, the drought time series is calculated here for only those drought events of 24 months or less. Correlation is weak (r = 0.36 for the smoothed data), but the year-to-year variation is similar, excepting the early 1950s, and 1979 and 1988, when the two time series diverge briefly. This similarity is consistent with reported covariabilities between ENSO and soil moisture indices (modelled soil moisture or PDSI), although here the relationship with the number of contiguous large-scale dry extremes of soil moisture is somewhat less distinct. At smaller scales, the correlations are likely to be stronger in regions known to be influenced by ENSO, such as Oceania and Southeast Asia, as shown in the PCA loadings in Plate 6.5, and as suggested by the high correlations found by Sheffield and Wood (2008) between ENSO and soil moisture in these regions. Variations in the tropical Pacific can also weaken the summer East Asian monsoon and lead to drought in eastern China (Li et al, 2010). For the US, it is likely that individual drought events, such as those in the 1930s and 1950s, were linked to SST conditions in the tropical Pacific Ocean, but particularly La Niña conditions (Schubert et al, 2004; Seager et al, 2005).

The link between the ENSO and large-scale drought may also be investigated using a non-linear,

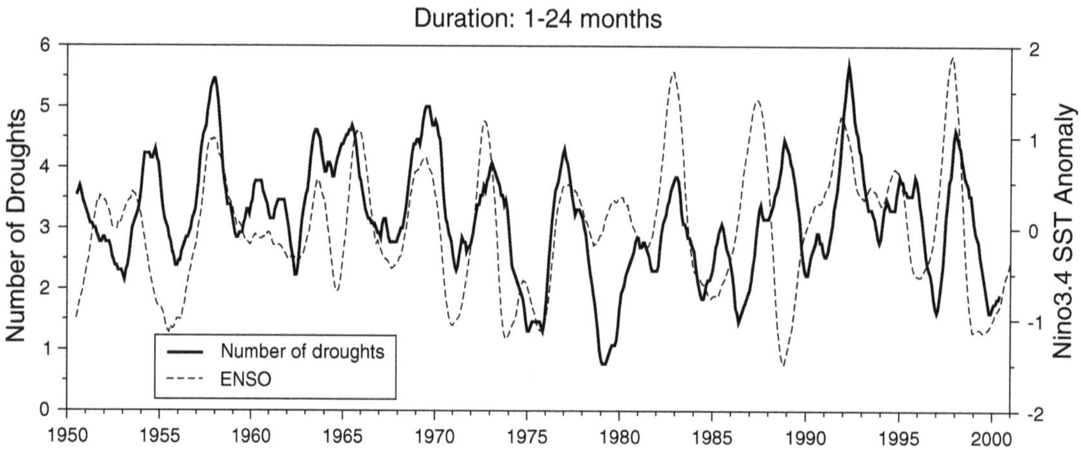

Figure 6.11 *Monthly time series of number of droughts of less than or equal to 24 months (black line) and Nino3.4 SST anomalies (dashed line)*

Note: The time series has been smoothed using a 13-month running mean filter.

Source: adapted from Sheffield et al (2009)

categorical-type relationship, rather than a linear correlative one. Figure 6.12 shows composites of soil moisture and two drought characteristics (contiguous area in drought and number of droughts) over a 36-month window that straddles the year of an El Niño or La Niña event, helping to show the potentially

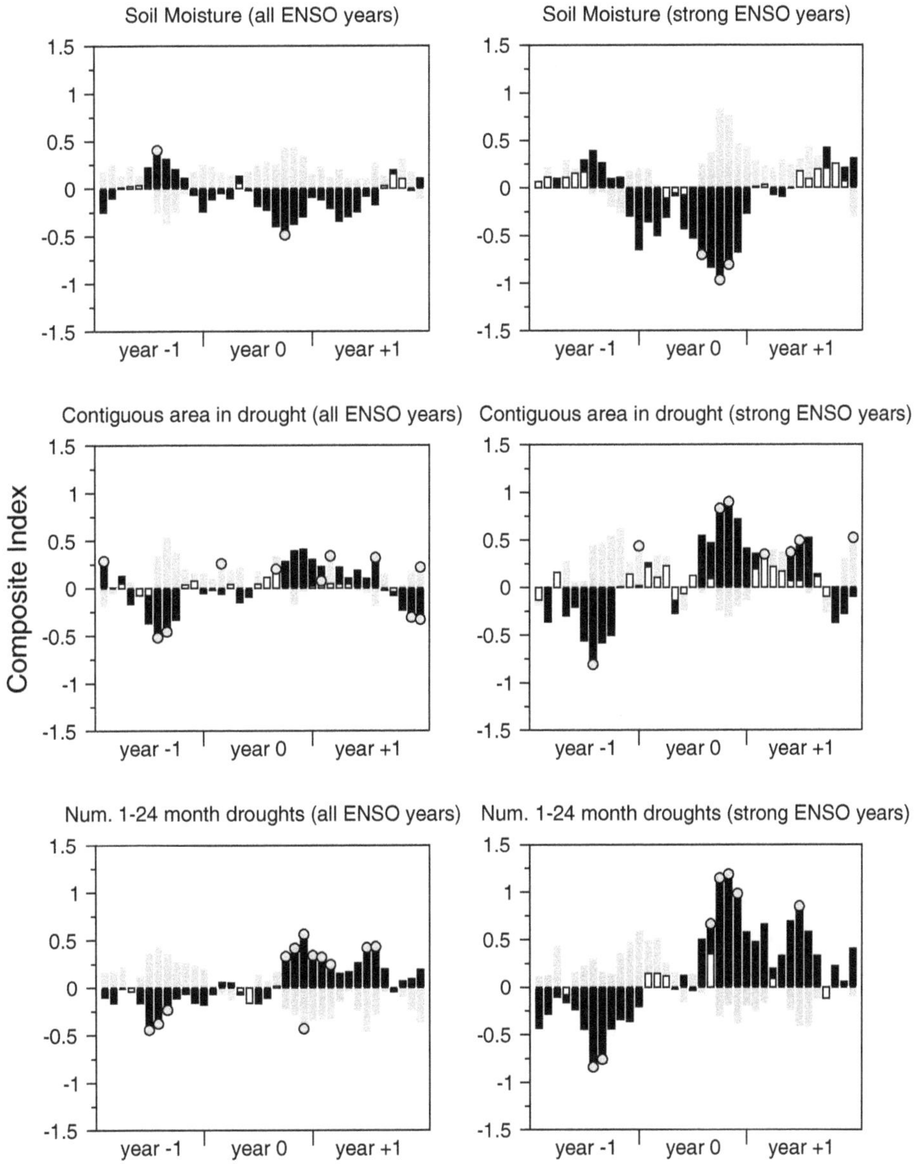

Figure 6.12 *Composite average soil moisture, contiguous area in drought and the number of large area droughts occurring globally for a 36-month window*

Note: This is centred on positive (El Niño, black bars) and negative (La Niña, grey bars) anomalies of ENSO SSTs. White bars indicate the overlap of the positive and negative phase composite values when they are of the same sign. Results are shown for all ENSO years (left panels), defined as years with five consecutive overlapping SST anomalies of greater than ±0.5°C and for strong ENSO years (right panels), similarly defined but for ±1°C. Statistical significance of the index values, as determined from the hypergeometric distribution at the 0.05 level, is indicated by a grey circle at the top of the bars.

Source: adapted from Sheffield et al (2009)

lagged response of the land surface to the ENSO event. The soil moisture data are averaged over all terrestrial land regions, excluding Antarctica and Greenland. The contiguous area is derived from the cluster analysis. The number of droughts is again filtered for droughts of less than 24 months. The three characteristics are first standardized to z-scores (subtract the mean and divide by the standard deviation) and then averaged for each month in the 36-month window over all El Niño or La Niña years. ENSO years are taken from the CPC Oceanic Niño Index (ONI), which is the three-month running mean of SST anomalies over the Niño 3.4 region. The figure shows results for two sets of ENSO years: (1) all ENSO years, defined as years with five consecutive overlapping anomalies of greater than ±0.5°C; (2) strong ENSO years, similarly defined but for a threshold of ±1°C. Statistical significance of the composite index was determined using the hypergeometric distribution by representing the index's departure from zero as a binomial variable with values 'success' and 'failure' (Dracup and Kahya, 1994). A 'success' is defined as the occurrence of an index value above (below) zero. The hypergeometric distribution was then used to determine the cumulative probability that at least m successes (the number of positive index values occurring in La Niña or El Niño years) are obtained in n trials (the number of La Niña or El Niño years) from a finite population of size N (the total number of years) containing k successes (the total number of positive values). A cumulative probability value of less than 0.05 (95 per cent confidence level) indicates that the index value is significant, and would only happen by chance 5 per cent of the time.

There is a clear tendency for all drought characteristics to be more pronounced in the second half of the ENSO year and at the beginning of the following year. During El Niño years, soil moisture is more likely to be drier globally, with a contemporaneous increase in the number of large-scale drought events. During La Niña years, the opposite is likely to be true. The signal is weaker for all ENSO years (the composite index stays below 0.5), but is much more robust for strong ENSO years. The response for contiguous area is not as strong, especially for all ENSO years, with the La Niña signal being particularly weak. The peak composite index for all characteristics falls at the end of the ENSO year (September–November) and the anomalies persist to about May of the next year, with only the larger El Niño index values being statistically significant. This is consistent with the impacts of ENSO on the global terrestrial climate, and highlights that large-scale drought occurrence is also impacted, with global drought increasing during El Niño events. The soil moisture response is shorter because the global average value reflects changes all over the world, including those unaffected by ENSO variability or subject to drought in La Niña years, and thus may cancel out the signal derived from ENSO-affected regions.

Figure 6.13 shows the time series of long-term droughts (longer than over 12 months; the number of droughts longer than 24 months were too few to discern their temporal variation). We compare this with the AMO and PDO, which are the main modes of decadal climate variability in the North Atlantic and Pacific respectively (see Chapter 3). All time series in Figure 6.13 are smoothed using a 121-month (ten-year) running mean filter. Correlations of the smoothed and filtered number of droughts with the smoothed climate indices are 0.62 for the AMO and 0.57 for the PDO. As the smoothed time series is serially correlated (each value is dependent on the previous value), we determined the statistical significance of the correlations using Monte Carlo simulation (McCabe and Palecki, 2006). A set of 10,000 pairs of time series (one for the number of droughts and one for the climate oscillation, AMO or PDO) were generated, which had the mean, variability and lagged (one-month) correlation of the original data. The distribution of the correlations of each smoothed pair was then used to determine whether the observed correlation was statistically significant for a two-tailed test at the 95 per cent level. The results show that the neither correlation is significant at this level, although the AMO correlation is significant at the 90 per cent level.

Many studies (e.g. Kitzberger et al, 2007) have noted the regional influence of decadal variations in the northern Pacific and Atlantic oceans, as characterized by the PDO and AMO. Other studies have shown that different combinations of modes of the PDO and AMO (along with ENSO) are associated with multiple regions (McCabe et al, 2004; Kitzberger et al, 2007), and may have a wide impact across the northern hemisphere (e.g. Sutton and Hodson, 2005). Our results reflect this, with the majority of results of the identified long-term drought events (85 per cent) occurring in the northern mid- to high latitudes, indicating that these oceanic oscillations have an appreciable influence on global drought occurrence.

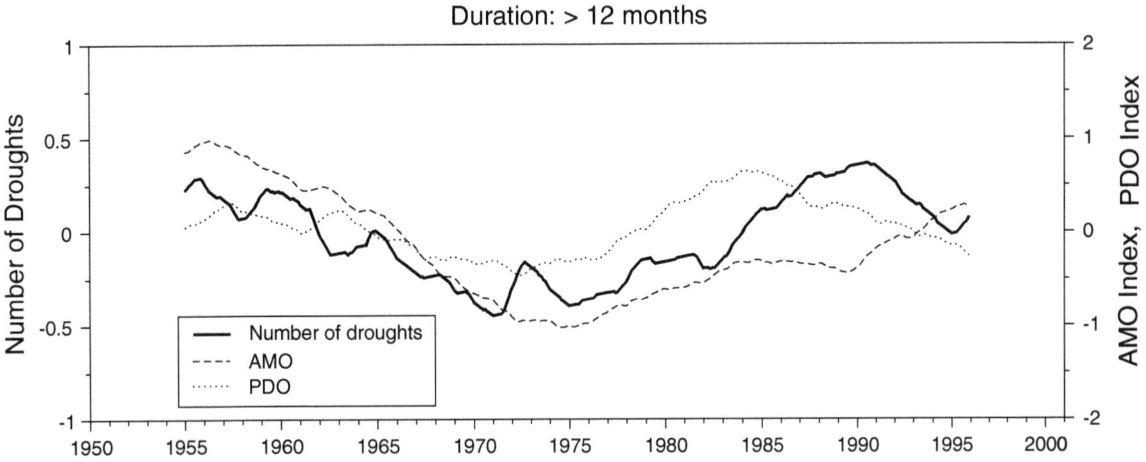

Figure 6.13 *Monthly time series of number of droughts longer than 12 months (black line) and the AMO (dashed line) and PDO indices (dotted line)*

Note: The time series is smoothed using a 121-month running mean filter.

Source: adapted from Sheffield et al (2009)

Summary

The abundance of instrumental data available over the past 100 years allows us to reconstruct the variations in hydrology across the world in reasonable detail. Of course, as we go back further in time, the density of data from weather stations and other sources is sparser or non-existent in some regions, and the reconstructions from the earlier periods must be viewed with caution. This is especially the case prior to the 1950s in regions such as the Amazon and central Africa. There are also many uncertainties in the hydrological modelling used to bring these sparse observations together to generate continuous depictions of hydrological variables in time and space (Box 4.1). In fact, it is unlikely that the simulated variability from day to day, or even from year to year, in parts of central Africa or the Amazon in the early part of the 20th century bear much resemblance to actual events, at least on small spatial scales. Nevertheless, the data can give us a broad picture of regional changes on decadal timescales and provide detailed information on seasonal and local scales in regions of dense observations.

This chapter has provided an overview of the variation in climate, hydrology and drought over the 20th century. Globally, there may have been a decreasing trend in soil moisture since the mid-20th century, when records are most reliable. However, this is highly dependent on the timescale over which the trend is calculated, because of decadal variation of precipitation. For example, Sheffield and Wood (2008) estimated a small but increasing trend in soil moisture during 1950–2000, on the basis of a previous version of the VIC dataset. However, drier conditions in many parts of the world have caused a drying trend over 1950–2008. The continuation of decadal variability in precipitation in upcoming years may yet bring further changes in the sign of the trend. Nevertheless, regional trends over the past 50 years or so are distinct, most obviously over the western hemisphere, and especially in North America, where increasing precipitation has increased soil moisture and decreased drought occurrence. There has also been significant drying over Africa, driven by falling levels of precipitation, especially over West Africa. Elsewhere, Europe appears to have been experiencing drying of soil moisture in recent years, most notably in the Mediterranean region, with an increase in drought extent. East Asia shows positive trends in drought extent, whereas Central Asia and the Tibetan Plateau indicate decreasing trends, but with sharp rises in recent years because of recent long-term drought. Drought extent over Australia has fallen in relation to the dry conditions of the 1940s and 1960s.

The analysis shows the complexity of drought occurrence over large time and space scales. Within the long-term trends, interannual and decadal variations in soil moisture and drought characteristics are apparent in many regions. Globally, large-scale drought events were more frequent in the 1960s to early 1970s, and 1980s to 1990s. In terms of the largest events, extending more than 10 million km^2, the 1950–1960s and 1980s suffered the most. Variations at interannual timescales are driven by ENSO variability, although the AMO appears to play an important role in many regions across the globe, such as West and East Africa, Central Asia and the high latitudes of North America. High correlations have been documented between the Mediterranean and the NAO, and between Southeast Asia, the Amazon basin and the ENSO. These types of connections hold implications for the prediction of drought, since modes of climate variability are often linked to regional changes in precipitation and temperature several months in advance. Given the dominance of ENSO on climate and drought throughout the tropics, and in many parts of the mid-latitudes, there is urgent need to improve forecasting of ENSO in terms of SSTs in the tropical Pacific. In the next chapter, we look at the regional occurrence of drought and explore the individual major drought events that took place over the 20th and early 21st centuries.

References

Aguilar, E. et al (2005) 'Changes in precipitation and temperature extremes in Central America and northern South America, 1961–2003', in *Journal of Geophysical Research*, vol 110, D23107, p15 doi:10.1029/2005JD006119

Alley, W. M. (1984) 'The Palmer Drought Severity Index: Limitations and assumptions', in *Journal of Climate and Applied Meteorology*, vol 23, pp1100–1109

Andreadis, K. M. and Lettenmaier, D. P. (2006) 'Trends in 20th century drought over the continental United States', in *Geophysical Research Letters*, vol 33, L10403, p4, doi:10.1029/2006GL025711

Barlow, M., Cullen, H. and Lyon, B. (2002) 'Drought in central and southwest Asia: La Nina, the warm pool, and Indian Ocean precipitation', in *Journal of Climate*, vol 15, no 7, pp697–700

BOM (2010) Australian Bureau of Meteorology website, www.bom.gov.au/lam/climate/levelthree/c20thc/drought.htm

Boulanger, J. P., Leloup, J., Penalba, O., Rusticucci, M., Lafon, F. and Vargas, W. (2005) 'Observed precipitation in the Parana-Plata hydrological basin: Long-term trends, extreme conditions and ENSO teleconnections', in *Climate Dynamics*, vol 24, no 4, pp393–413

Briffa, K. R., van der Schrier, G. and Jones, P. D. (2009) 'Wet and dry summers in Europe since 1750: Evidence of increasing drought', in *International Journal of Climatology*, vol 29, no 13, pp1894–1905

Brohan, P., Kennedy, J. J., Haris, I., Tett, S. F. B. and Jones, P. D. (2006) 'Uncertainty estimates in regional and global observed temperature changes: A new dataset from 1850', in *Journal of Geophysical Research*, vol 111, D12106, p21, doi:10.1029/2005JD006548

Dai, A., Trenberth, K. E. and Qian, T. (2004) 'A global data set of Palmer Drought Severity Index for 1870–2002: Relationship with soil moisture and effects of surface warming', in *Journal of Hydrometeorology*, vol 5, pp1117–1130

Dracup, J. A. and Kahya, E. (1994) 'The relationships between US streamflow and La Niña', in *Water Resources Research*, vol 30, no 7, pp2133–2141

Foley, J. A., Botta, A., Coe, M. T. and Costa, M. H. (2002) 'El Nino-Southern Oscillation and the climate, ecosystems and rivers of Amazonia', in *Global Biogeochemical Cycles*, vol 16, no 4, p27

Giorgi, F. and Francisco, R. (2000) 'Uncertainties in regional climate change predictions: A regional analysis of ensemble simulations with the HADCM2 GCM', in *Climate Dynamics*, vol 16, pp169–182

Herweijer, C. and Seager, R. (2008) 'The global footprint of persistent extra-tropical drought in the instrumental era', in *International Journal of Climatology*, vol 28, pp1761–1774

Herweijer, C., Seager, R., Cook, E. R. and Emile-Geay, J. (2007) 'North American droughts of the last millennium from a gridded network of tree-ring data', in *Journal of Climate*, vol 20, pp1353–1376

Hulme, M. (1992) 'Rainfall changes in Africa: 1931–1960 to 1961–1990', in *International Journal of Climatology*, vol 12, pp685–699

IPCC (2007) S. Solomon, D. Qin, M. Manning, Z. Chen, M. Marquis, K. B. Averyt, M. Tignor and H. L. Miller (eds) *Climate Change 2007: The Physical Science Basis. Contribution of Working Group I to the Fourth Assessment Report of the Intergovernmental Panel on Climate Change*, Cambridge University Press, Cambridge and New York

Jiang, T., Qiang, Z., Deming, Z. and Yijin, W. (2006) 'Yangtze floods and droughts (China) and teleconnections with ENSO activities (1470–2003)', in *Quaternary International*, vol 144, pp29–37

Kitzberger, T., Brown, P. M., Heyerdahl, E. K., Swetnam, T. W. and Veblen, T. T. (2007) 'Contingent Pacific-Atlantic Ocean influence on multicentury wildfire synchrony over western North America', in *Proceedings of the National Academy of Sciences,* vol 104, no 2, pp543–548

L'Hôte, Y., Mahé, G., Somé, B. and Triboulet, J. P. (2002) 'Analysis of a Sahelian annual rainfall index from 1896 to 2000; The drought continues', in *Hydrological Sciences Journal–Journal des Sciences Hydrologiques,* vol 47, pp563–572

Li, H., Dai, A., Zhou, T. and Lu, J. (2010) 'Responses of East Asian summer monsoon to historical SST and atmospheric forcing during 1950–2000', in *Climate Dynamics* vol 34, no 4, pp501–514

Liverman, D. M. (1999) 'Vulnerability and adaptation to drought in Mexico', in *Natural Resources Journal,* vol 39, no 1, pp99–115

Lloyd-Hughes, B. and Saunders, M. A. (2002) 'A drought climatology for Europe', in *International Journal of Climatology,* vol 22, pp1571–1592

McCabe, G. J. and Palecki, M. A. (2006) 'Multidecadal climate variability of global lands and oceans', in *International Journal of Climatology,* vol 26, pp849–865

McCabe, G. J., Palecki, M. A. and Betancourt, J. L. (2004) 'Pacific and Atlantic Ocean influences on multidecadal drought frequency in the United States', in *Proceedings of the National Academy of Sciences,* vol 101, pp4136–4141

Mitchell, T. D. and Jones, P. D. (2005) 'An improved method of constructing a database of monthly climate observations and associated high-resolution grids', in *International Journal of Climatology,* vol 25, pp693–712

Monteith, J. L. (1965) 'Evaporation and environment', in G. E. Fogg (ed) *The State and Movement of Water in Living Organisms,* Society for Experimental Biology Symposia, vol 19, pp205–234

Nicholls, N., Lavery, B., Frederiksen, C., Drosdowsky, W. and Torok, S. (1996) 'Recent apparent changes in relationships between the El Nino–Southern Oscillation and Australian rainfall and temperature', in *Geophysical Research Letters,* vol 23, no 23, pp3357–3360

Rajagopalan, B., Cook, E., Lall, U. and Ray, B. K. (2000) 'Spatiotemporal variability of ENSO and SST teleconnections to summer drought over the United States during the twentieth century', in *Journal of Climate,* vol 13, pp4244–4255

Rouault, M. and Richard, Y. (2005) 'Intensity and spatial extent of droughts in southern Africa', in *Geophysical Research Letters,* vol 32, no 15, p4

Schubert, S. D., Suarez, M. J., Pegion, P. J., Koster, R. D. and Bacmeister, J. T. (2004) 'On the cause of the 1930s Dust Bowl', in *Science,* 303, pp1855–1859

Seager, R., Kushnir, Y., Herweijer, C., Naik, N. and Miller, J. (2005) 'Modelling of tropical forcing of persistent droughts and pluvials over western North America: 1856–2000', in *Journal of Climate,* vol 18, pp4065–4088

Seager, R., Burgman, R., Kushnir, Y., Clement, A., Cook, E., Naik, N. and Miller, J. (2008) 'Tropical Pacific forcing of North American medieval megadroughts: Testing the concept with an atmosphere model forced by coral-reconstructed SSTs', in *Journal of Climate,* vol 21, no 23, pp6175–6190

Sheffield, J. and Wood, E. F. (2007) 'Characteristics of global and regional drought, 1950–2000: Analysis of soil moisture data from off-line simulation of the terrestrial hydrological cycle', in *Journal of Geophysical Research,* vol 112, D17115, p21, doi:10.1029/2006JD008288

Sheffield, J. and Wood, E. F. (2008) 'Global trends and variability in soil moisture and drought characteristics, 1950–2000, from observation-driven simulations of the terrestrial hydrological cycle', in *Journal of Climate,* vol 21, pp432–458

Sheffield, J., Goteti, G., Wen, F. H. and Wood, E. F. (2004) 'A simulated soil moisture based drought analysis for the United States', in *Journal of Geophysical Research-Atmosphere,* vol 109, D24, p19

Sheffield, J., Goteti, G. and Wood, E. F. (2006) 'Development of a 50-yr high-resolution global dataset of meteorological forcings for land surface modelling', in *Journal of Climate,* vol 19, pp3088–3111

Sheffield, J., Andreadis, K. M., Wood, E. F. and Lettenmaier, D. P. (2009) 'Global and continental drought in the second half of the twentieth century: Severity-area-duration analysis and temporal variability of large-scale events', in *Journal of Climate,* vol 22, no 8, pp19621981

Sutton, R. T. and Hodson, D. L. R. (2005) 'Atlantic Ocean forcing of multidecadal variations in North American and European summer climate', in *Science,* vol 309, pp115–118

Trenberth, K. and Branstator, G. (1992) 'Issues in establishing causes of the 1988 drought over N. America', in *Journal of Climate,* vol 5, pp159–172

van der Schrier, G., Briffa, K. R., Jones, P. D. and Osborn, T. J. (2006) 'Summer moisture variability across Europe', in *Journal of Climate,* vol 19, no 12, pp2818–2834

Verdon-Kidd, D. C. and Kiem, A. S. (2009) 'Nature and causes of protracted droughts in southeast Australia: Comparison between the Federation, WWII, and Big Dry droughts', in *Geophysical Research Letters,* vol 36, L22707, p6, doi:10.1029/2009GL041067.12

Visbeck, M. H., Hurrell, J. W., Polvani, L. and Cullen, H. M. (2001) 'The North Atlantic Oscillation: Past, present, and future', in *Proceedings of the National Academy of Sciences,* vol 98, no 23, pp12,876–12,877

Wells, N., Goddard, S. and Hayes, M. J. (2004) 'A self-calibrating Palmer Drought Severity Index', in *Journal of Climate,* vol 17, pp2335–2351

Willmott, C. J. and Matsuura, K. (2009) 'Terrestrial air temperature and precipitation: Monthly and annual time series (1900–2008)', V2.01, Center for Climatic Research, University of Delaware, Newark, DE

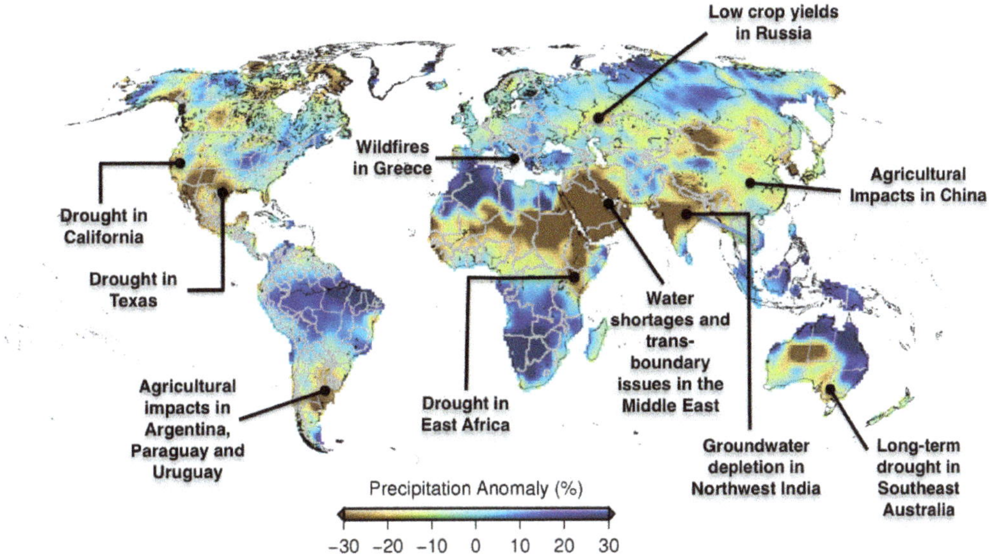

Plate 1.1 *The global impacts of drought in 2008–2009*

Note: The underlying shading represents the 2008–2009 precipitation anomalies above (blue colours) or below (brown colours) the long-term mean precipitation.

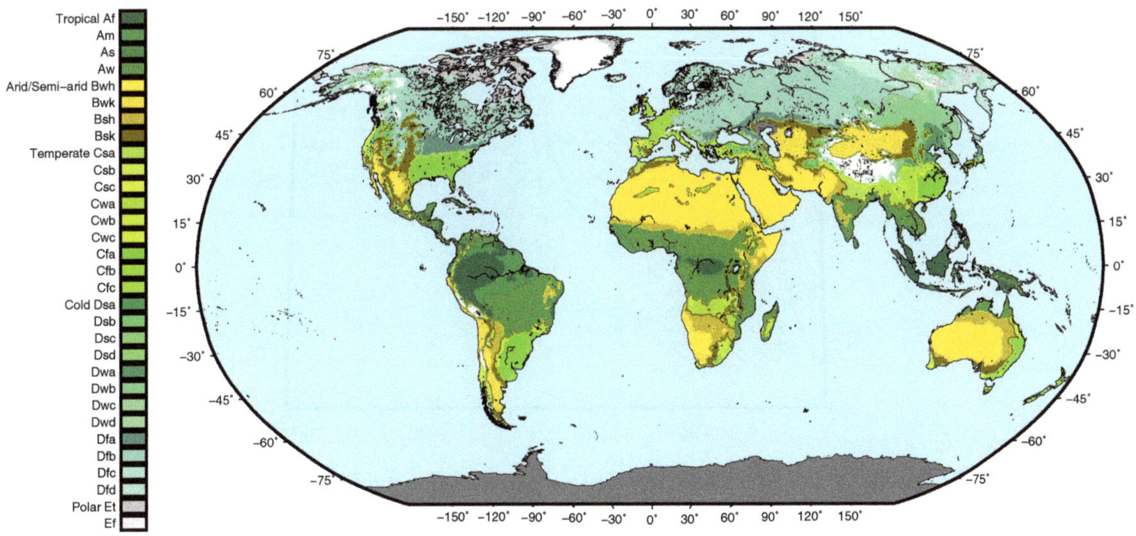

Plate 3.1 *Climate classification of the world, based on the Köppen-Geiger system*

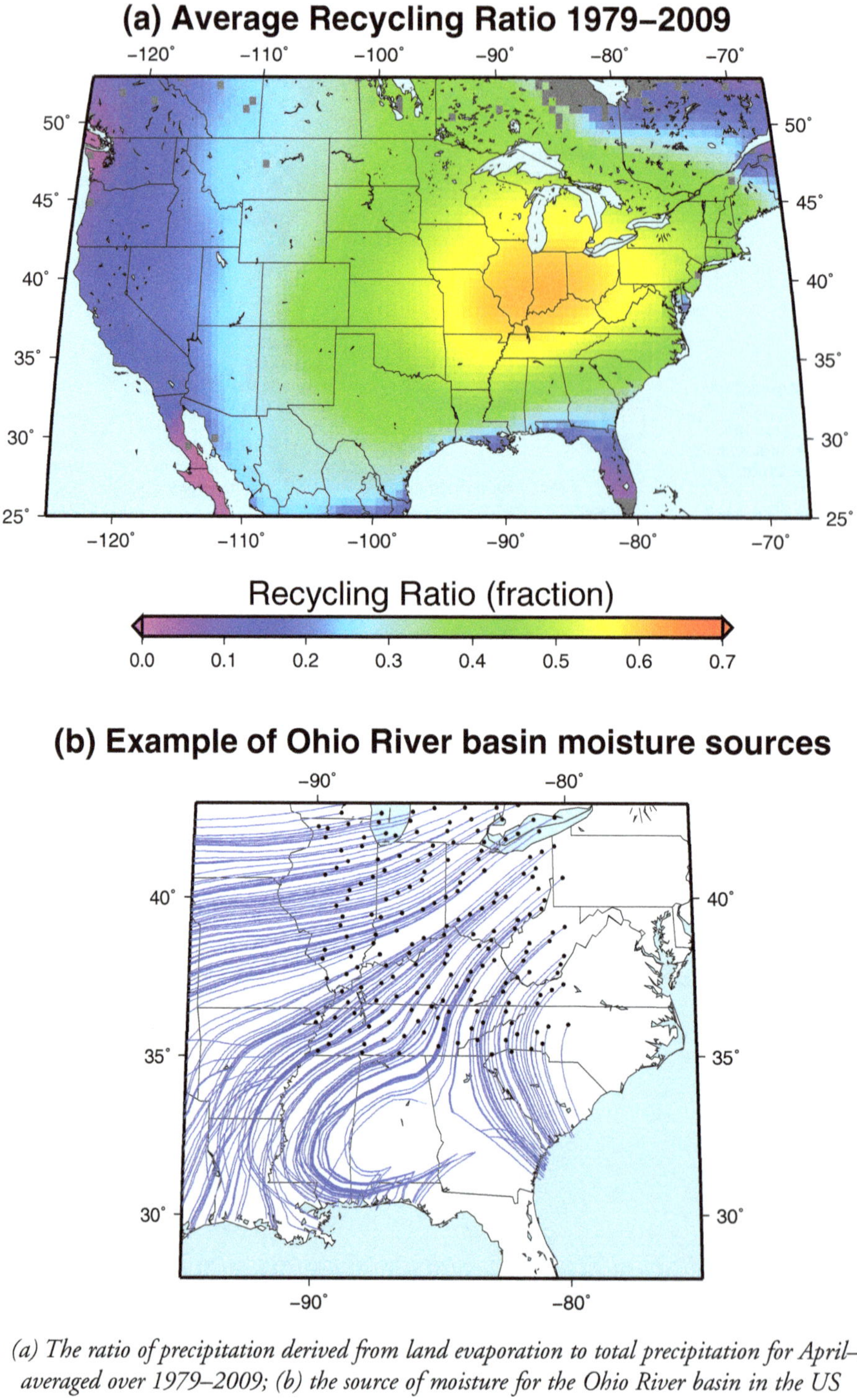

Plate 3.2 *(a) The ratio of precipitation derived from land evaporation to total precipitation for April–September averaged over 1979–2009; (b) the source of moisture for the Ohio River basin in the US*

Note: The lines represent the atmospheric flows that bring moisture from downwind to the region, part of which falls as precipitation. The source of the moisture is a combination of evaporation from the oceans and evaporation from the land of previously fallen precipitation.

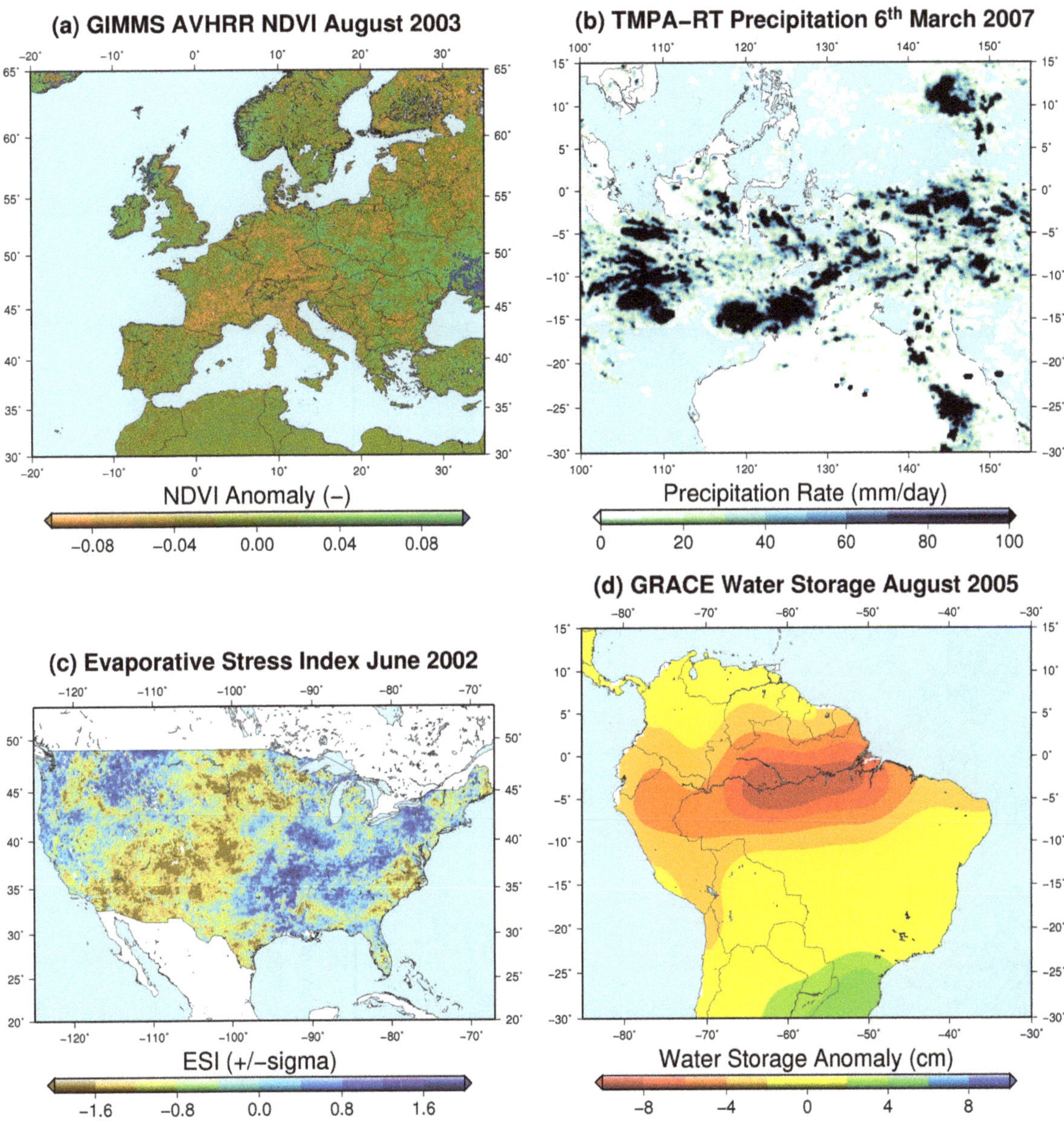

Plate 4.1 *Examples of remote-sensing data used to assess large-scale drought and changes in the hydrological cycle*

Note: (a) Vegetation condition in Europe during the 2003 drought, as derived from GIMMS AVHRR NDVI data. Green shows where vegetation growth is above average relative to 1982–2006. Brown colours indicate below-average growth. (b) Precipitation over Southeast Asia for March 2007, as derived from the TRMM Multisatellite Precipitation Analysis (TMPA) showing detailed storm patterns. (c) Eight-week clear-sky composite for May–June 2002 of the ALEXI Evaporative Stress Index (ESI), from Anderson et al (2007), which is based on thermal infrared remote-sensing data. (d) Total terrestrial water storage (includes surface water, soil moisture and groundwater, plus snow and ice in colder regions) from the GRACE satellite for August 2005. The data are anomalies calculated relative to the mean over 2003–2010. The drought in 2005 over the Amazon is clearly shown as an anomalous signal spanning the inundation area of the Amazon River and its tributaries.

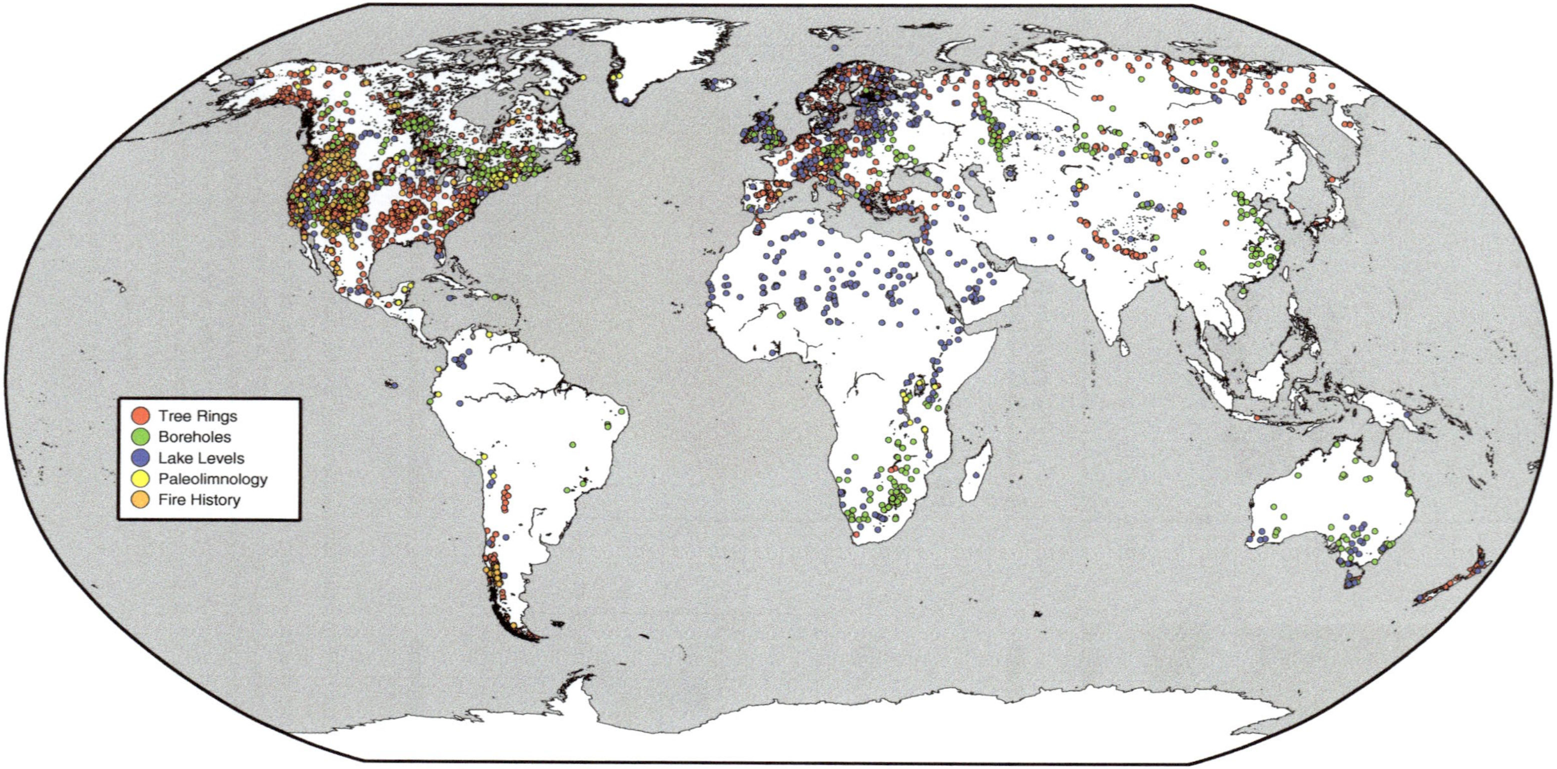

Plate 5.1 *Location of land paleoclimate records that have been used to reconstruct climate over the past hundred to several thousand years. These include tree rings, boreholes, lake levels and paleolimnology records*

Source: Data are taken from the US National Climatic Data Center World Data Center for Paleoclimatology at www.ncdc.noaa.gov/paleo/paleo.html

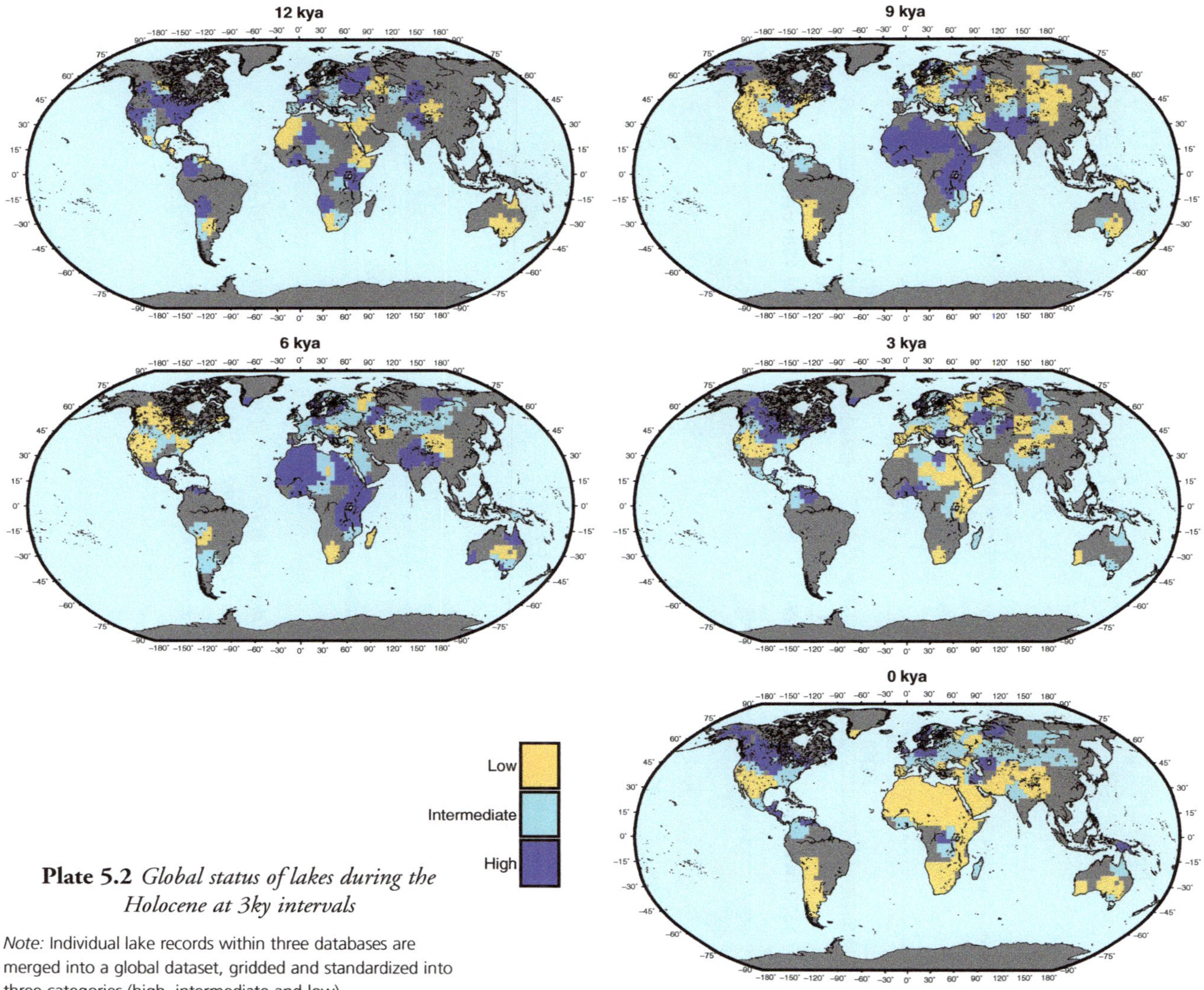

Plate 5.2 *Global status of lakes during the Holocene at 3ky intervals*

Note: Individual lake records within three databases are merged into a global dataset, gridded and standardized into three categories (high, intermediate and low).

Source: Derived from the dataset of Viau and Gajewski (2001)

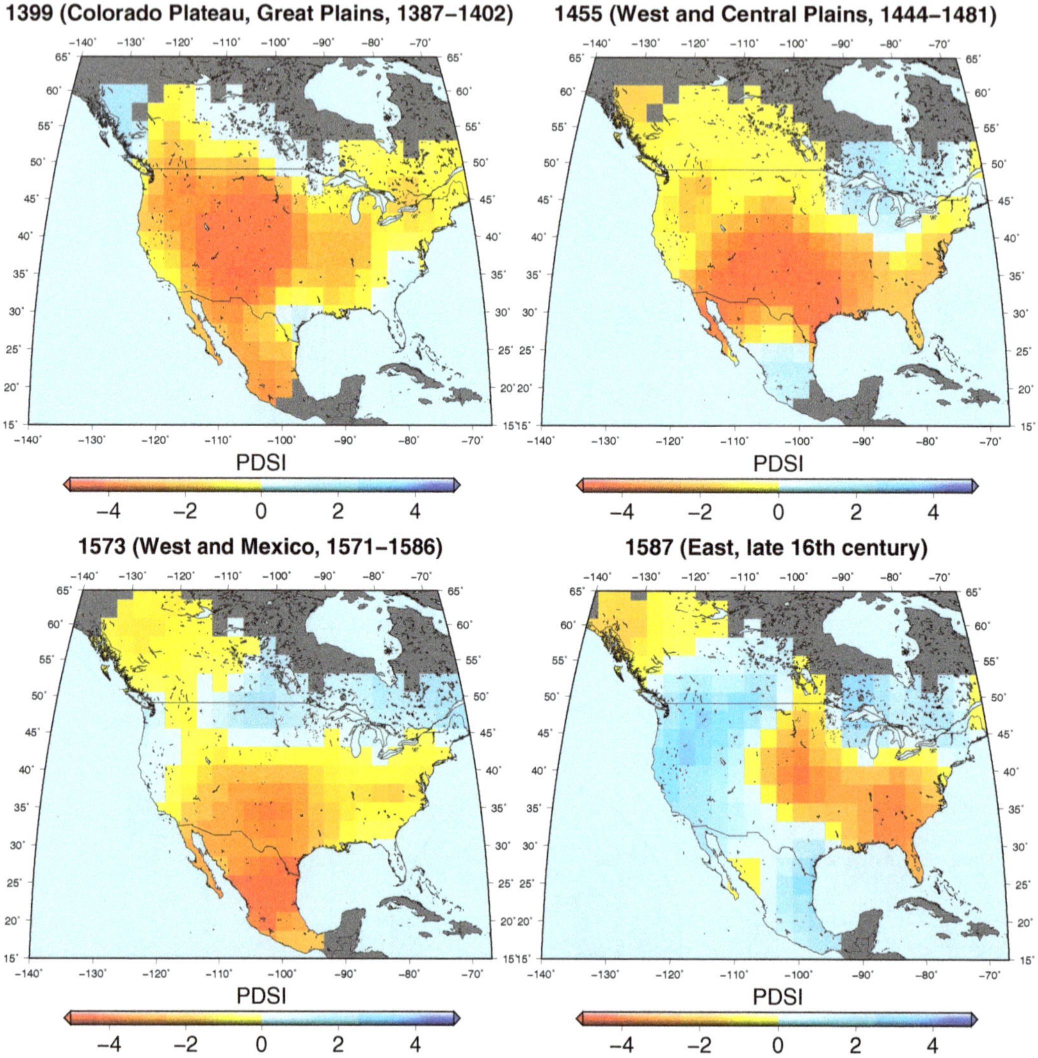

Plate 5.3 *North American drought during the peak of the series of megadroughts in the 14–15th centuries, as quantified by the PDSI*

Note: The last map shows drought conditions during the short-term drought in the eastern US, linked with the collapse of the Jamestown colony.

Source: Data taken from the *North American Drought Atlas*, Cook and Krusic, 2004

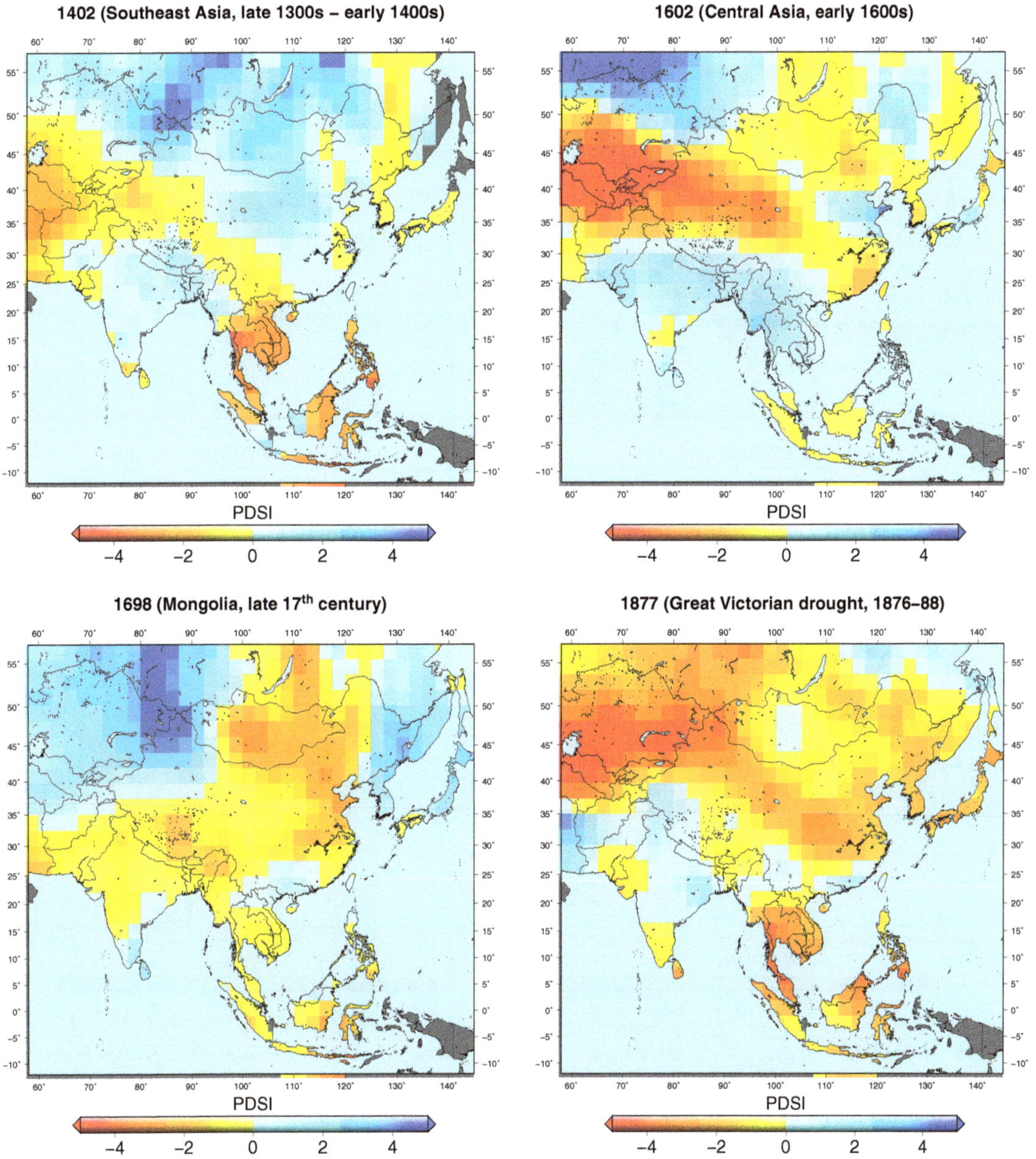

Plate 5.4 *Examples of droughts in Asia during the monsoon (July–August)*

Note: The event in 1402 was part of series of droughts across Southeast Asia that was associated with the demise of the Angkor civilization in what is now Cambodia. The 1602 drought was part of a wider event that was particularly severe across Central Asia. Dry conditions in Mongolia in the later 17th century were particularly severe in the central-west region. The Great Victorian drought of 1876–1888 was widespread and is associated with famine and colonial revolt.

Source: Data are taken from the *Monsoon Asia Drought Atlas* (MADA) of Cook et al (2010)

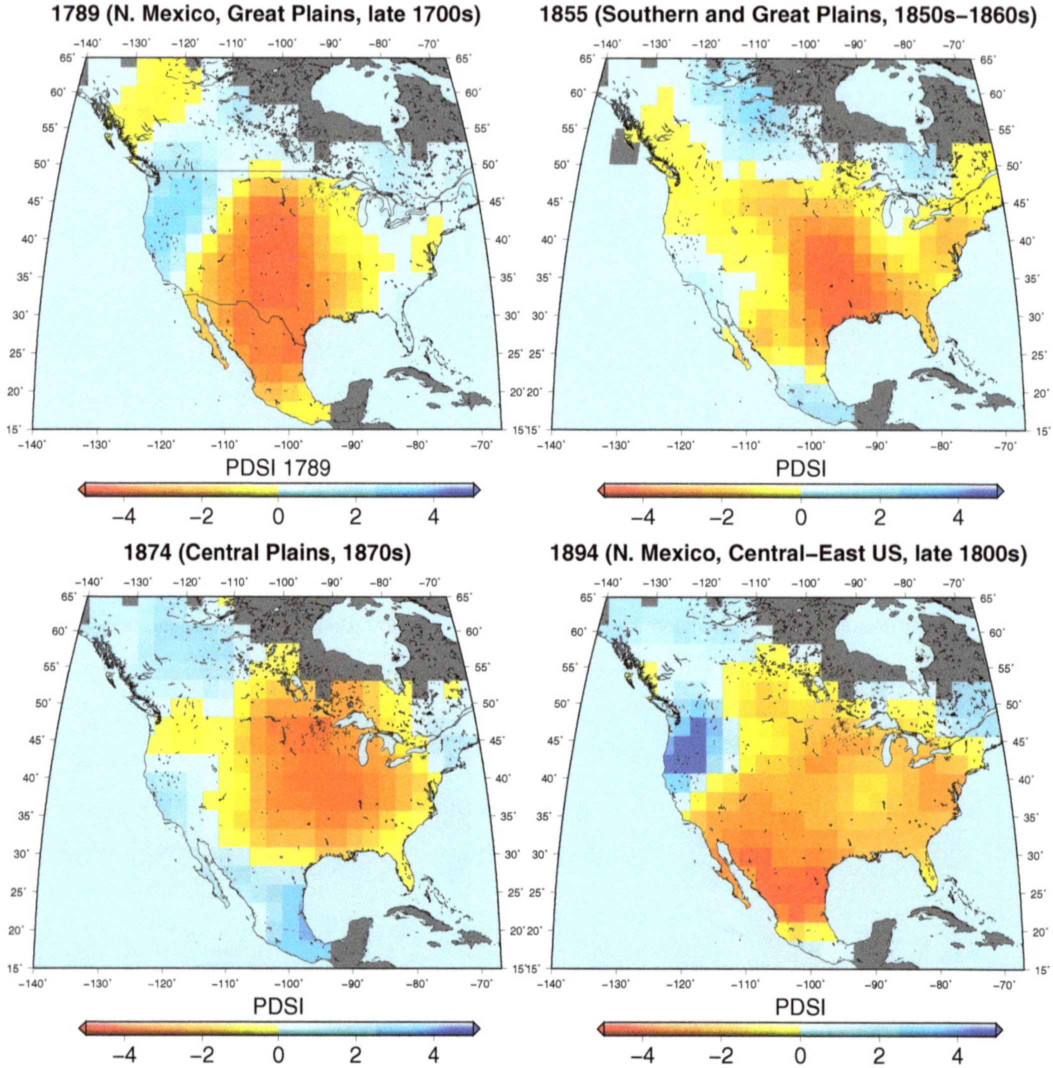

Plate 5.5 *North American drought during the peak of the severe drought events of the late 1700s (which impacted on the indigenous peoples of northern Mexico), the 1850s to 1860s (known as the Civil War Drought), the 1870s (which instigated massive locust swarms) and the 1890s (which prompted federal action to help settlers in the West)*

Source: Drought is quantified by PDSI data from the *North American Drought Atlas*, Cook and Krusic (2004)

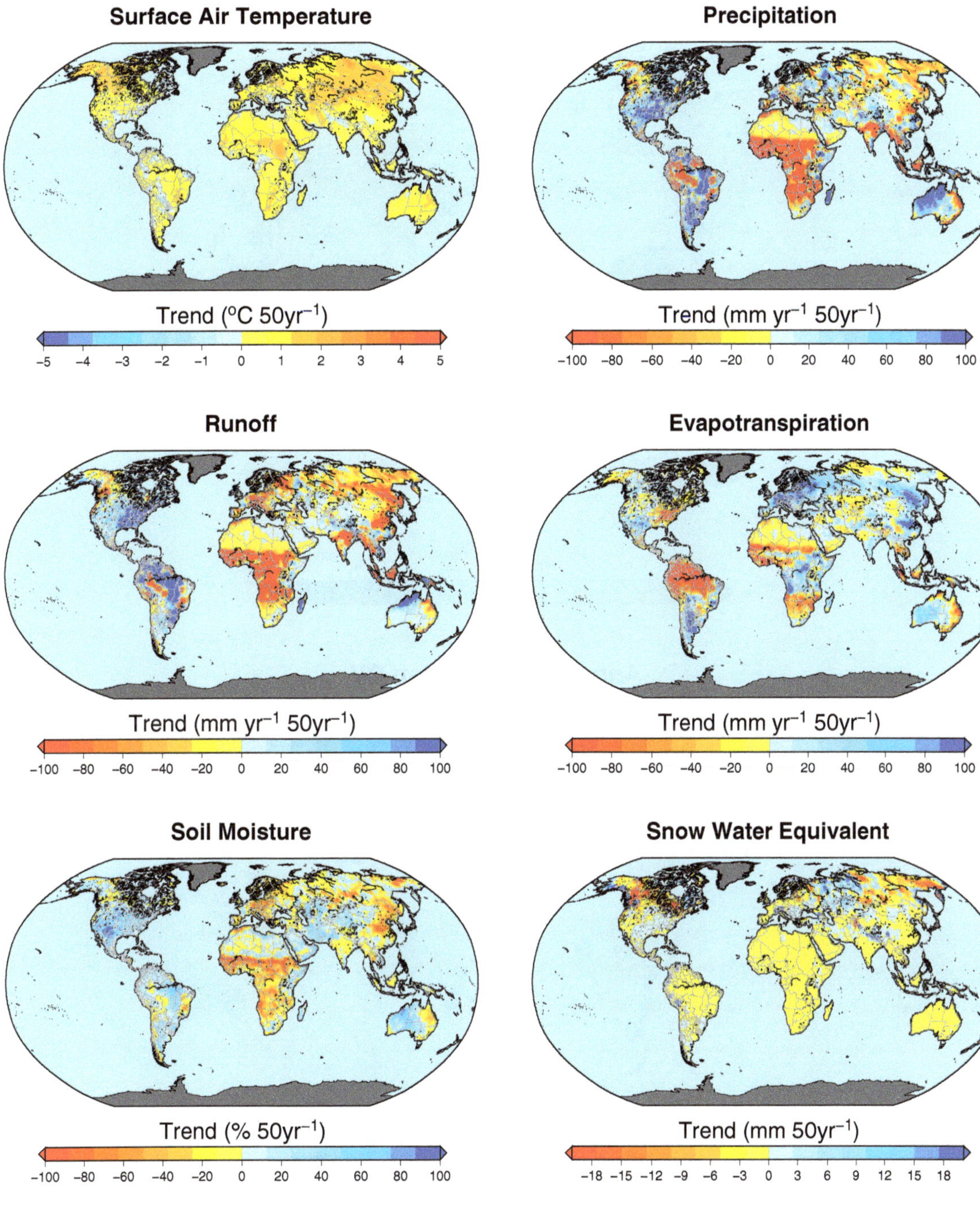

Plate 6.1 *Annual trends in air temperature, precipitation, run-off, evapotranspiration, soil moisture and snow water equivalent for 1950–2000*

Sources: The precipitation and temperature data are taken from the updated dataset of Sheffield et al (2006), which merges the CRU TS3.0 dataset of Mitchell and Jones (2005) and the Willmott and Matsuura (2009) dataset. Runoff, evapotranspiration, soil moisture and SWE are taken from the updates to the VIC model simulations of Sheffield and Wood (2007)

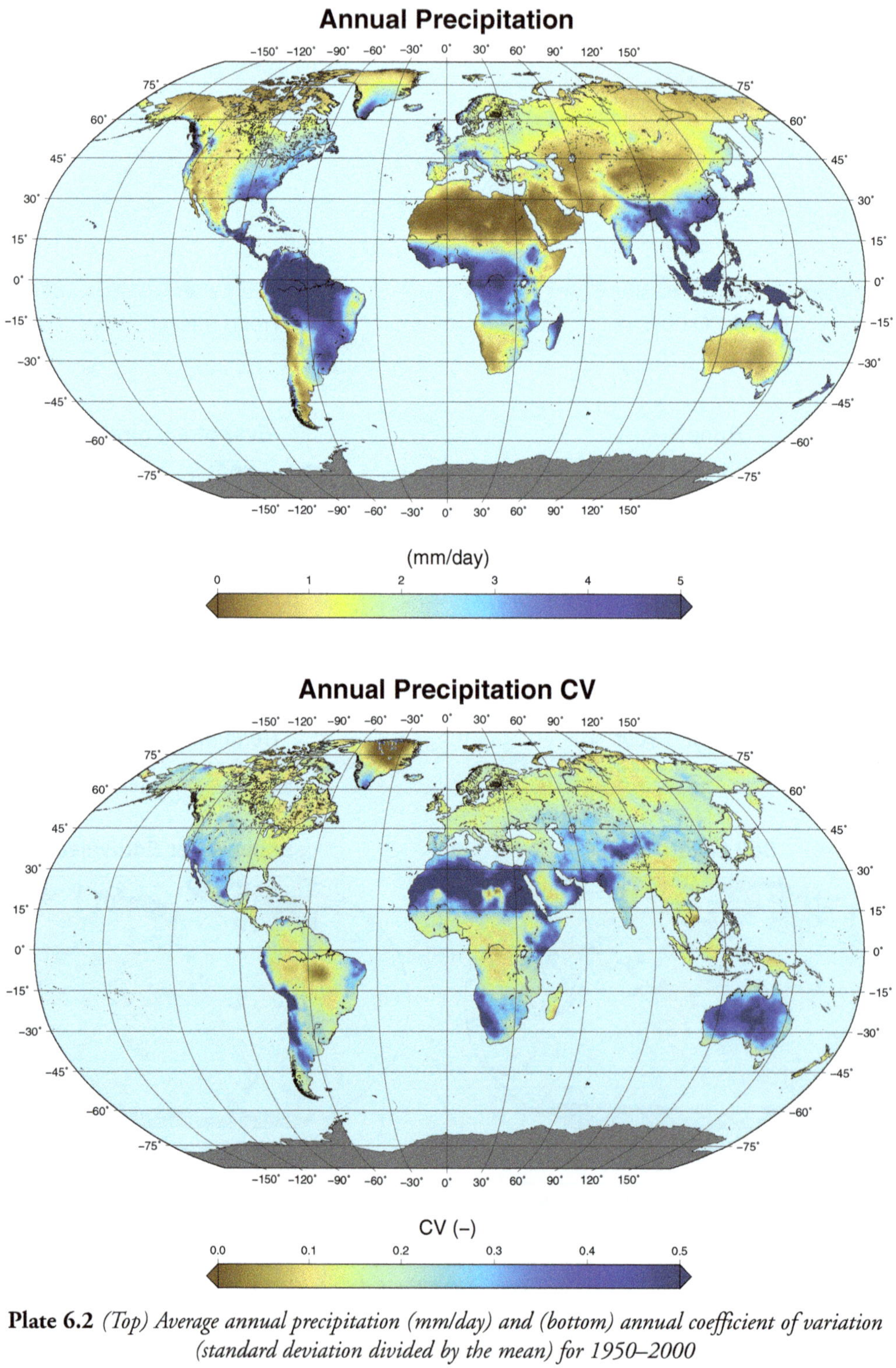

Plate 6.2 *(Top) Average annual precipitation (mm/day) and (bottom) annual coefficient of variation (standard deviation divided by the mean) for 1950–2000*

Note: No data are plotted for Antarctica.

Source: CRU TS3.0 dataset

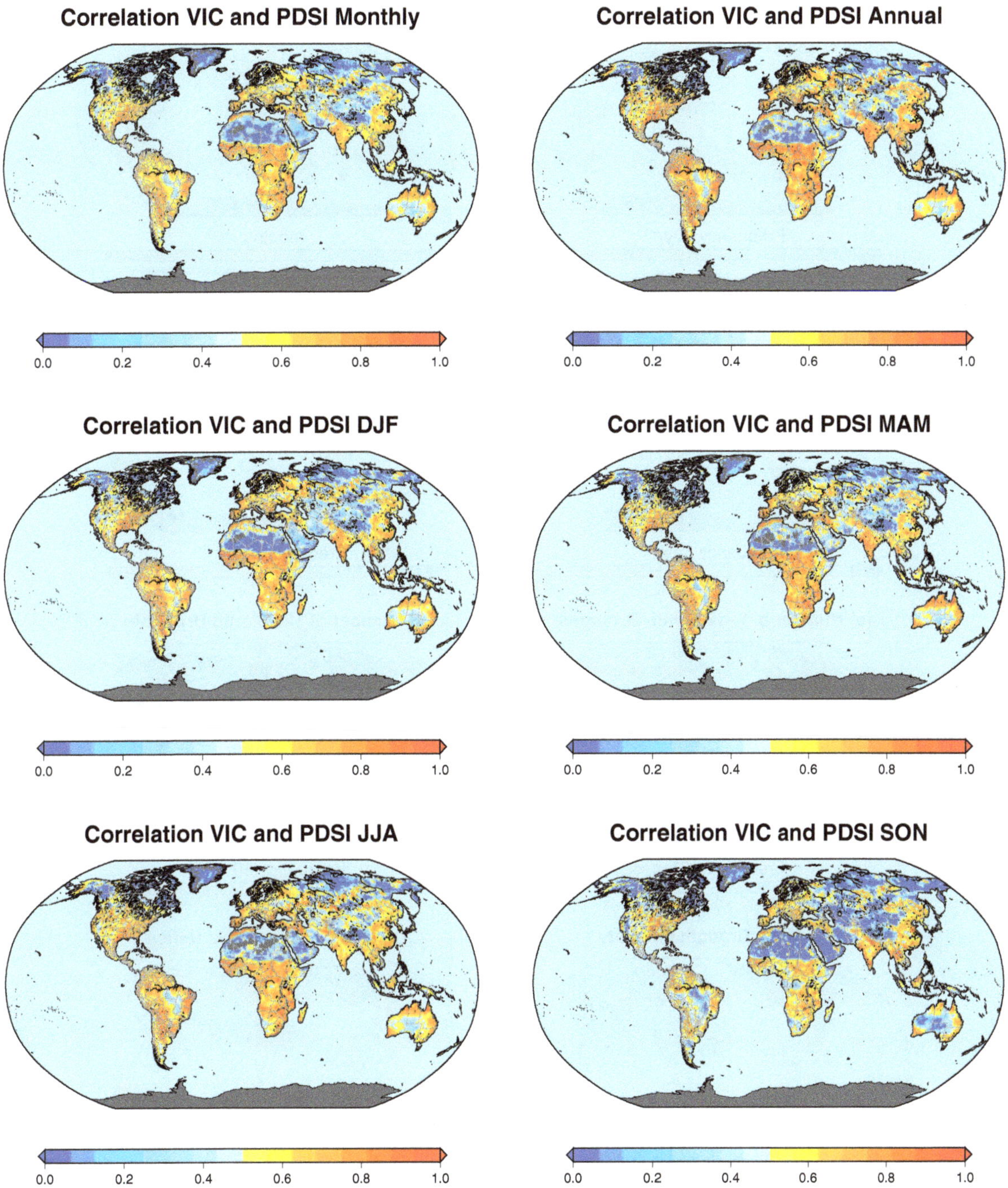

Plate 6.3 *Correlation between the VIC soil moisture index and the PDSI at monthly, annual and seasonal timescales*

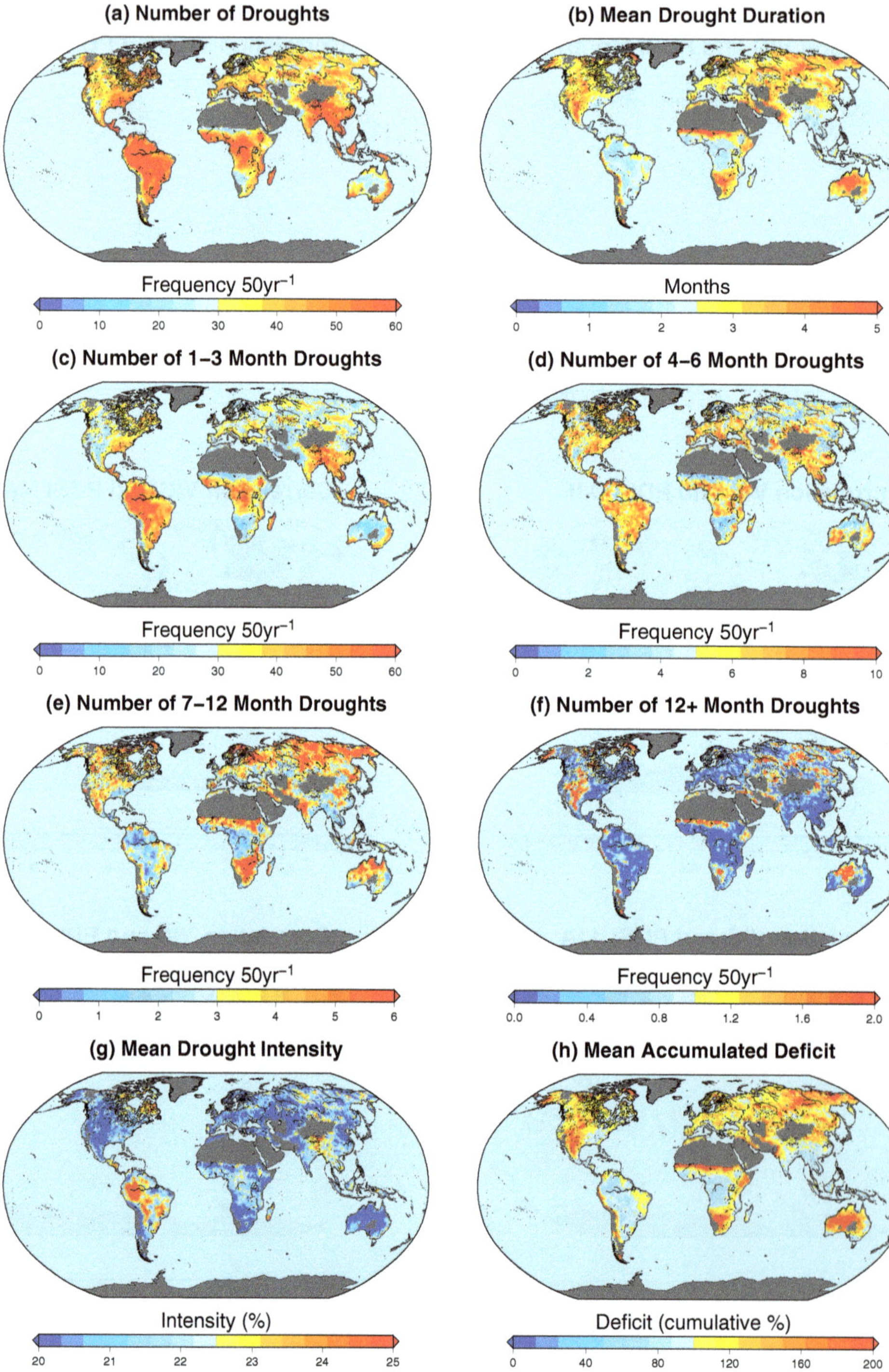

Plate 6.4 *Global variation, 1950–2000, of (a) frequency of droughts; (b) mean drought duration; frequency of occurrence of (c) very short (1–3 month), (d) short (4–6 month), (e) medium (7–12 month) and (f) long-term (>12 months) droughts (threshold, $q_0(\theta)$ = 20 per cent); (g) mean drought intensity; and (h) mean accumulated deficit/severity for $q_0(\theta)$ = 50 per cent*

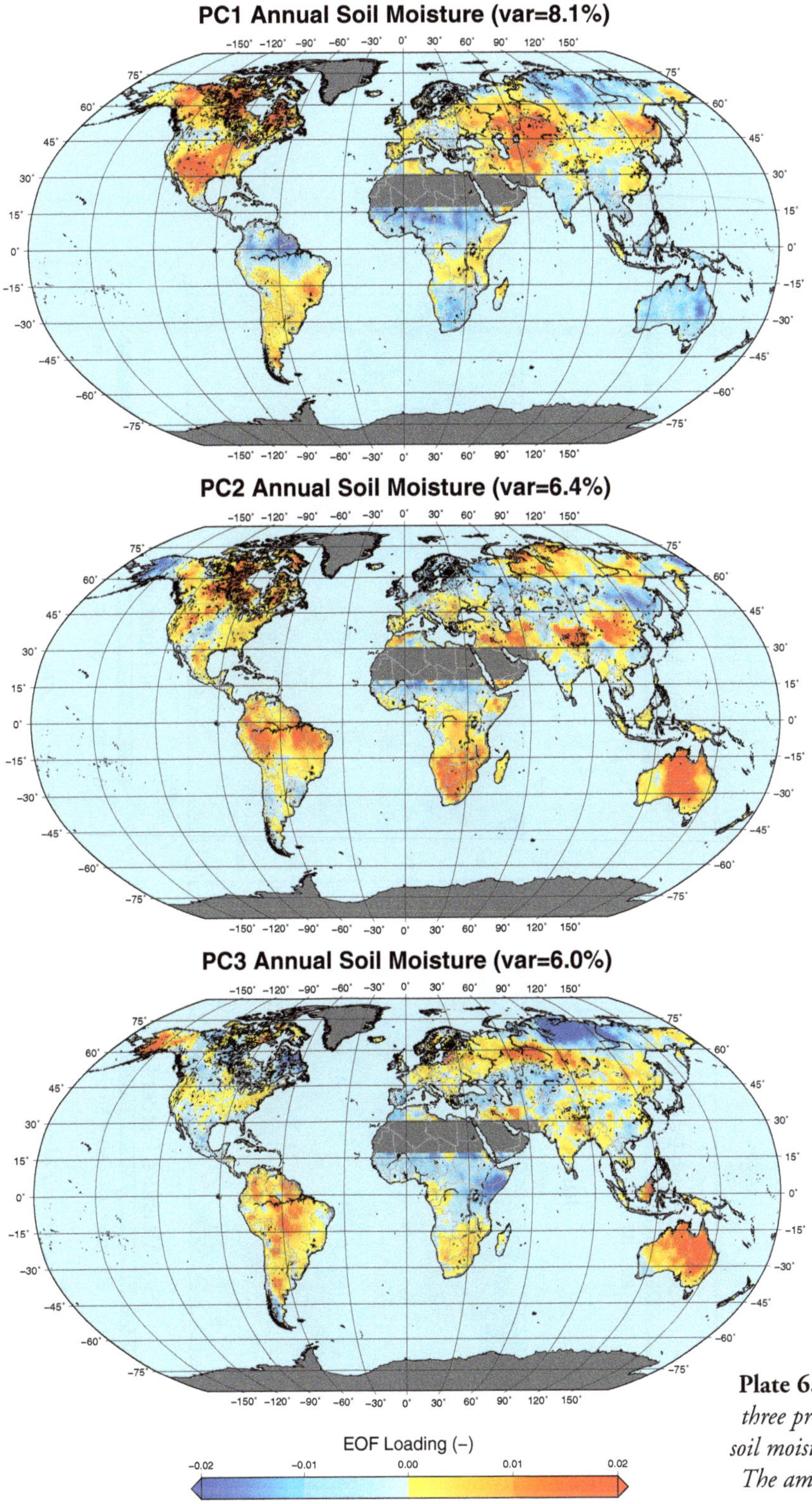

Plate 6.5 *Spatial loadings of the first three principal components of annual soil moisture percentile for 1950–2000. The amount of variance explained by each component is also given*

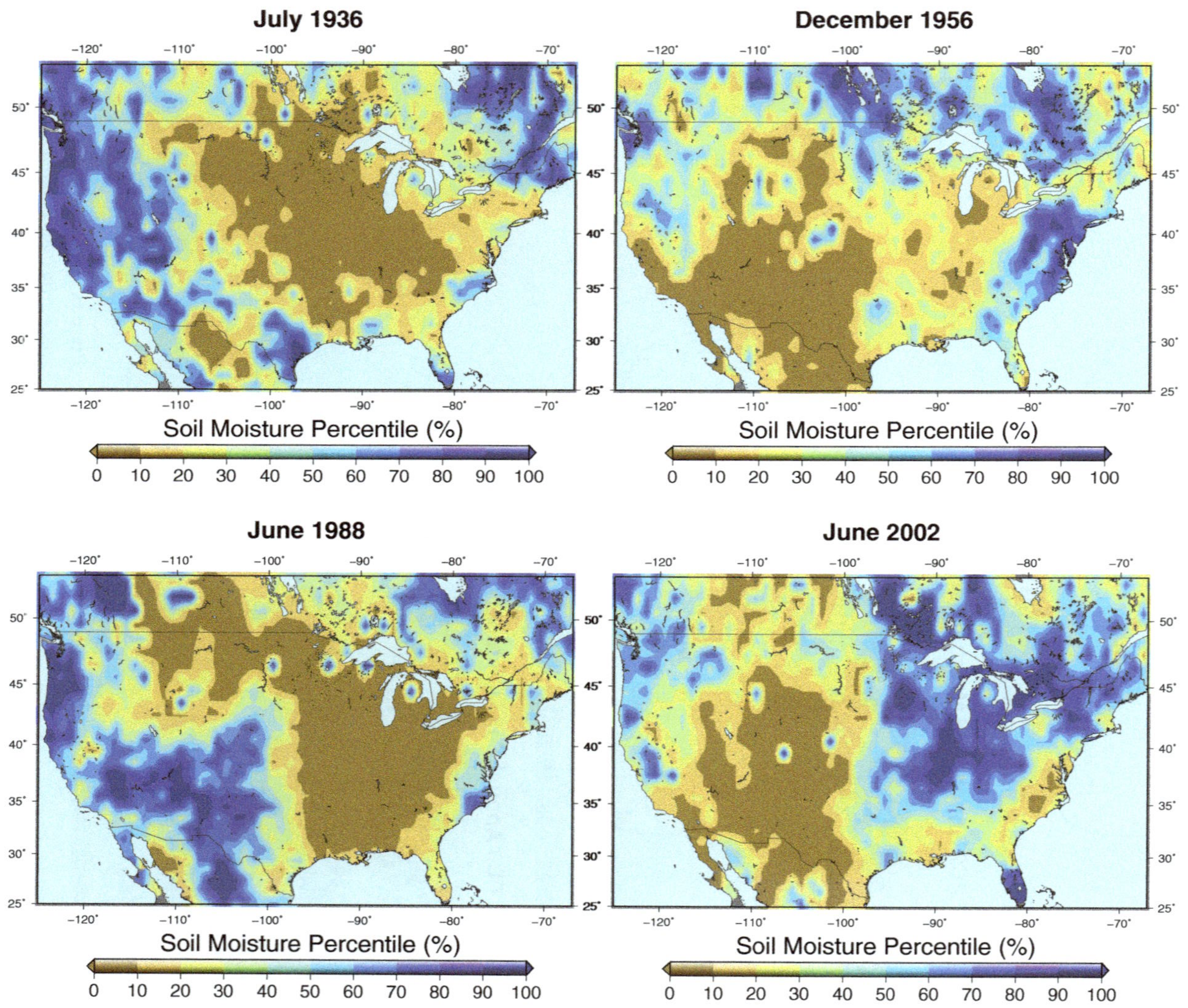

Plate 7.1 *Maps of the peak extent of major droughts of North America during the 20th and early 21st centuries, showing the 1930s Dustbowl drought, 1950s southwest drought, 1988 central US drought and the 2000s drought in the West*

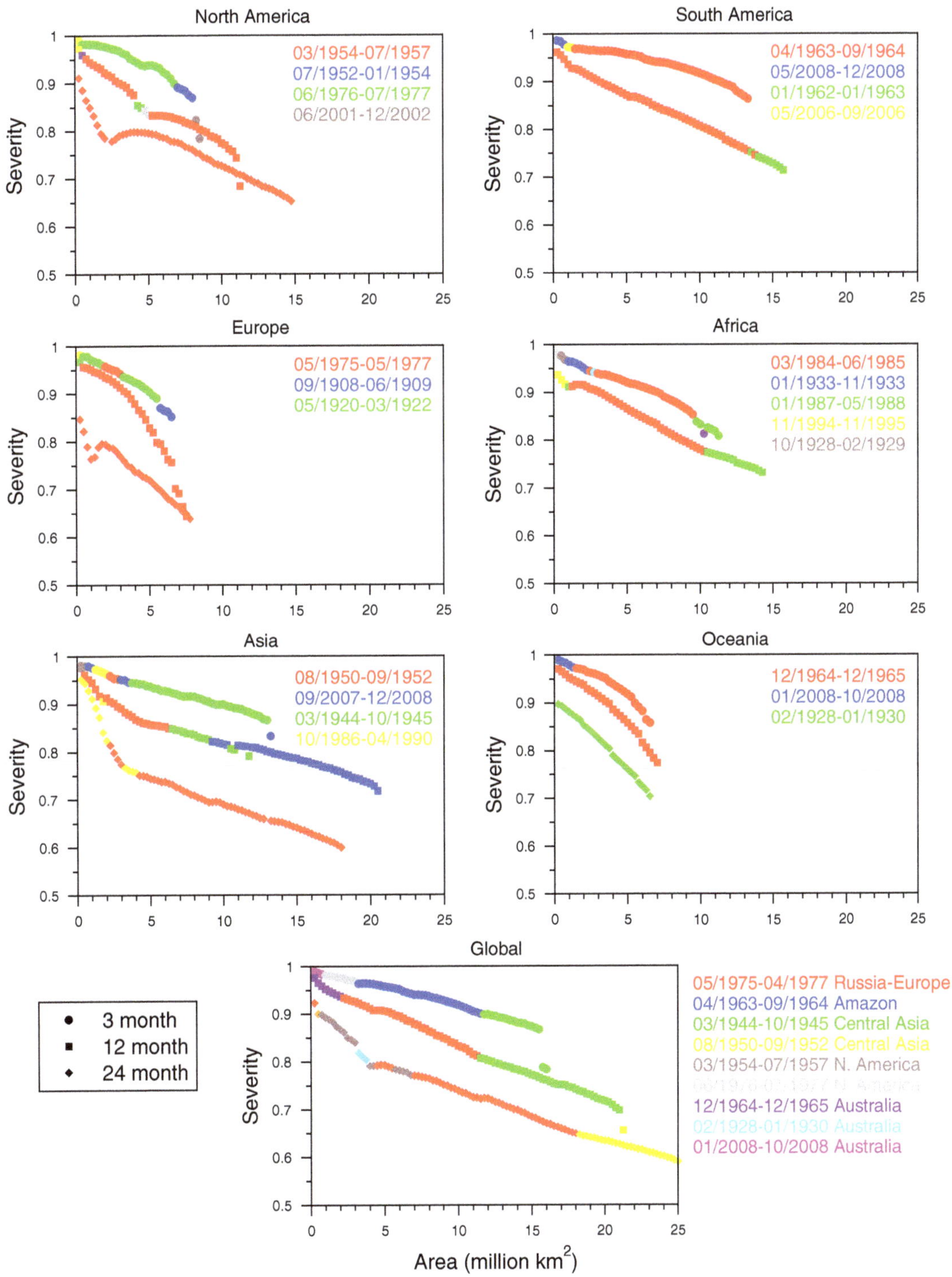

Plate 7.2 *Continental and global drought severity–area–duration (SAD) envelope curves for 3-month, 12-month and 24-month durations*

Note: Each curve represents the maximum bound, or envelope, of severities from the set of all droughts that occurred over the continent. They are generated by selecting the drought with the maximum severity for each incremental drought area. The curves may be made up of data from several different drought events, as some events are more severe for small areas while others are more severe for large areas.

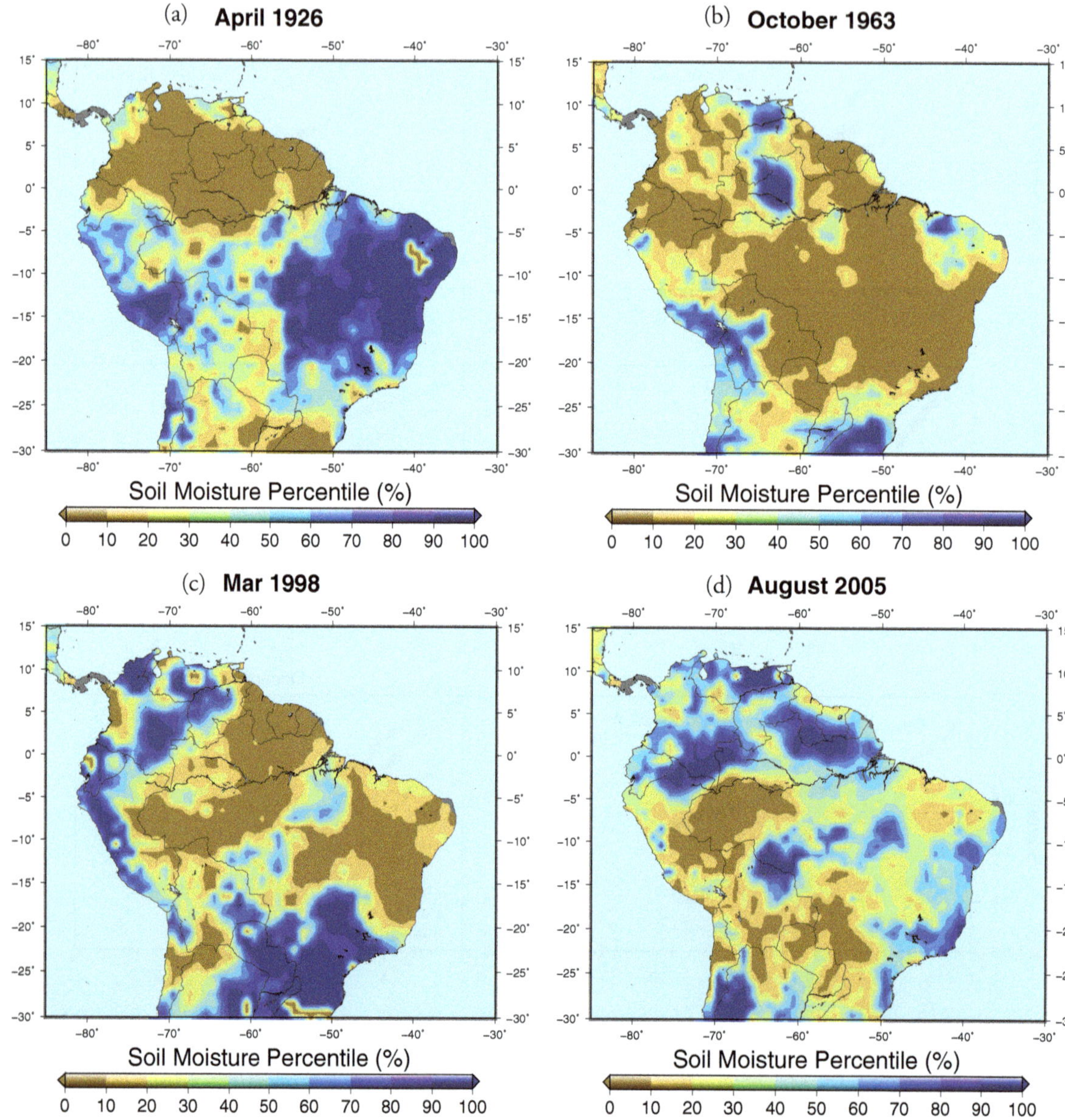

Plate 7.3 *Maps of the peak extent of major droughts of the Amazon and northern South America during the 20th and early 21st centuries*

Note: Events shown are the 1926 drought in northern Amazonia and the Rio Negro, the 1963–1964 drought that affected most of northern South America, the 1997–1998 drought that was forced by the strong El Niño of the time, and the recent 2005 drought.

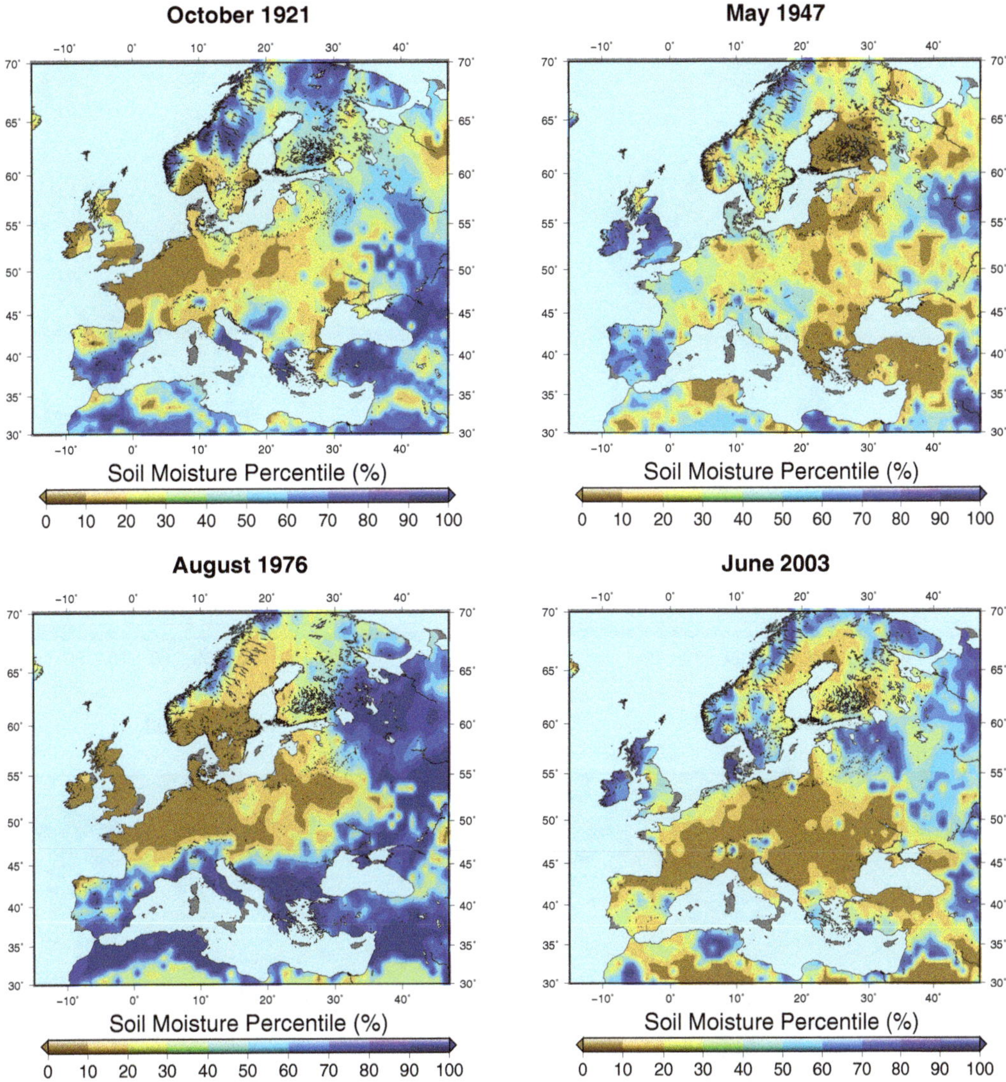

Plate 7.4 *Maps of the peak extent of major droughts of Europe during the 20th and early 21st centuries*

Note: The events shown are the 1921 drought, which affected much of northwestern Europe, the central European drought of 1947, which led to food shortages across much of the continent, the northwestern drought of 1975–1976 and the central European drought of 2003, which was characterized by an exceptional heat wave.

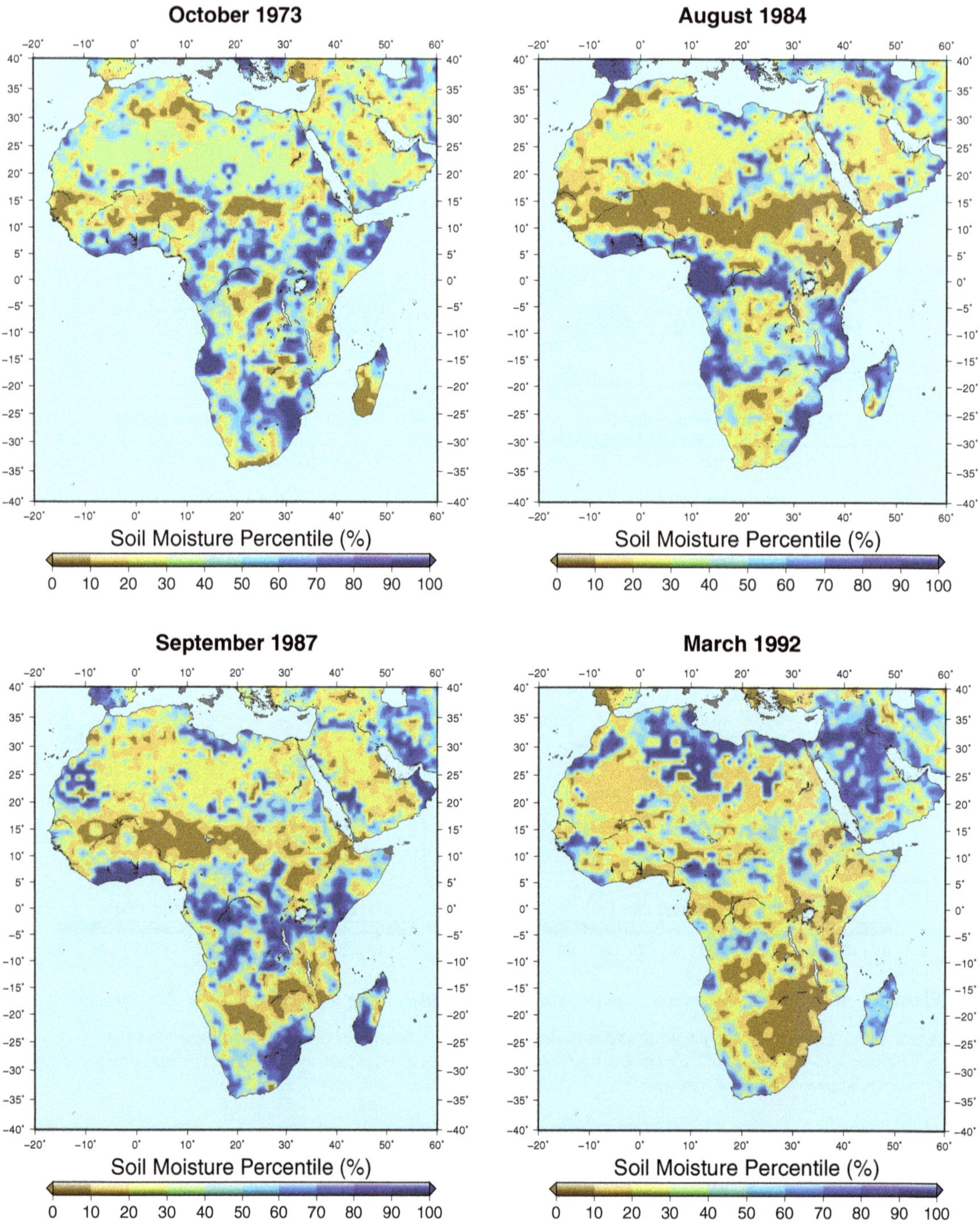

Plate 7.5 *Maps of the peak extent of major droughts of Africa during the 20th century*

Note: Shown are the devastating droughts in the Sahel in the early to mid-1970s and the 1980s, and the drought in southern Africa in 1991–1992, which was one of a series of droughts in the 1990s that affected the region.

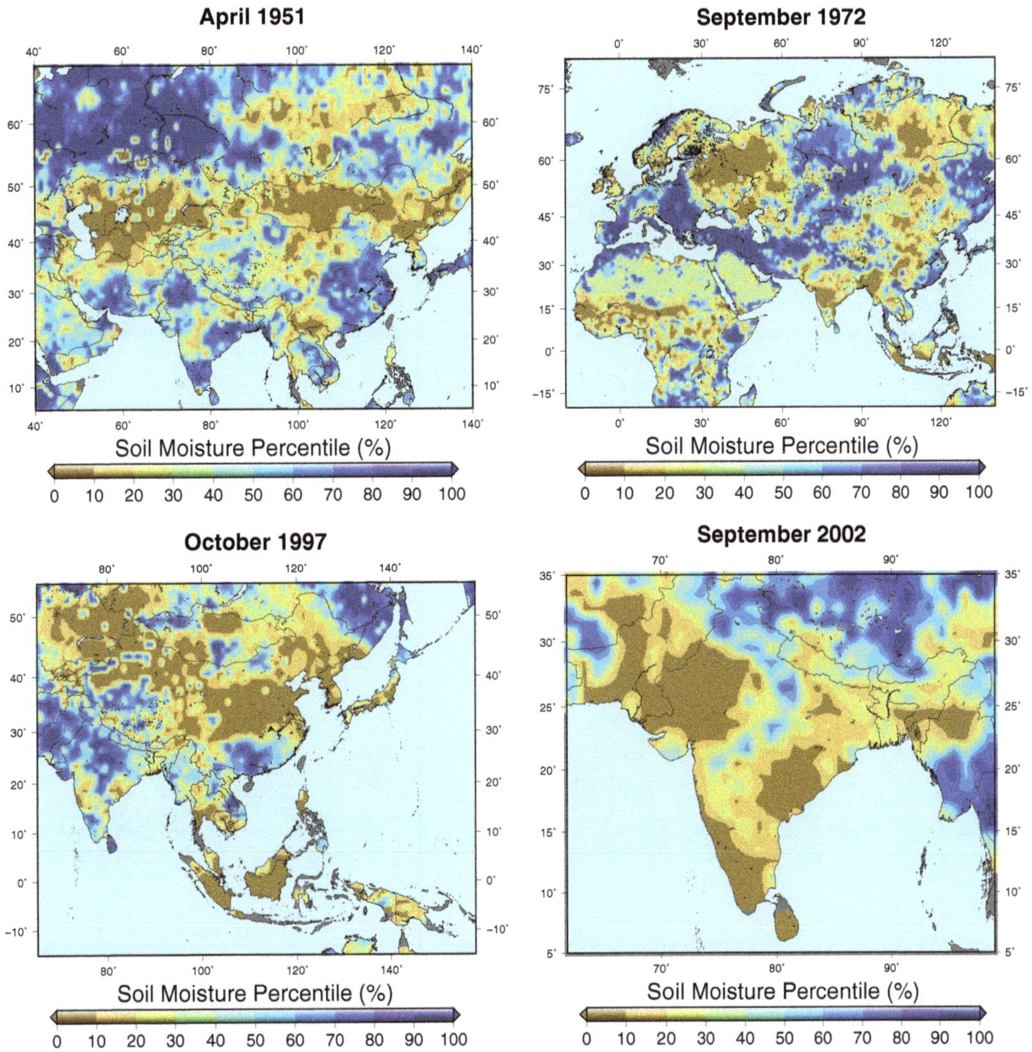

Plate 7.6 *Maps of the peak extent of major droughts in Asia during the 20th and early 21st centuries*

Note: The events shown are the early 1950s drought, which affected much of central Asia, and the drought of 1972 which impacted across much of the continent, including western Russia, northern Eurasia, and the monsoonal regions of southern and Southeast Asia, as well as West Africa, as driven by the strong El Niño of 1972–1973. The drought in 1997 extended over much of China and across into central Asia. Drought in the late 1990s to early 2000s was severe in western Asia. At the same time drought in northern China has been part of a drying trend in that region over the past decade. The drought in India in 2002 was one of the severest on record, although the El Niño that year was only moderate.

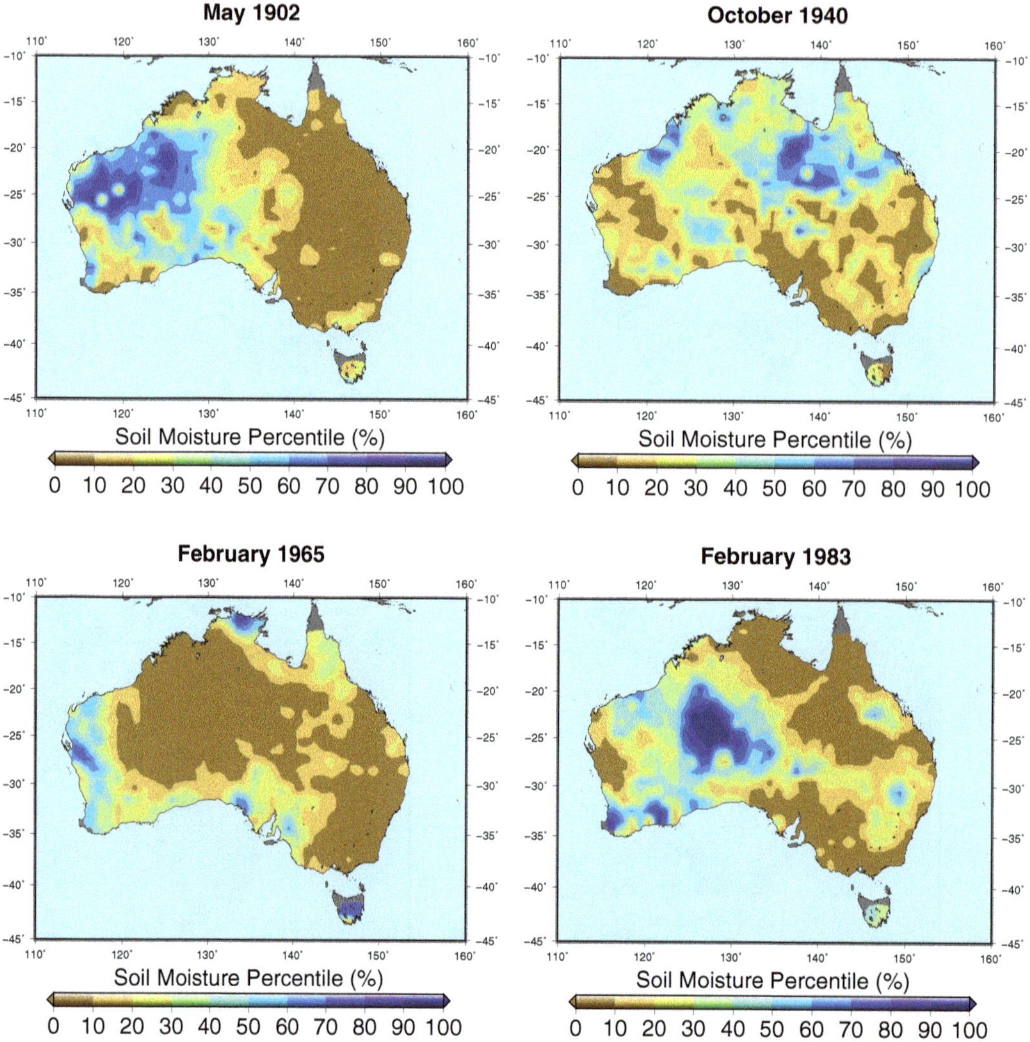

Plate 7.7 *Maps of the peak extent of major drought events for Australia*

Note: The 1902 event was part of the 'Federation drought' of 1895–1902. The 'WWII drought' affected most of southern Australia throughout 1939–1945. The peak extent of drought for Australia during 1901–2008 occurred in 1965, part of a short but intense event. The 1982–1983 El Niño-forced drought mainly affected eastern Australia, and the Murray-Darling River basin in particular.

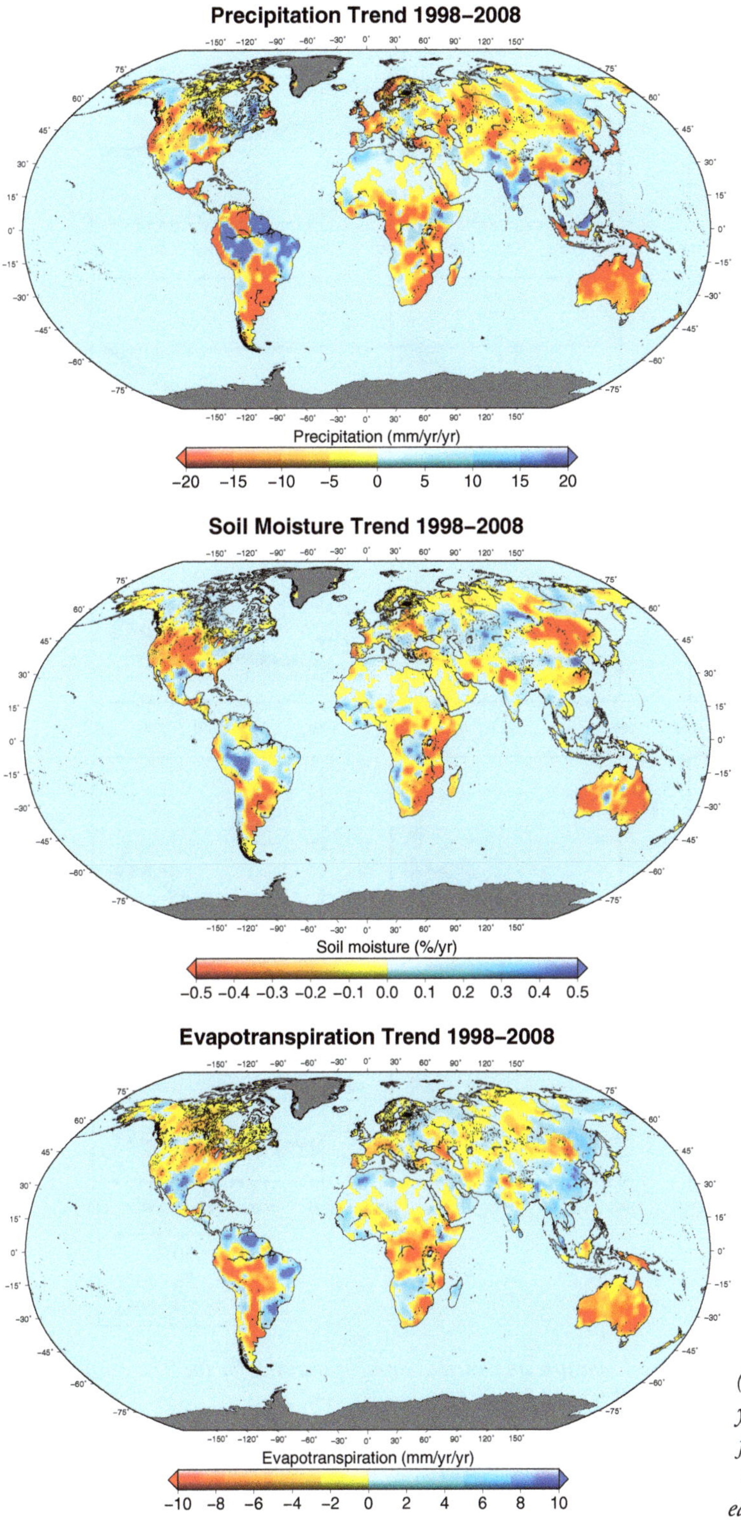

Plate 8.1 *Trends in precipitation (mm yr^{-2}), soil moisture (% volumetric yr^{-1}) and evapotranspiration (mm yr^{-2}) for 1998–2008, showing drying in the western US, southeast South America, eastern Africa, northern China/Mongolia and Australia*

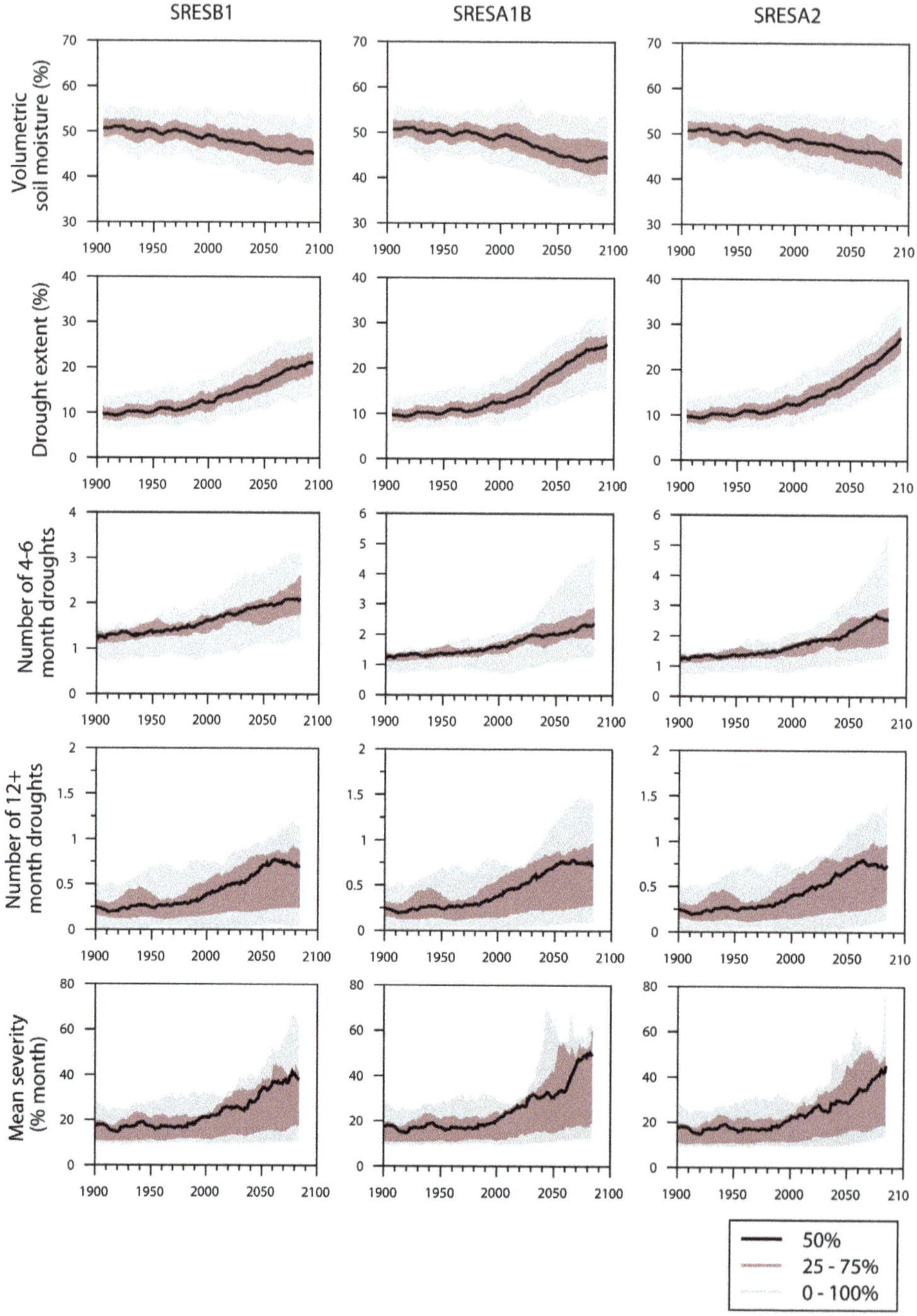

Plate 8.2 *Global average time series of soil moisture and various drought statistics for the 21st century for the SRES B1, A1B and A2 scenarios*

Note: Monthly soil moisture and drought extent ($q_0(\theta)$ = 10.0 per cent) is shown as the 121-month running mean. The frequency of 4-month to 6-month duration droughts ($q_0(\theta)$ = 10.0 per cent), 12+ month duration droughts ($q_0(\theta)$ = 50.0 per cent) and the maximum severity (defined as the drought intensity times the drought duration) are calculated over a 30-year moving window. The data are shown as the 0, 25, 50, 75 and 100 per cent percentiles of the distribution of the multimodel ensemble of soil moisture and drought statistics.

Source: reproduced with permission from Sheffield and Wood (2008a)

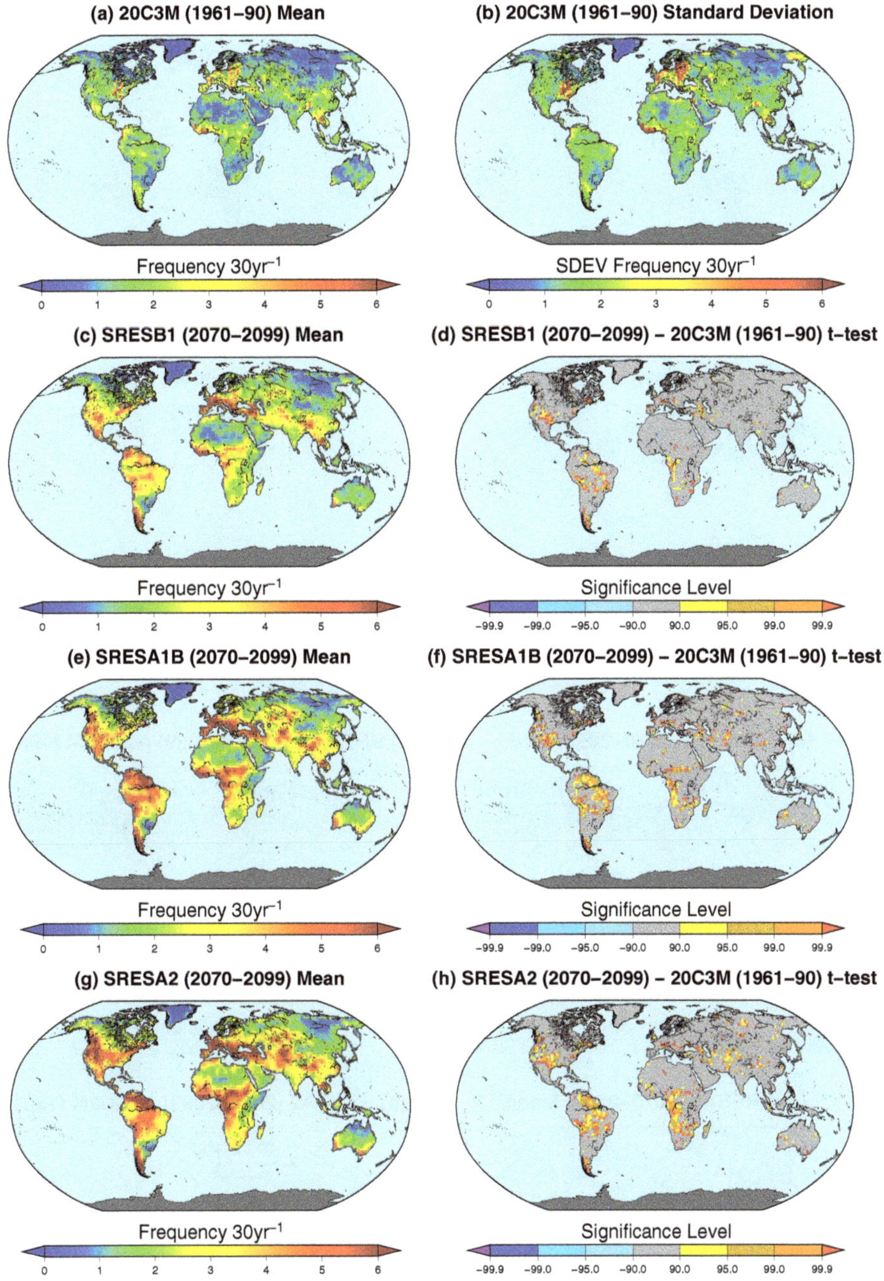

Plate 8.3 *Multimodel ensemble mean of frequency of $D_{4\text{-}6}$ droughts in the 20C3M (1961–1990) simulation and for 2070–2099 in the SRES B1, A1B and A2 future climate scenarios; also shown are the standard deviation of the 20C3M (1961–1990) multimodel ensemble and the statistical significance of the difference in ensemble means between the 20C3M simulation and the future climate predictions*

Note: Statistically significant changes were estimated by calculating t-test statistics under the null hypothesis that the mean of distribution of 20C3M period drought frequency equals the mean of distribution for the future climate period. Results are shown for significance levels of 90, 95, 99 and 99.9 per cent, with negative values indicating a decrease in the frequency of drought.

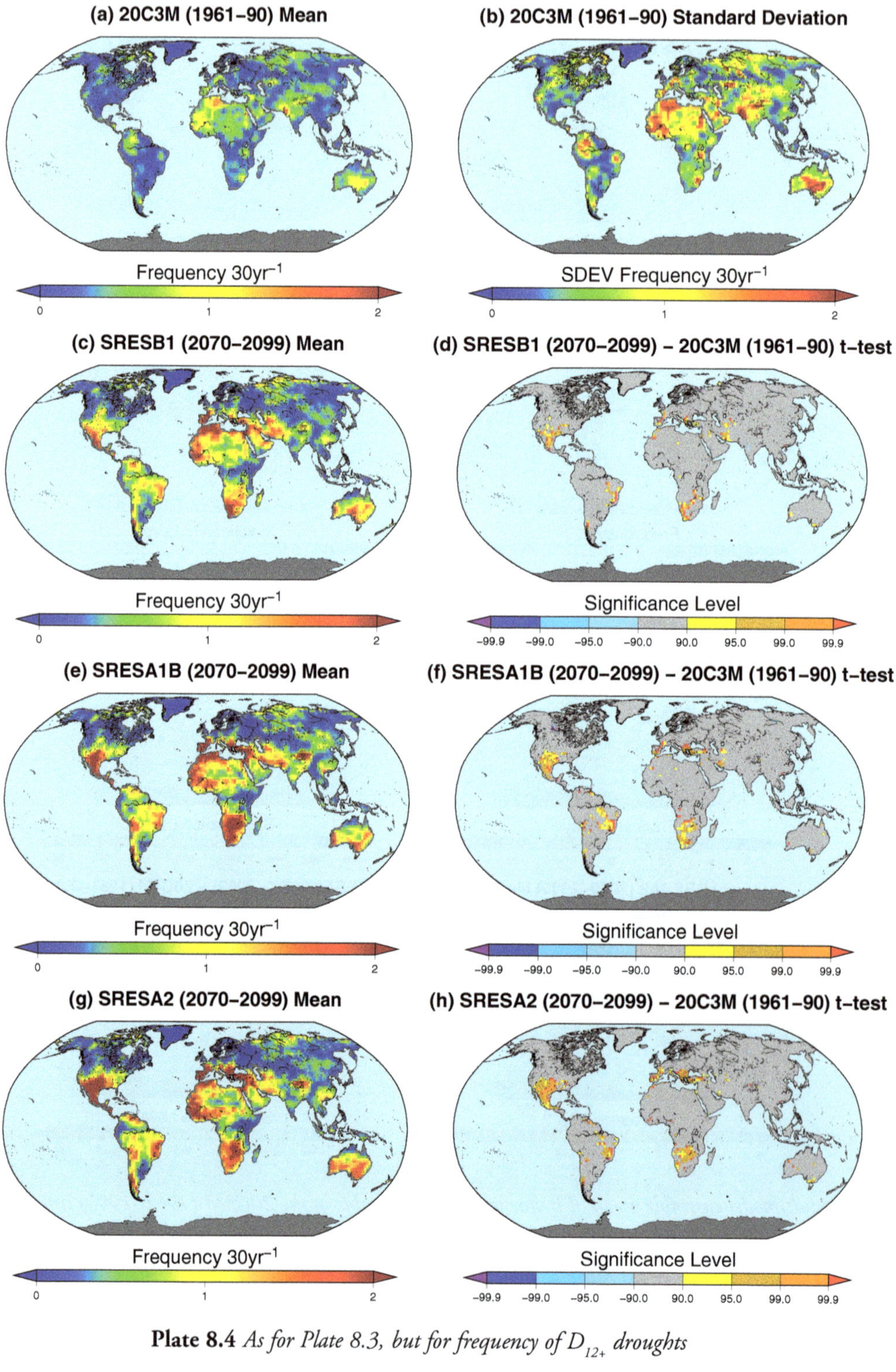

Plate 8.4 *As for Plate 8.3, but for frequency of D_{12+} droughts*

7

Major Drought Events of the 20th Century

Introduction

The last chapter described drought over the 20th century in terms of the variations of hydrology and climate, and soil moisture and large-scale drought in particular, without specific attention to individual drought events. This chapter delves into the regional occurrence of drought and individual drought events, their characteristics, causes and impacts. It concludes by assessing the most severe drought events of the 20th century. Individual drought events are identified objectively using a cluster algorithm to track the extent of spatially contiguous regions of low soil moisture. The soil moisture data are taken from the long-term modelling simulations using the Variable Infiltration Capacity (VIC) model (see Box 4.1). Where possible, these events are linked to drought reports and reviews of their causes and characteristics in the scientific literature. The severity of an event is quantified using a severity–area–duration (SAD) analysis (see Box 4.2). In this approach, the severity of the drought is recorded at regular intervals during its lifetime and over incremental areas within the event. The severity values can be used to evaluate events against each other in a region or across the globe, in synchrony or at other times. This provides a quantitative ranking of individual drought events in terms of their space-time characteristics. The set of identified droughts is not exhaustive, but offers a picture of events which were significant in terms of their spatial extent, duration and severity, and therefore their impacts. Such characteristics are used to identify events that are less well known, because they have occurred in more remote regions, or in the earlier part of the 20th century. The events are discussed in detail in sub-sections for continents and regions of the world.

North America

Figure 7.1 shows the time series of the monthly soil moisture index, averaged over the US and sub-regions from 1901 to 2008. Also shown is the monthly time series of the percentage area of each region that experienced severe drought conditions, where a severe drought is defined through an index value of less than 20 per cent. North America, and the US in particular, has undergone a number of major droughts during the last 100 years, many of which have been severe in terms of their impacts to agriculture and society. A number of significant dry events can be identified in the time series of average values that correspond to these major historic events. The time series of the percentage area affected by severe drought conditions highlights the extent of these significant events. For example, the 1930s and the mid-1950s show dry spells (low soil moisture percentile values) spanning a number of years, which reflect the multiyear 1930s Dustbowl drought and the 1950s US Southwest drought, respectively.

Two moderate events occurred during the 1960s, which correspond to the drought conditions in the Midwest (Diaz, 1983) and Northeast (Namias, 1966, 1967; Diaz, 1983). During 1988, there was a short but severe drought period that affected a large area of the country, peaking in the spring and summer of 1988 (Namias, 1991; Mo et al, 1991; Trenberth and Branstator, 1992; Lyon and Dole, 1995; Sud et al, 2003). The maximum total area of the US affected during the Dustbowl drought in the 1930s was 53.8 per cent in June 1934, 50.1 per cent in July 1936 and 58.2 per cent at the end of 1939. During the 1950s drought, 64.3 per cent was affected in October 1952,

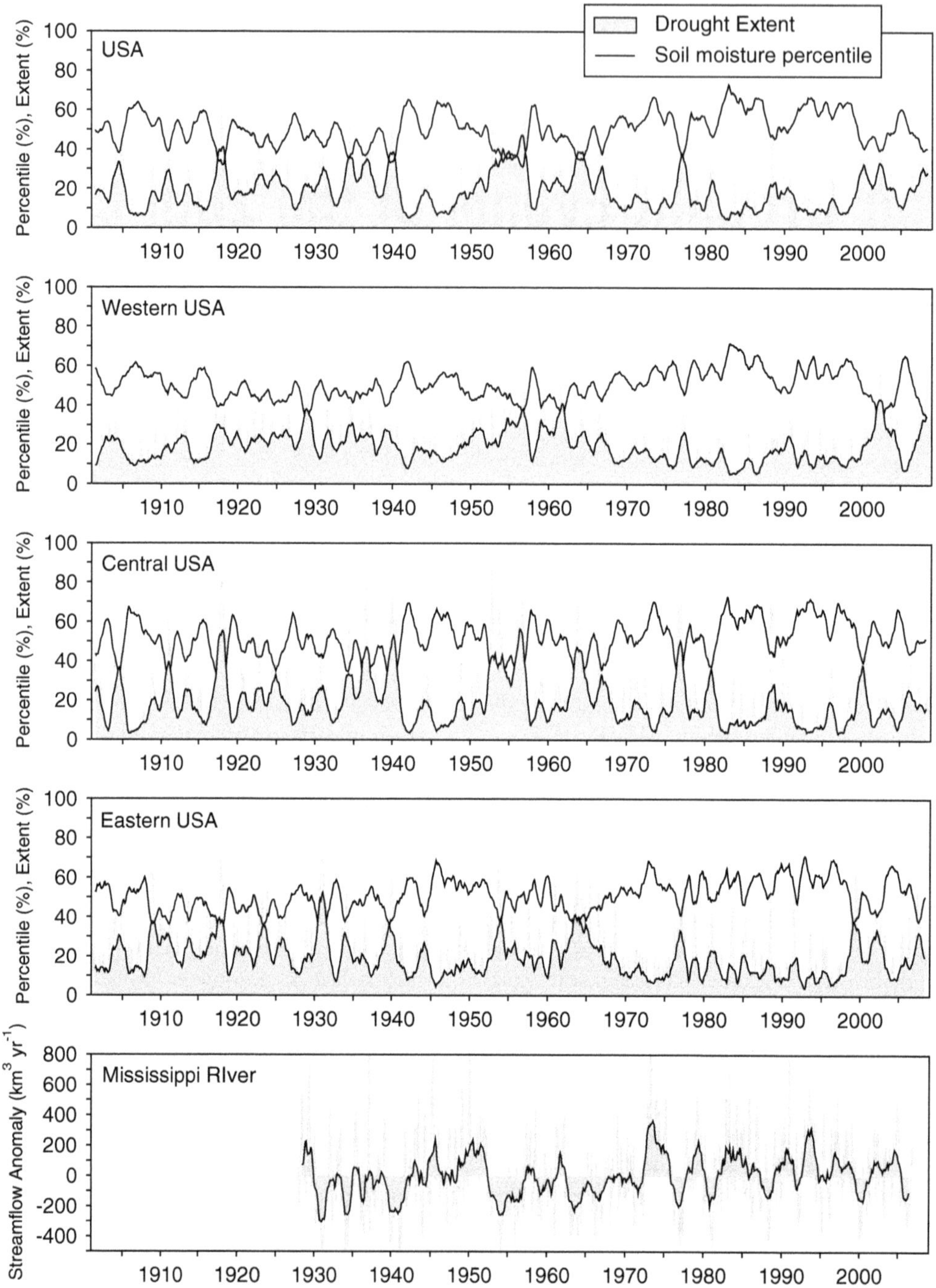

Figure 7.1 *Time series of monthly regional soil moisture index and the percentage area in drought for the US and three sub-regions*

Note: The data are smoothed using a 13-month moving average filter. The original unsmoothed area in drought is shown by the grey lines. The bottom panel shows the monthly streamflow anomaly for the Mississippi River, as measured at Vicksburg.

Source: streamflow data from the Global Runoff Data Centre, Koblenz, Germany at http://grdc.bafg.de

although there are similarly high values in October 1956 (56.7 per cent areal extent) and January 1957 (56.2 per cent), with overall values greater than 40 per cent during 1952–1957. Significant peaks in March 1964 (46.8 per cent) and July 1966 (47.3 per cent) indicate that dry conditions also prevailed in the 1960s. The spike in early 1977 (53 per cent areal extent) shows the peak of the moderate drought of 1976–1977, which mostly affected the western third of the United States throughout this period and the north-central states during the spring of 1977 (Diaz, 1983). For the drought of 1988, the peak extent during June 1988 was 48.3 per cent, but as with the 1977 drought, the duration of the event was small compared with the 1950s drought, although the impacts were significant (Trenberth and Branstator, 1992).

Plate 7.1 shows snapshots of some of the major droughts of the 20th century to hit North America. The two most severe were the 1930s Dustbowl and 1950s Southwest drought, but the 1988 central US drought and the recent drought in the West that peaked in 2002, were also severe, but less extensive. We discuss these events in more detail next.

The Dustbowl drought

The Dustbowl drought of the 1930s is one of the worst natural disasters to hit the US, impacting mainly across the central US Great Plains. The drought lasted up to eight years in some places, but peaked in 1934, 1936 and 1939–1940. It covered up to 60 per cent of the US and extended from the southern states to the Canadian prairies and from the eastern US coast to the Rocky Mountains in the west (Plate 7.1). The drought was particular devastating, since it compounded the existing economic problems of the Great Depression and led to the mass migration of people from the worst-affected areas to California and other regions of the western US (Worster, 1979). The drought was made worse by poor land management, which led to soil erosion and the great dust storms that gave the drought its name. The economic cost of the drought is difficult to estimate, given its extensive coverage, long duration and that it occurred many years ago. It is also difficult to discern the impacts of the drought itself, when taking into account the underlying depressed economic conditions of the time. One gauge of the costs is reflected by the amount of government financial assistance, which is equivalent to US$13 billion today (Warrick, 1980), indicating that the total costs were much higher. Agricultural production was estimated to have fallen by 17 per cent across the region of Missouri, Iowa, Nebraska and Kansas (Rosenberg, 1993). Yet much was learned, including better understanding of the interactions of climate and agriculture, and land management to control erosion and maintain fragile soils in dry-land environments. This led to reduced impacts in subsequent severe drought episodes, such as the 1950s Southwest drought.

Climate model recreations of the conditions at the time have enabled the identification of possible reasons for the drought. These point to the existence of ocean conditions that were ideal for inducing weather patterns that would bring about drought across the US. These were cooler than normal sea surface temperatures in the tropical Pacific – La Niña mode of the El Niño Southern Oscillation (ENSO) – and warmer than normal conditions in the tropical Atlantic. These conditions shifted the jet stream to a more southerly location than normal, and reduced the amount of moisture usually transported from the Gulf of Mexico into the central US and the Great Plains (Schubert et al, 2004a). It is thought that the increase in dust in the atmosphere may have also exacerbated the dry conditions (Cook et al, 2009). Overall, there were a number of factors that contributed to the severity and impacts of the Dustbowl drought, but many uncertainties still remain.

The 1950s drought

The next major drought to hit North America occurred during the 1950s. At the time, economic prosperity was returning to the US after the Second World War and technological advances were finding their way into everyday life. Yet the drought caused great suffering, mainly over the southwest US and Texas, and may have lasted for up to ten years in some places (Namias, 1982; Diaz, 1983; Barlow et al, 2001; Schubert et al, 2004b). It was at its height most notably in Texas from 1951 to 1956 (Diaz, 1983), with most of Texas declared as a federal drought disaster area. At its peak, the drought reached from one side of the country to the other side (Plate 7.1). The drought began in the southwest in 1950 and spread north and westwards over the next few years to cover ten states by 1954, bringing heat waves to major cities such as Dallas. In the southern Great Plains, agriculture was decimated, with crop yields down by up to 50 per cent and extreme temperatures drying out grazing lands, forcing ranchers to feed their cattle prickly pear cactus. The drought reached its peak at the end of 1956 and was finally dissipated in most regions with the advent of spring rains of 1957.

Although the Southwest drought was comparable to the Dustbowl drought in its severity, its socio-economic impacts were less because of lessons learnt from the Dustbowl, in terms of agricultural and water management such as irrigation, as well as a much stronger prevailing economy, driven by the post-war boom and technological advances. The 1950s drought is also linked with La Niña conditions, coupled with pan-Pacific cooling across the tropics and extra-tropics, which acts on decadal timescales (Barlow et al, 2001) similar to the Pacific Decadal Oscillation (PDO). There is also evidence to indicate that feedbacks with soil moisture were partly the cause for the persistence of low precipitation (Schubert et al, 2004b).

The 1988 central US and Canadian prairies drought

The 1988 drought was the first major drought in the US and Canada since the 1950s and although it was shorter it had major impacts, mainly because of its location over the major grain- and corn-growing regions of the US. It is estimated to have had direct and indirect costs of US$39 billion in the US and US$1.8 billion in western Canada, taking into account the impacts on the energy, water and agricultural sectors, as well as the impacts on the ecosystem. Grain surpluses and federal aid may have reduced these impacts somewhat. The drought was not only the costliest drought in US history up until that time, but also compared to all following disasters, until recently, when it was overtaken by the devastation of Hurricane Katrina, which had estimated costs of up to US$156 billion (Burton and Hicks, 2005).

The drought began in the western US and spread northwards, and then across the northern Great Plains, where it had it the greatest agricultural impacts. The drought initiated damaging wildfires across the West, including the devastation of much of Yellowstone National Park. Thousands of deaths are estimated to have been due to the ensuing heat waves in many metropolitan areas in the Midwest. Further south, the drought affected the navigation of the upper Mississippi River. At its peak, the drought covered almost 50 per cent of the US (Plate 7.1). This is smaller than the Dust Bowl drought (60 per cent) at its peak, but the 1988 drought covered major parts of the corn- and soy-growing regions of the mid-western US, and so had major agricultural impacts, estimated to amount to US$30 billion.

It is generally agreed that a complex combination of SST anomalies, persistent stationary atmospheric circulation and soil moisture–precipitation feedbacks were key factors in the development and severity of the drought (Sud et al, 2003). The spring of 1988 was characterized by dry conditions over much of the western and mid-western US (Janowiak, 1988). Extensive drought conditions throughout the central US were initiated during May, forced by changes to atmospheric circulation during April–June (Trenberth and Branstator, 1992), which led to below-normal precipitation across the Central Plains. The maximum extent of negative precipitation anomalies occurred in June across a large region, extending from the upper Midwest to the East Coast (Ropelewski, 1988). Precipitation returned to near-normal levels during July and August for large parts of the drought-affected area, first in the east in July and then through the upper Midwest, and much of the Central and Southern Plains region in August (Ropelewski, 1988; Namias, 1991). Despite the end of meteorological drought through July and August, record high surface air temperatures and accompanying significant evapotranspiration ensured that severe hydrological and agricultural drought conditions prevailed throughout the summer (Trenberth and Branstator, 1992). There was a return to wetter soil moisture conditions from September onwards, across the lower Mississippi and Ohio river basins.

Recent drought in the US West

Long-term drought in the West has persisted since about 1999, but reached its peak in the summer of 2002, covering about 41 per cent of North America, including the Canadian prairies and parts of northern Mexico, as seen in the time series in Figure 7.1 and the map in Plate 7.1. The drought particularly affected the Colorado River basin, where it covered more than 90 per cent of the area. This was driven by near record-low precipitation (Lawrimore and Stephens, 2003), and caused severe water shortages, exacerbated by increased human demand. Figure 7.2 shows the long-term storage of Lake Powell, which is one of the largest reservoirs in the US. The reservoir dams the upper Colorado River, supplying water to the states of Colorado, Utah, Wyoming and New Mexico, under the Colorado River Compact of 1922. There has been a steady decline since the previous full level in about 1999, with some recovery in recent years to about 50 per cent of capacity. At the height of the drought in the upper Colorado in the winter of 2005, the reservoir was at 45m below pool level, or about 30 per cent of capacity. Similar deficits happened in almost all western states with Arizona, New Mexico, Nevada, Utah and Wyoming being particularly hard hit.

This drought peaked in 2002, and despite being as severe and widespread as previous recent droughts, it was estimated to have cost much less (US$10 billion) and caused no fatalities. This stands in contrast to the drought of 1988, which caused huge economic damages and thousands of deaths in the related heat waves. The 1988 drought impacted across regions of intensive agriculture over the northern and eastern Great Plains, during the spring and summer, whereas the Western drought in 2002 affected regions of low population density and relatively little agriculture. Nevertheless, major impacts included wildfires, tree die-off and insect outbreaks. Fires during the summer of 2002 were the second worst in the last 50 years, with some of the largest fires spreading across Arizona, Colorado and Oregon. In Arizona and New Mexico, about 12,000 km^2 of ponderosa and piñon pine trees were killed off by the drought, combined with the devastating effects of insect outbreaks (Raffa et al, 2008).

Severity–area–duration analysis

The recent drought in the West, the 1988 drought and other smaller droughts, such as the 1980 drought in the southern US and 2007 drought in the Southeast, all had major impacts that were exacerbated by their location in terms of effects on population and agriculture. They do, however, pale into insignificance compared to the droughts in the 1930s and 1950s, whose impacts were compounded by their lengthy duration and spatial extent. The presence of these earlier events and even more lengthy ones in the palaeoclimatic record also indicates the potential for another Dustbowl-size event in the future.

The longest droughts identified for North America are in the 1950s (Cook et al, 1999; Stahle and Cleaveland, 1988; Palmer, 1965; Karl and Quayle, 1981), with only the 2001–2002 drought (Cook et al, 2004) of comparable duration (see Table 7.1). It should be noted that some of these events covered high latitudes of Canada and Alaska (although they joined temporarily with US droughts during their lifetime), where freezing temperatures prolong the anomalously low soil moisture values through the winter (Sheffield and Wood, 2007). The most spatially extensive droughts were the events of 1954–1957 (central US and much of Canada), 2001–2002 (most of the western parts of the

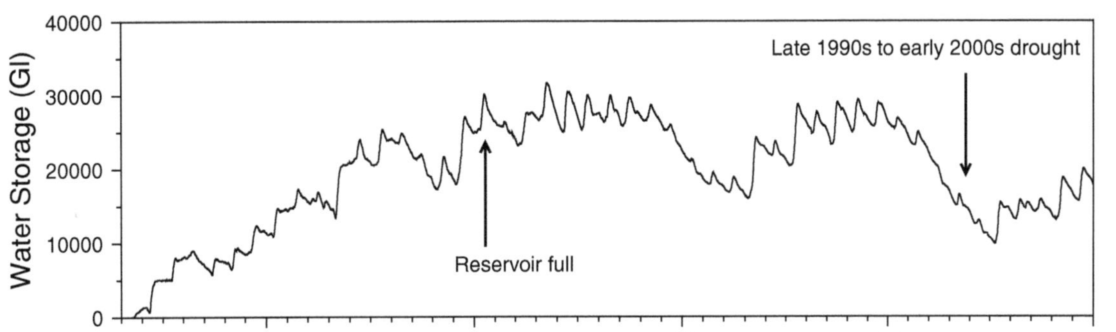

Figure 7.2 *Water storage of Lake Powell, 1963–2010, showing the long-term drought during the late 1990s to early 2000s*

Note: The lake stores about 30km^3 (30,000 gigalitres) of water when full. Filling of the reservoir began in 1963 and was completed in 1980. The drought in the late 1990s and early 2000s is clear and the recovery up until 2010 has only been slight.

Source: US Bureau of Reclamation (www.usbr.gov/uc)

US and Canada), and 1976–1977 (northern US and central Canada) (Cook et al, 1999; Keyantash and Dracup, 2004). The drought of 1917–1918 was short-lived but peaked in the winter of 1918, encompassing the eastern half of North America, from Mexico up to Canada. The 1976–1977 event, which spanned the US, is interesting because it was at its peak during the winter and so had relatively little impact nationally, although regionally it was the drought of record in California and the Columbia river basin.

The SAD envelope curves in the upper left panel of Plate 7.2 identify the most severe droughts for different combinations of duration and spatial extent which occurred over North America during the 20th century (the other panels refer to other continents and are discussed later on). The 1954–1957 drought dominates the SAD envelope curves for durations of 12 and 24 months. For shorter-term (three-month) durations, the 1976–1977 drought was the severest event for areas of up to about 7 million km^2. Note that despite the 1988 drought being considered the worst US drought on record in economic terms (Trenberth and Branstator, 1992), it does not appear on the envelope curve. This is because other drought events were more severe for all

Table 7.1. *List of top five ranked drought events for each continent in terms of duration and maximum spatial extent*

	Duration (months)	Spatial extent (million km^2)
Africa	1992–1993 (21)	1984–1985 (8.2)
	1987–1988 (17)	1993–1994 (7.8)
	1984–1985 (16)	1987–1988 (7.7)
	1985–1986 (14)	1992–1993 (7.3)
	1994–1995 (13)	1983 (7.2)
Asia	1986–1990 (42)	1944–1945 (10)
	1950–1952 (26)	1987–1988 (9.8)
	1982–1984 (23)	1968–1969 (9.7)
	1992–1994 (21)	1997–1998 (8.7)
	1984–1986 (20)	1982–1983 (8.5)
Europe	1975–1977 (25)	1920–1922 (5)
	1920–1922 (23)	1994 (4.9)
	1946–1948 (18)	1992–1993 (4.2)
	1980–1981 (14)	1975–1977 (4)
	1906–1907 (14)	1945 (3.9)
N. America	1954–1957 (41)	1954–1957 (7.5)
	1963–1964 (20)	2001–2002 (7.5)
	2001–2002 (19)	1952–1954 (6.9)
	1952–1954 (19)	1961 (6.9)
	1939–1941 (19)	1917–1918 (6.7)
Oceania	1928–1930 (24)	1964–1965 (5.9)
	1951–1953 (15)	2005 (5.2)
	1964–1965 (13)	1986 (4.8)
	1963–1964 (13)	1961–1962 (4.6)
	1944–1945 (13)	1935 (4.6)
S. America	1982–1983 (19)	1963–1964 (10.7)
	1963–1964 (18)	1995 (8.7)
	1967–1968 (16)	1967–1968 (8.5)
	1958–1959 (15)	1962–1963 (8.0)
	1988–1989 (15)	1961 (6.8)

Note: The duration (months) and extent (millions km^2) are given in brackets after the time period of each event.

combinations of duration and extent, yet their actual impacts may have been lower because of their location and/or timing.

Andreadis et al (2005) in a similar study of drought for the US only, found the 1950s drought to be the most severe for large areas and long durations. Although our chosen cluster threshold (see Box 4.2) splits this drought into two events (1952–1954 and 1954–1957), we also find that these are the most severe for large extents and durations (Plate 7.2). Figure 7.3 shows that the 1954–1957 event is unique in terms of severity beyond 7 million km² in extent. Both studies identify the 1976–1977 drought that covered much of the US and southern Canada as one of the most severe for short durations. Andreadis et al (2005) also find that the drought in the West (1998–2003) is less severe but comparable to the droughts of the 1950s (and 1930s), especially for smaller spatial extents. We find that neither the recent western drought nor the 1930s drought are on the SAD envelope curves, although they are ranked highly in terms of duration and spatial extent, and are relatively severe at three-month duration (Figure 7.3). For both studies, the important 1988 drought (1988–1989 in this book; 1987–1993 in Andreadis et al, 2005) is considered a severe event, but does not contribute to the SAD envelope curves. However, Figure 7.3 shows that this is comparable to many other events, and is generally more severe than the 1930s series of droughts at three-month duration.

South America

South America contains extremes of climate from the humid Amazon rain forest to the relative dryness of the northeast coasts of Brazil, to the Pampas grasslands of Argentina, the deserts of the Atacama and the cooler Patagonian Steppe in the south. Along the spine of South America run the Andes Mountains, which play a role in shaping climate across the continent. Overall, the continent is rich in freshwater resources, although this is mediated by the spatial and temporal variability that can ultimately lead to the occurrence of drought (Chapter 3). ENSO is a major cause of this variation, with La Niña driving drier conditions in central Argentina and Chile, and El Niño reducing water supply in many countries, including Columbia, and affecting hydropower – which accounts for the majority of electricity generation in many South American countries.

Amazonian droughts of the 20th century

The Amazon River basin is a crucial component of the climate of the region and the world, returning copious amounts of water via evapotranspiration back to the atmosphere and streamflow to the Atlantic, and acting as a giant carbon sink, assimilating carbon from the atmosphere to grow the forests. As such it is thought to

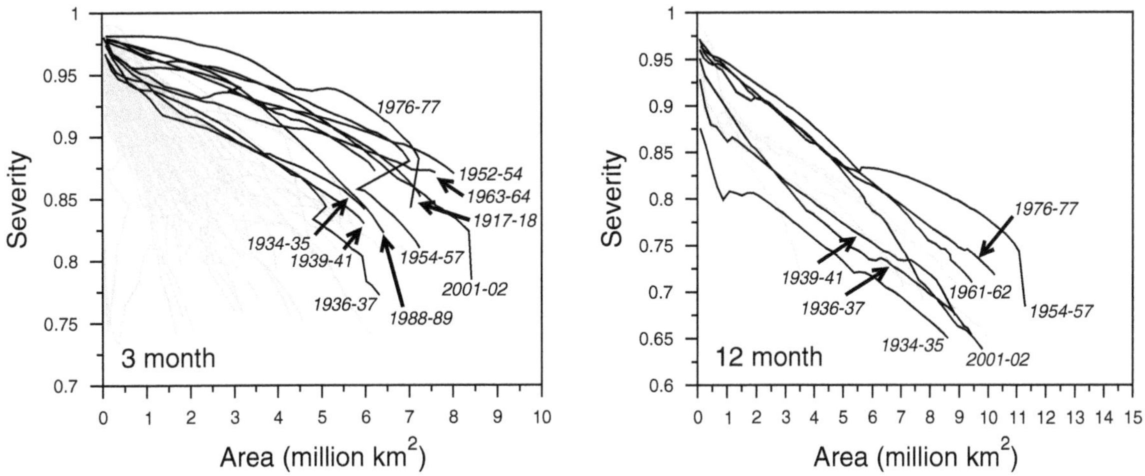

Figure 7.3 *Drought severity–area–duration curves for North America at 3-month and 12-month durations*

be a mediator of greenhouse warming, providing a buffer against anthropogenic emissions. Some of the best records of drought can be found in streamflow measurements along the Amazon, especially at the beginning of the 20th century, when rain gauges were non-existent across much of the region. Figure 7.4 shows the water level record back to 1903 at Manaus on the Rio Negro, which is a northern tributary of the Amazon that drains almost 700,000km^2 of the western Amazon River basin, including part of the Colombian highlands. The most pronounced anomaly is the minimum water level in 1926 (Williams et al, 2005), which is 50 per cent less than the climatological mean, when the high water was 6m below the mean of 27.8m. This drought was compounded by dry conditions in the previous year. Corroborating evidence includes that of forest fires in the region and high temperatures in Venezuelan cities. Local reports indicated that the 1926 drought was so severe that it killed vast numbers of fauna. Furthermore, fires killed thousands of rubber workers (Meggers, 1994). The spatial extent of the drought is shown in Plate 7.3a. From SST reconstructions, this was known to be an El Niño year. The El Niño was not particularly strong, but warm temperatures persisted in the tropical Pacific from May 1925 for 16 months, replicating conditions that had occurred in 1911–1912. Interestingly, the Nordeste (northeast) region of Brazil was wet at this time, when normally an El Niño would bring dry conditions. However, drought is also controlled by sea surface temperatures (SSTs) in the South Atlantic Ocean, which in 1926 were greater than normal, inducing low pressure and higher levels of rainfall over the region. This is also shown further south in the records for the Parana River (see below). The influence of the Atlantic on South American drought is discussed further in the case of the 2005 Amazon drought.

After a relatively long-term wet period during the 1940s–1950s, major droughts occurred in the Amazon in 1963–1964, 1970, 1983, 1988 and 1997–1998, as shown in the soil moisture and streamflow records for the Amazon River basin in Figure 7.5. The spatial extent of the 1963–1964 and 1997–1998 droughts are shown in Plate 7.3. Most of these are associated with El Niño events, and appear also in the Manaus water level record (Figure 7.4) of the Rio Negro (Marengo, 1995). Figure 7.5 shows the time series of ENSO activity based on anomalies of SSTs in the Niño3.4 region of the tropical Pacific, and indicates a reasonable match with the soil moisture record. Large drought events occurred in the Rio Negro during 1925–1926, 1935–1936, 1966–1967, 1979–1980, 1983, 1992 and 1997 (Marengo, 1995). The very strong 1982–1983 El Niño brought 70 per cent less precipitation in Amazonia during January to February, with river discharge in the lower Amazon falling to less than 2 per cent of normal levels. Rainfall deficits were severe from southern Venezuela up to Panama. Then one of the strongest El Niño events of the 20th century occurred in 1997–1998, causing widespread drought mainly over the eastern Amazon (Plate 7.3c), and inducing fire in secondary forests. Primary forests are not normally affected but

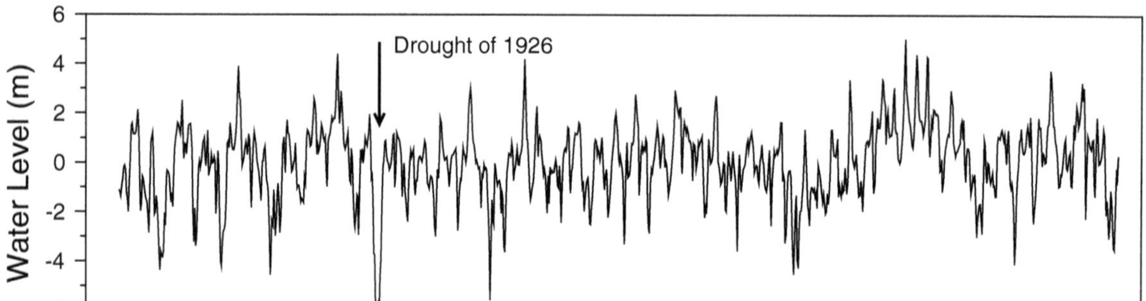

Figure 7.4 *River water level for 1903–1992 for the Rio Negro (a tributary of the Amazon River) at Manaus*

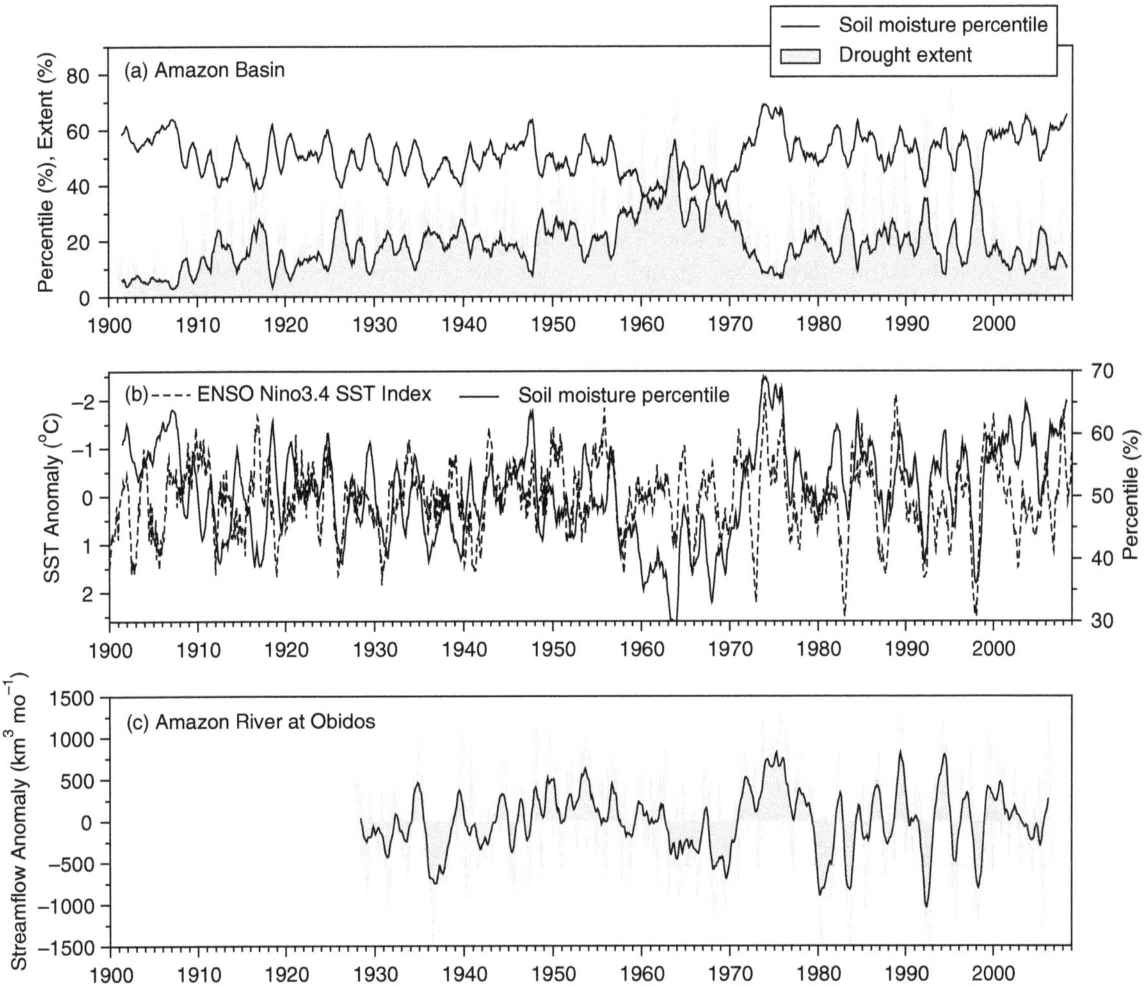

Figure 7.5 *(a) Soil moisture percentile and spatial extent of drought for 1901–2008 for the Amazon River basin. (b) Niño3.4 SST anomalies based on the HadSST1 dataset (available from www.esrl.noaa.gov/psd/gcos_wgsp/Timeseries/Nino34/) compared to the soil moisture index. The y-axis for the SST anomalies has been inverted. The positive anomalies indicate El Niño events. (c) Streamflow anomalies for the Amazon River measured at the Obidos gauging station*

Source: streamflow data from the Global Runoff Data Centre, Koblenz, Germany at http://grdc.bafg.de

the severity of this drought caused significant impacts in the northern state of Roraima during the dry season, burning almost a quarter of cultivated lands, with 50 per cent more fires in 1997 over the whole Amazon compared to 1996. Tree mortality rose during the drought but soon returned to normal levels afterwards, hinting at the resilience of the rain forest (Williamson et al, 2000). Overlying these short-term but severe events is evidence of decadal variations across the basin, although the north is out of phase with the south (Marengo, 2009).

Amazon drought, 2005

Most recently, the drought in 2005 that covered much of the south and western parts of the Amazon River basin had dramatic impacts, which made newspaper headlines across the world. In some places, such as at the confluence of the Amazon and Tapajós rivers, the water levels were 15m lower than usual. Newspaper photographs depicted boats stranded on dried-up floodplains, far away from the course of the river and the ubiquitous river transport replaced by pedestrian traffic. Impacts were felt across central and southwestern Amazonia (Plate 7.3d), reducing soybean and maize production by 65 and 55 per cent respectively (Magrin et al, 2007). This drought appears to have been the worst in the last 100 years since streamflow records began on tributary rivers such as the Rio Negro at Manaus in the heart of the rain forest. In the east it was the worst in 50–60 years. Despite the relatively small spatial extent of the drought at its peak during July–September, streamflow for the whole Amazon River basin was anomalously low (Figure 7.5), probably because the drought impacted the wetter western part of the basin.

The climate of the region is dominated by the annual north–south undulation of the Intertropical Convergence Zone (ITCZ; see Chapter 3), which is driven by the seasonal cycle of solar radiation, with the wet season following the ITCZ (December–February) and dry season (June–November) falling in its wake. This is also influenced by ENSO and conditions in the tropical Pacific. At the time of the 2005 drought, however, ENSO was in neutral mode (Figure 7.5), implying that the cause would be found elsewhere. The severity of the drought was nevertheless compounded by the El Niño-driven dry conditions in 2002–2003, followed by only a slight recovery in 2004 (Zeng et al 2008). Variations in the ITCZ undulation are also driven partly by north–south gradients in SSTs in the Atlantic, which are especially strong for northeastern Brazil (and the West African Sahel). However, connectivity with the Amazon interior where the drought manifested is less established. Nevertheless, a warmer northern tropical Atlantic during 2005 and the associated atmospheric patterns hint that the Atlantic was partly the cause. Higher temperatures induce higher evaporation and rainfall in low-pressure systems, bringing wetter weather to the US (and a devastating hurricane season in 2005), while causing neighbouring high pressure to the south and Amazonia, and lower rainfall. Overall, the cause of the drought is believed to have been warmer oceans, but also reduced recycling of precipitation within the Amazon because of lower evapotranspiration. Forest fires may also have caused decreased forest cover and therefore transpiration, released smoke particles that interfered with precipitation, and increased carbon emissions. The 2005 Amazon drought was also interesting because of the implications for future change. This and other major droughts were associated not with El Niño, which affects the central and eastern Amazonia, as in 1926, 1983 and 1998, but with the warming of the tropical North Atlantic, which also was the likely cause of the prolonged drought in the mid-1960s. This is an important consideration because of the projection of continued warming in the North Atlantic in the future.

In addition to the severity and impacts of the drought, part of the reason for the heightened interest was the central role that rain forests, such as those in the Amazon, in central Africa and in Southeast Asia, have on the global carbon cycle and the possible feedbacks with climate change. The response of the rain forest to this drought has also been the subject of intense debate about whether forests in wet regions are resilient to drought and may actually benefit from clearer skies during dry periods (Saleska et al, 2007), although this may actually be a result of uncertainties in the remote-sensing data (Samanta et al, 2010). Typically the Amazon rain forest absorbs up to 2 billion tonnes of CO_2 every year and tropical forests are estimated to have absorbed one fifth of anthropogenic emissions over the past few decades. On-the-ground studies over the past 30 years have found that the Amazon forest is sensitive to drought, leading to tree and vegetation die-off (Phillips et al, 2009). This results in a lack of uptake of carbon in the form of CO_2 and therefore less anthropogenic emissions are offset. There are also releases from the Amazon carbon stores through natural emissions and decay. For the 2005 drought it is estimated that an extra 5 billion tonnes of CO_2 were added to the atmosphere, which is equivalent to the annual emissions of Europe and Japan. The worry is that climate change may exacerbate dry periods, leading to a positive feedback with the Amazonian carbon cycle and further drying. Model simulations have shown that the Amazon rain forest may disappear by the end of the 21st century because of this mechanism, coupled with continued deforestation and fire setting for land clearing (Cox et al, 2008).

Other droughts in South America

Marengo (1995) summarizes variability and changes in streamflow across South America during the 20th century. The associations between low flows and drought in rivers and ENSO is evident from the records for many rivers across the Amazon and northern South America in general. Streamflow from the Magdalena and Cauca rivers, which flow through Columbia to the

Caribbean Sea, is lower during El Niño events. The Orinoco River, which drains much of Venezuela and parts of Columbia, shows low streamflow in 1931–1939 and 1973–1976, and does not show much relationship with ENSO. The Trometas (eastern Amazonia) and Ji-Parana (western Amazonia) rivers are generally in drought when ENSO is in its warm El Niño phase and cool La Niña phase respectively. This is also the case for rivers across northern South America, from Venezuela to French Guiana. In southeastern Brazil, the streamflow of the Säo Francisco river was lower than normal during the 1950s, 1970s and during 1987–1992 (Marengo, 1995). Further south, in southern Brazil and Uruguay, La Niña also causes drought in the Uruguay and Negro rivers. The Parana and Uruguay rivers, which flow through southern Brazil, northern Argentina and parts of Paraguay, and make up the La Plata River basin, appear to be linked to the ENSO (Bischoff et al, 2000; Berri et al, 2002), with La Niña forcing lower flows than normal. This influence may vary with time, however (Boulanger et al, 2005). Figure 7.6 shows soil moisture and streamflow time series for the Parana River basin over the 20th century, indicating a reasonable correspondence in the occurrence of dry periods, but no clear contribution from the ENSO. The influence of SSTs in the South Atlantic may also play a role in climate variability and drought across southeast South America (Camilloni and Barros, 2000). The Parana was low during 1907–1918, 1926, 1935–1946, 1954–1958, 1962 and 1968–1974 (Marengo, 1995; Barrucand et al, 2007) and is aligned with dry periods in the soil moisture record. Since the late 1970s, the Parana, Paraguay and Uruguay rivers show increased streamflow (Figure 7.6), driven by increasing regional rainfall (Liebmann et al, 2004). The 1962 drought affected much of Argentina with no rain recorded at gauges in the Mendoza region for up to 110 days (Llano and Penalba, in press). Barrucand et al (2007) also report dry periods in 1985–1986 and 1995 in Argentina. The drought in 1985–1986 was restricted to the southern hemisphere summer and the northeast of the country, and was associated with high temperatures. The drought in 1995 was widespread, affecting the central and northeast regions.

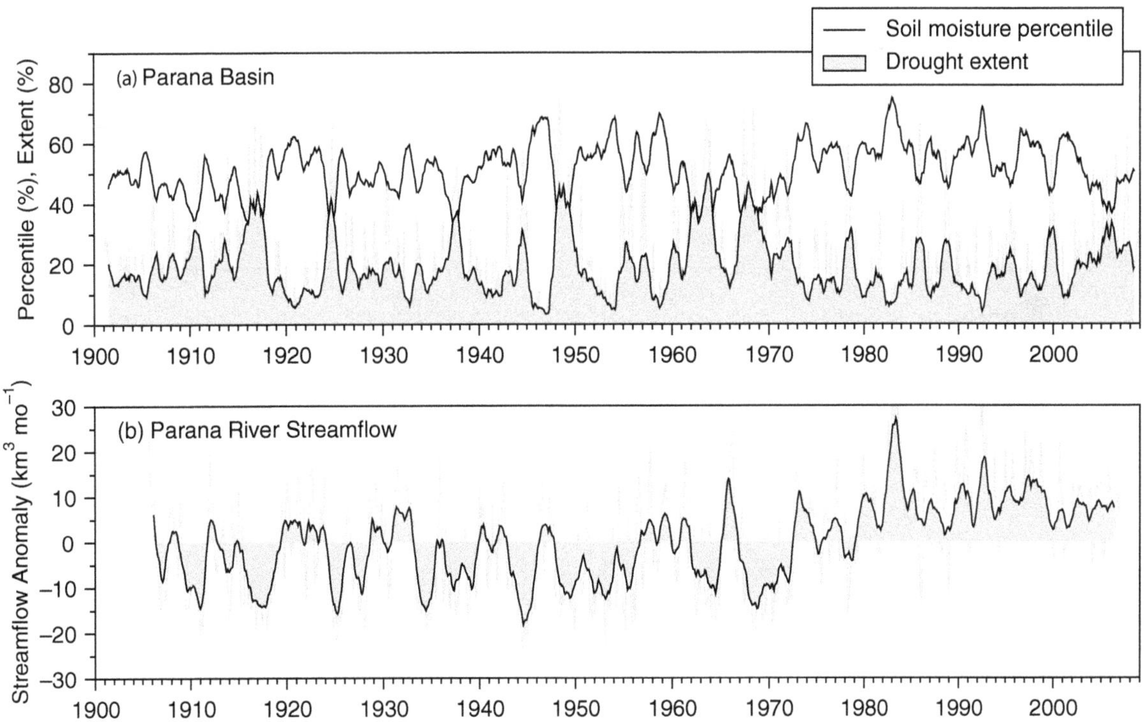

Figure 7.6 *(a) Soil moisture percentile and spatial extent of drought over 1901–2008 for the Parana River basin; and (b) streamflow anomalies for the Parana River at the Timbues gauging station*

Source: streamflow data from the Global Runoff Data Centre, Koblenz, Germany at http://grdc.bafg.de

Precipitation in north-central Chile is mainly controlled by the ENSO (Quintana, 2000), as well as SSTs and sea-ice concentration in the Amundsen Sea and Bellingshausen Sea to the south, but has been declining over the 20th century (Vuille and Milana, 2007). During La Niña events, precipitation falls to at least 35 per cent below normal, affecting agriculture, livestock and timber production. A very severe drought occurred during the La Niña event of 1998–1999, with pronounced atmospheric blocking from a stronger and more southerly subtropical high in the southeast Pacific. Rainfall during July in the capital Santiago was only 0.2mm, compared to normal rainfall of more than 80mm, and this drought persisted into 1999. During La Niña years between 1877 and 1999, the annual average precipitation in Santiago was 210mm, in comparison to the average of about 360mm (Quintana, 2000).

Severity–area–duration analysis

There are several references to drought in the Amazon in the literature, particularly those related to ENSO variability (e.g. the 1982–1983 drought identified here as one of the longest events) (Marengo et al, 1998). However, the drought of 1963–1964 has received little attention (Marengo et al, 2008), despite being identified here as having the largest peak extent and the most severe up to 12 month durations (Plate 7.2). In South America as a whole the longest droughts identified occurred during 1982–1983 (19 months) and 1963–1964 (18 months), and the most spatially extensive events were mainly in the 1960s (see Table 7.1). The SAD envelope curves (Plate 7.2) indicate that the 1963–1964 drought was the worst drought for durations up to 12 months, especially for larger spatial extents. This drought initiated in eastern Brazil in April 1963 reached its maximum extent and severity in October before moving northeast and dissipating in southern Amazonia the following spring. The 1962–1963 drought, which ranged from Patagonia up to southern Brazil, makes up the 12-month SAD curve for very large spatial extents. The droughts that ravaged Argentina in 2006 and 2008 contribute to the three-month curve but only for extents of less than 2 million km^2. Figure 7.7 shows the individual SAD curves for all events, and some of the major events are highlighted. It is remarkable that the series of droughts in the 1960s are dominant in terms of severity for 3- and 12-month durations, with the 1963–1964 drought being by far the most severe drought of a 3-month duration. The 1997–1998 drought is also prominent, being only second in severity to the 1960s droughts.

Europe

The effects of short-term summer hydrological drought has been tempered somewhat in Europe because of the extensive water resources management. Yet this can do

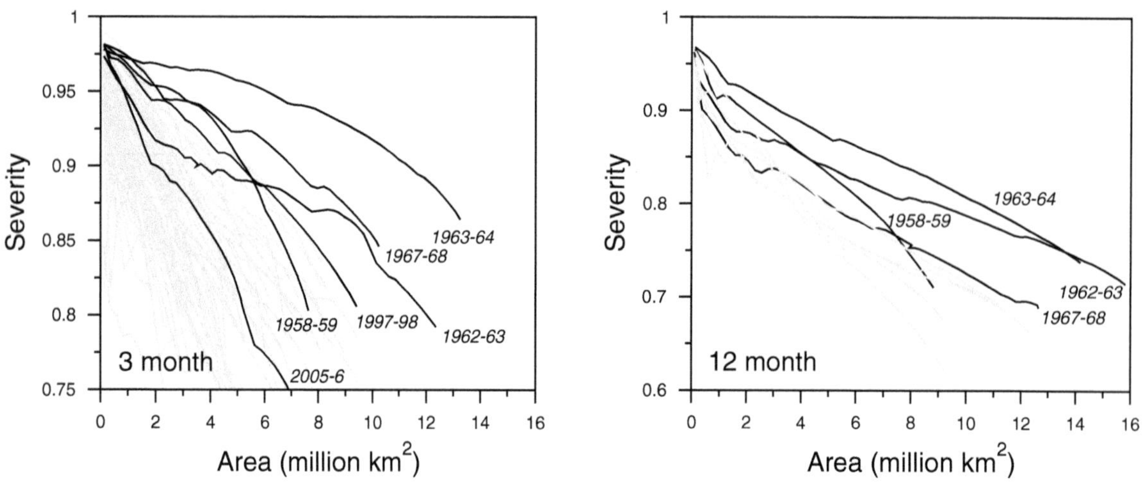

Figure 7.7 *Drought severity–area–duration curves for South America at 3-month and 12-month durations*

little to prevent large-scale and long-term drought from impacting on the environment, and especially under increasing demand from the population and agriculture sector. Some of the more southerly regions rely heavily on irrigation and so are susceptible to drought. The amount of agricultural water use is about 80 per cent of total water in Greece, more than 50 per cent in Italy, 68 per cent in Spain and 52 per cent in Portugal (Bradford, 2000). The value for northern Europe is about 10 per cent. As described in Chapter 3, persistent anticyclonic (high pressure) systems are often the cause of drought in Europe, although these may be accompanied by cyclonic (low pressure) systems in a neighbouring region, which experiences wetter conditions.

Over the 20th century there have many major drought events in Europe, in particular during 1921, the 1940s and 1950s, 1976, early 1990s and 2003. Bradford (2000) and Feyen and Dankers (2009) list droughts that have occurred over the past 30 years. These include droughts in northern and western Europe in 1976, most of Europe in 1989 and again in 1991. There was a lengthy event over much of western Europe during 2003, which had severe impacts through the associated heat wave. Figure 7.8 gives the soil moisture time series

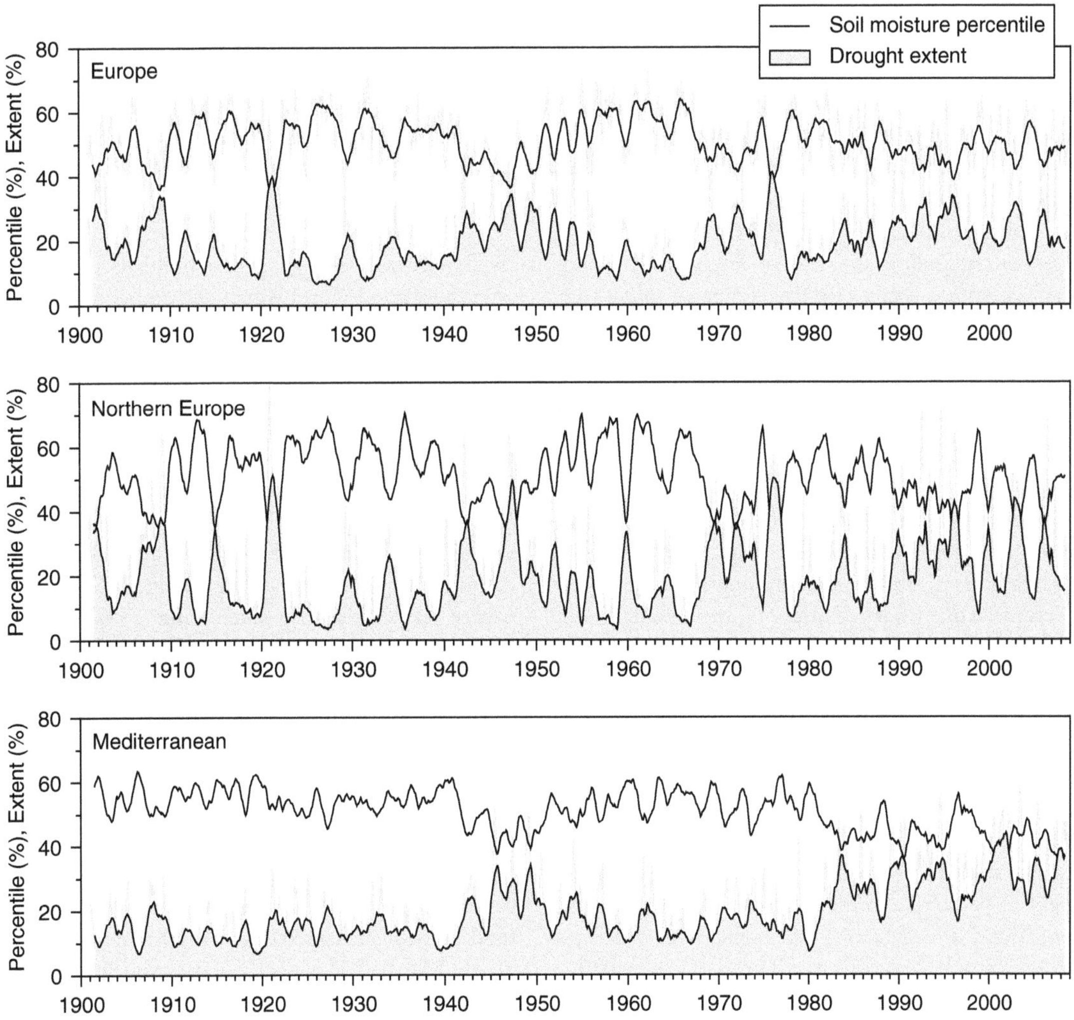

Figure 7.8 *Soil moisture percentile and spatial extent of drought for 1901–2008 for (top) Europe, (middle) northern Europe and (bottom) the Mediterranean region*

for Europe, northern Europe and the Mediterranean regions for 1901–2008 and clearly shows these drought periods. Of particular note are the droughts in 1921, the 1940s and 1950s, 1976 and recent years. The decreasing trend in Mediterranean soil moisture in the series in Figure 7.8 and concurrent increase in drought extent reflects not only the decreasing trend in precipitation across the region (e.g. Xoplaki et al, 2004), but may also be linked to increasing temperatures driving evaporation (e.g. Briffa et al, 2009). Individual countries that have been particularly affected by drought include Greece (1987–1988 to 1992–1993), Italy (1988–1990), Spain (1990–1996) and the UK (1988–1992 and 1995–1996). More recently, droughts struck across the Iberian Peninsula in 2005, which was perhaps the most serious event in 60 years and led to a 10 per cent reduction in total EU cereal yields (UNEP, 2006). On average, the annual economic cost of drought in Europe is estimated to be about €5.3 billion (US$7 billion) since 1991 (EC, 2007). As well as impacts to agriculture, water supply, sanitation, tourism and electricity supply, river transport is also a vital resource. With the EU showing increasing interest in expanding the use of rivers and inland waterways, since it is cost-effective compared to rail and road transport, and reduces greenhouse gas emissions, it is placing a high priority on understanding drought occurrence to ensure that this goal is achievable.

Drought of 1921

1921 appears to have been one of the severest drought years across Europe. This drought is linked with famine in eastern Europe and Russia, although for the latter this was more likely a result of poor governmental food and land policy (see later section on Russia). Brooks and Glasspoole (1922) analysed UK precipitation data, which were available back to the late 18th century, and concluded that 1921 was the driest year in England and Wales since 1854 and only 1887 was drier for the entire British Isles. Although the rainfall data previous to 1850 are problematic, they also hint that 1921 was the driest year since 1758. Modern analyses have found this drought to be unique in the 20th century. Santos et al (2000) found it to be a 100-year event (with a chance of occurring only once every 100 years on average) in terms of meteorological drought across central Europe, and Briffa et al (1994) identified it as the most spatially extensive summer drought across Europe (Plate 7.4). The time series in Figure 7.8 and the maximum spatial extent in Table 7.1 confirm this. The map in Plate 7.4 indicates that the drought mainly affected northwestern Europe, but stretched to central Europe and parts of the Mediterranean. Pekarova et al (2006) looked at records for 18 major European rivers, the earliest of which started in 1807 (Goete River in Sweden), and found the driest year of the period 1850–1997 was 1921. Our own analysis of selected rivers shown in Figure 7.9 also highlights this year as one of the driest on record, affecting the Danube and Rhine, two major European rivers, and the Thames in the England. Bonacina (1923) reported shortly after the event that the drought practically affected the whole of Europe to varying degrees. Rainfall was down to below 40 per cent of normal levels, from England through to Italy, with much of the deficit happening in the warmer summer months in the west and extending from the late summer through to the following year in Italy.

Late 1940s central European drought

As Europe was struggling to rebuild its infrastructure after the Second World War, it succumbed to devastating drought and crop failures in 1947. Much of Europe was already suffering from a lack of food, especially in the east, which was adjusting to post-war reform and failed harvests in 1945. Western Europe could no longer depend upon eastern agriculture and suffered likewise. Poor harvests also occurred in 1946 in much of European Russia, including the Ukraine, Moldavia, and the lower and middle Volga, leading to near-famine conditions by the end of the summer. The following winter was particularly severe, and brought much of Europe to a standstill, which compounded the problems, as well as the transport of coal. By the spring, the harsh winter led to flooding. This was then followed by one of the driest and hottest summers on record during 1947 (Plate 7.4). Much of central Europe had then suffered lower than normal harvests for three years. In some parts of eastern Europe, such as northeastern Romania, more than 90 per cent of the population was suffering from hunger, after crop failures. This led to food aid being delivered from the US and the instigation of the Marshall Plan to rebuild European economies. Ironically, although crops were devastated across much of central Europe, 1945 and 1947 were vintage wine years in France.

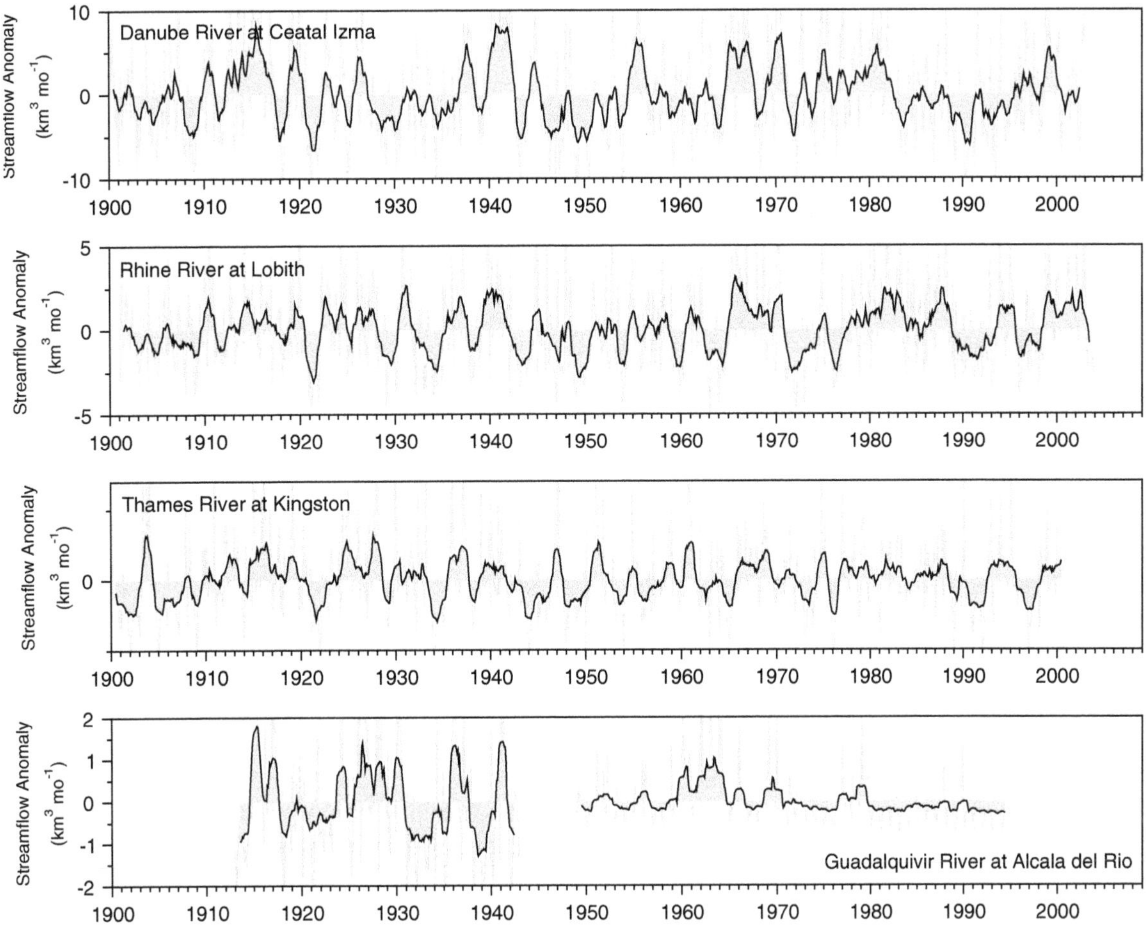

Figure 7.9 *Monthly streamflow anomalies for 1901–2008 of the Danube, Rhine, Thames and Guadalquivir rivers*

Note: The dark lines are the 13-month moving average. The anomalies for the Guadalquivir are calculated separately for before and after the building of a major dam in the 1930s, which reduced the streamflow mean and variability dramatically.

Source: streamflow data from the Global Runoff Data Centre, Koblenz, Germany at http://grdc.bafg.de

Tree-ring reconstructions and instrumental data over Slovakia, analysed by Büntgen et al (2009), show that the driest summers were 1947, 1948 and 1964, and overall long-term drier conditions prevailed during 1940–1960. The Czech Republic also suffered during 1947 (Trnka et al, 2007), which was the severest drought year in the 20th century, with less severe ones in 1976 and 1935. The main impacts were on agricultural production. Agriculture is particularly susceptible to drought in temperate zones such as central Europe during early spring, when plant survival is low and can lead to complete crop failure, and yields can be severely affected during flowering and grain-filling, which occurs in the late summer and autumn. This occurred in some regions in 1947 (Brázdil et al, 2009), when the harsh preceding winter, and spring and summer drought combined to reduce yields and bring failed harvests in some parts. The drought impacts were most severe in the southeast of the Czech Republic, where yields fell to 70 per cent of average for cereals and forage crops. Dry conditions affected over 80 per cent of Bulgaria in terms of precipitation during the 1940s (Tran et al, 2002), mainly occurring during the winter, and so affected snow accumulation and subsequent spring melt. The time series of streamflow anomalies for the Danube River (Figure 7.9), which drains much of central Europe, indicates the drought of 1947, and this lies within a series of dry years from the late 1940s to the early 1950s. Further west, 1947 was one of the severest drought years in southern Germany (Demuth and Kulls, 1997; Fink et al, 2004). Calanca (2007) looked at the Swiss Alpine region and similarly found 1940–1960 to be drier than normal, with 1945, 1947, 1949, 1952, 1959 and 1962 among the driest individual years, and on a par with the pan-European drought years of 1976 and 2003. Across Europe summer moisture, as defined by the Palmer Drought Severity Index (PDSI), was lowest during the 1940s and early 1950s (van der Schrier et al, 2006).

1975–1977 northwestern Europe

The drought during 1975–1977 in northwestern Europe is considered the worst event of the 20th century in this region. During the summer of 1976, it was exceptionally widespread (Plate 7.4), affecting much of northwestern Europe, spreading from Scandinavia to France, and reaching as far as eastern Europe subsequently (Bradford, 2000). The UK was particularly hit, especially in the southeast, where the drought lasted for up to 16 months, with 45 days without rain in some places. May 1975 to August 1976 endured the lowest rainfall on record (Zaidman et al, 2001). Around £500 million (US$780 million) in agricultural costs were encountered, as topsoil was eroded and huge numbers of buildings suffered from subsidence costing £60 million (US$95 million). Lack of air-conditioning meant huge problems at workplaces. The drought began in April 1975, followed by a dry and cool winter of 1975, leading to serious problems by the spring, with no rainfall recorded in many regions for a month. An unprecedented heat wave lasted from 22 June to 16 July 1976, when temperatures were consistently above 27°C, peaking at above 32°C for two weeks in that period. In many places rainfall was less than half of the normal level from October 1975 to August 1976, and equivalent to semi-arid regions. There were widespread heath and woodland fires in southern England, but no water to put them out. Various water-saving measures were introduced, such as hose-pipe bans, a campaign to share baths of no more than 5 inches (12.5cm) in depth, and communal standpipe supplies in some cities. Northern France, Denmark, Germany and even Poland were also affected. In France, the drought reached its peak between May and August 1976. Agricultural impacts were very high in France, with milk yields down by 25 per cent, and 20,000 pigs had to be slaughtered because of the lack of water. The heat wave also caused the deaths of more than 6000 people (Poumadère et al, 2005). Rains finally came in August and normal rainfall patterns returned by October of 1976, although water restrictions carried on for months afterwards.

Early 1990s: Spain

Long-term drought of varying severity happened through much of the early 1990s across Europe, with economic and environmental costs. This is reflected in the time series of soil moisture and drought extent in Figure 7.8, particularly in the Mediterranean. The period from 1992 to 1995 was one of the driest in the century for the Iberian Peninsula, and especially Spain. Many other significant drought events have affected sub-regions, but events during the 1940s, 1950s and 1990s affected a major part of the peninsular (Vicente-Serrano, 2006). The drought in Spain in the early 1990s was estimated to have cost more than €3.7 billion (US$4.9 billion), mostly in agricultural damage (Garrido and Gómez-Ramos, 2000). Losses caused by drought on cereal crops in 1992 were put at over 50 million pesetas (47 per cent of the total production value) and almost 28 million pesetas in 1995. The most affected region was the Guadalquivir River basin, the second longest river in southern Spain, where precipitation was almost half of the normal level (600mm) during 1992–1995, causing irrigation allotments to drop from half the mean in 1992 to zero in 1995. 500,000ha of irrigated land was affected, which was devastating, given that about 75 per cent of water use in the region goes to agriculture. Water shortages and reduction of water quality affected supply to cities, such as in Seville. The bottom panel of Figure 7.9 shows the time series of available streamflow data for the Guadalquivir River over the past 100 years. First notice the difference in variability between the early and later periods, which reflects the construction of the reservoir on the river during the 1930 and 1940s, and the subsequent use of the reservoir water for irrigation. The streamflow anomalies are calculated with respect to the mean seasonal cycle of each period separately, in order to account for the change. The drop in flows is clear after the construction of the dam, which led to restricted migration of sturgeon to spawning grounds and subsequent extinction of the local population. Since the 1970s, the flows have fallen even further due to the drought conditions.

Western European heat wave, 2003

The fatal heat wave and drought of 2003 in Europe was the warmest summer on record and was associated with the deaths of more than 37,000 mostly elderly people across western Europe, and especially France. The drought was caused by a persistent blocking high-pressure pattern that lingered over western Europe. Economic losses were estimated at several million euros, mainly due to decreased agricultural production and environmental and ecosystem damage, such as from forest fires (De Bono et al, 2004). There was also a net loss of carbon across the European biosphere (Ciais et al, 2005). Temperatures above 40°C broke records across several countries. The heat wave was extremely unusual for Europe, but was restricted to Europe only, and was not unusual compared to the last 30 years for the globe as a whole. In fact, more regional temperature anomalies of the same magnitude occurred during 1998 after the strong El Niño (Chase et al, 2006). The extreme heat and lack of water drove grain, vegetable, fruit and wine yields down to between 4 and 11 per cent, with uninsured agricultural losses of more than US$12 billion (Schär and Jendritzky, 2004). Wheat and corn yields were estimated to be down by 15 and 28 per cent, respectively, with costs of several billion euros. In the UK, 2003 provided the driest February to October period since 1921.

The year 2003 was part of a wave of droughts throughout central Europe, which fell also in 2000 and 2001 (Trnka et al, 2007), during which the 2000 drought was short-lived but coincided with springtime planting and so had huge agricultural impacts. The 2003 drought was much more widespread, reaching to Germany, Switzerland and Austria (Plate 7.4). The drought was particularly problematic for activities dependent on river water. River levels were decreased to such an extent that navigation was reduced on the Danube, Elbe and Rhine rivers. For example, the Danube was at its lowest level in 160 years, and the Rhine at Lobith along the Dutch–German border was at its lowest ever recorded level. The drop in water level began to rot dykes in the Netherlands. Hydroelectric dams in Spain were operating at reduced capacity and water temperatures in excess of 30°C were causing power plants that relied on river water for cooling to be shut down. The excessive temperatures were also causing fish to die and driving widespread forest fires. These impacted on tourism, not only directly through lack of water and excessive heat, but also through degradation to the landscape.

Two years later, widespread drought and heat waves happened again across much of central, western and southern Europe (Figure 7.8), leading the EU to start to take drought far more seriously, especially in the context of climate change. For 2003, the heat wave was so strong that it is construed from climate modelling that such an event was twice as likely to happen because of human-induced climate change (Stott et al, 2004). In 2008, the city of Barcelona in Spain was suffering so badly from lack of water that it took extraordinary measures of importing water from Marseilles, with obvious ramifications for political and national pride. Briffa et al (2009) have recently shown that drought is becoming more widespread, and this trend can be attributed to increasing temperatures and the impact of evaporation.

Severity–area–duration analysis

For Europe, Lloyd-Hughes and Saunders (2002) and van der Schrier et al (2006) analysed Standardized Precipitation Index (SPI) and PDSI data, respectively, and found the 1940s and 1990s to be the most drought-prone eras. This is reflected for the 1940 to the 1950s in the drought time series (Figure 7.8) and streamflow records (Figure 7.9). Our analysis shows that the 1990s is the decade with the most events (12), with the 1940s ranking a close second (11). However, the SAD envelope curves (Plate 7.2) are dominated by two major events, namely the 1975–1977 and 1920–1922 events, with a minor contribution from an event in 1908–1909. The 1975–1977 drought dominates the SAD curves at 12-month and 24-month durations, and forms part of the 3-month curve for areas between 2 and 3 million km^2. The remainder of the 3-month curve is made up of the 1920–1922 drought and the 1908–1909 event at very large spatial extents. The longest droughts identified include events during

1975–1977, 1920–1922 and 1946–1948 (Table 7.1), whereas the 1980–1981 and 1906–1907 events were of long duration, but were limited in areal extent.

The SAD curves for all events at 3-month and 12-month durations are plotted in Figure 7.10. Recall that the envelope curves in Plate 7.2 are calculated from the upper limit of all curves shown in Figure 7.10. The events in the late 1940s and early 1950s approach the severity of the 1920–1922 and 1975–1977 events, especially for 12-month durations. The 1953–1954 drought highlighted in the figure covered most of central Europe and spread west into France and Germany, whereas two early 1950s droughts (1950 and 1951–1952) reached their peak in European Russia, the former more southerly than the latter. The 1953–1954 drought is very close to the 1920–1922 and 1975–1977 events, in terms of three-month SAD curves. It is also notable that events in the 1990s are comparably severe for 3-month duration and the 2003 (identified as 2002–2003 by the cluster analysis) event is similarly severe for 12-month duration.

Africa

African droughts are dominated by events during the mid-1970s and 1980s in West Africa (Hulme, 1992; Oba et al, 2001; L'Hôte et al, 2002; Tarhule and Lamb, 2003; Dai et al, 2004), and during the early 1990s, especially in southern Africa (Rouault and Richard, 2005) (Table 7.1). Figure 7.11 summarizes the variation in drought over the 20th and early 21st centuries across the African continent, in terms of soil moisture and drought extent, and a selection of the major events are mapped in Plate 7.5. In southern Africa, droughts occur recurrently, forced by high interannual variability in precipitation, which impacts on agriculture during the growing season. Although there is no evidence for changes in climate and drought over the 20th century, droughts were particularly severe in the 1980s and early 1990s (Richard et al, 2001), as shown in the soil moisture data (Figure 7.12) and reflected in the streamflow, for example for the Orange River in South Africa. Summertime drought is linked with El Niño events, but this may be tempered by the Quasi-Biennial Oscillation (QBO). The QBO is a mode of climate variability that manifests within the stratosphere and is influenced by changes in SSTs in the neighbouring South Atlantic and Indian Oceans. For example, the strong El Niño of 1997–1998 appears to have not affected southern Africa because of the overriding dynamics of the coupled Indian Ocean–atmospheric system (Webster et al, 1999).

Northeastern Africa is dominated by the Nile River, which runs through ten countries and supplies water to about 160 million people. The Nile exemplifies the sharing of river waters between nations through a

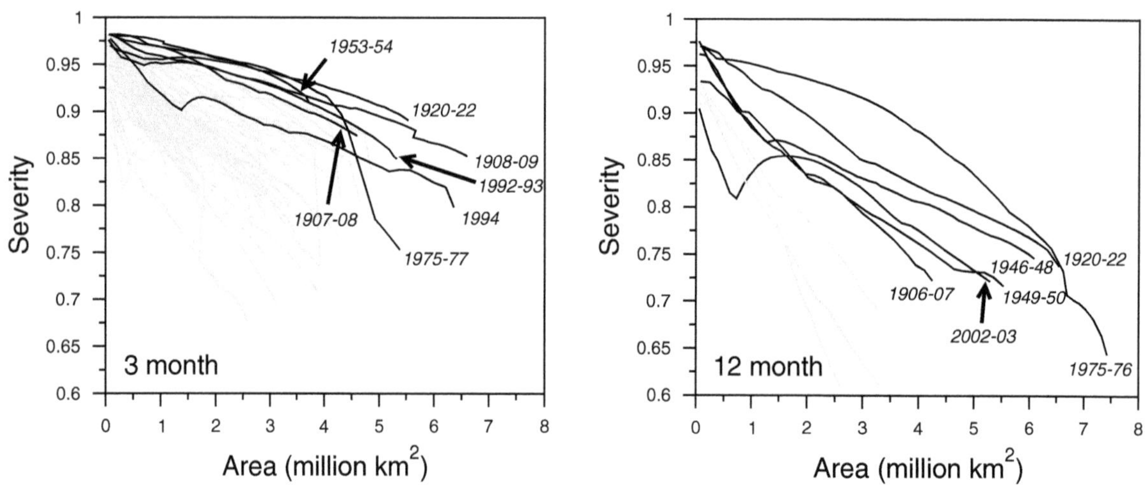

Figure 7.10 *Drought severity–area–duration curves for Europe at 3-month and 12-month durations*

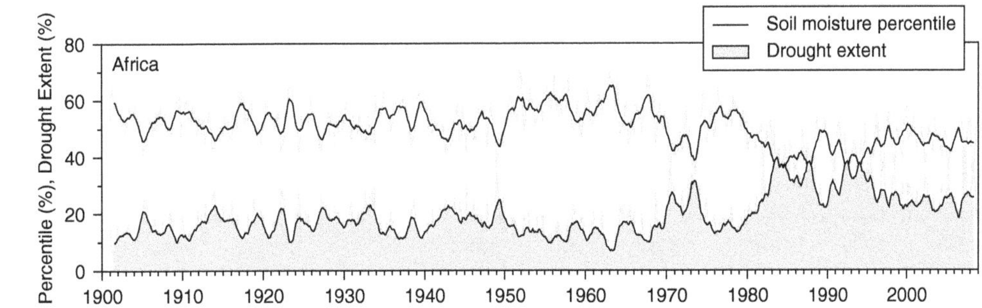

Figure 7.11 *Time series of soil moisture and drought extent for 1901–2008 for Africa (excluding the Sahara Desert)*

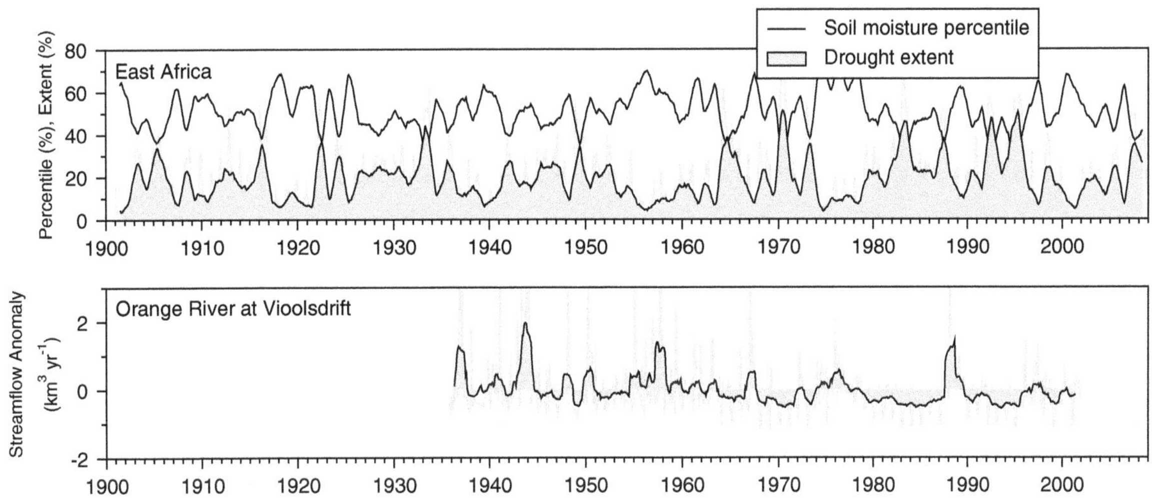

Figure 7.12 *Time series of soil moisture and drought extent for 1901–2008 for Southern Africa; also shown are streamflow anomalies for the Orange River in South Africa*

Source: streamflow data from the Global Runoff Data Centre, Koblenz, Germany at http://grdc.bafg.de

treaty, in this case the 1959 agreement between Egypt and Sudan over water allocations. However, such treaties are usually held in contention by other nations, and are especially put to the test in times of drought. The upstream source of the Nile derives mainly from the Blue Nile tributary. Figure 7.13 shows the 20th-century drought extent for East Africa and the Blue Nile region, and the streamflow anomalies for the river at Khartoum. This river drains part of the Ethiopian highlands and contributes 60 per cent of flows to the Nile from 10 per cent of the total drainage area. The streamflow corresponds with the ENSO (Abtew et al, 2009), and the El Niño forces dry years. One of the largest El Niño years was 1888, which marked the beginning of the great Ethiopian famine of 1888–1892 (Pankhurst, 1966). More recently, droughts in 1965, 1972–1973, 1983–1984, 1987–1988 and 1997, which correspond to El Niño events, caused reduced agricultural production, which is crucial for a predominantly rain-fed agricultural system, and low streamflow in the Blue Nile, with implications for downstream nations. Overall, the Blue Nile River shows a decline in flows from the end of the 1960s, although the available record ends in the mid-1980s.

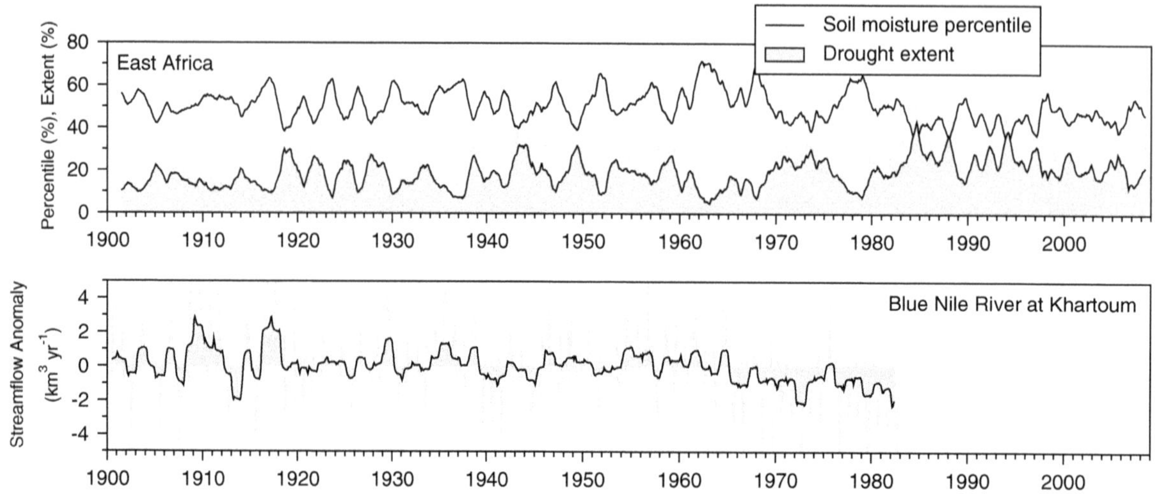

Figure 7.13 *Time series of soil moisture and drought extent for 1901–2008 for East Africa; also shown are streamflow anomalies for the Blue Nile River, which supplies almost 60 per cent of the flow of the Nile River*

Source: streamflow data from the Global Runoff Data Centre, Koblenz, Germany at http://grdc.bafg.de

This decline is consistent with overall dry conditions across the Sahel and sub-Saharan Africa, which are discussed in more detail below.

The famines of the Sahel in the 1970s and 1980s

The Sahel region of Africa, which forms the transition zone between the tropical climate of the Guinea coast and the Saharan desert, has suffered multiple long-term dry periods over the 20th century, including during 1910–1914, 1940–1944, and more recently across the 1970s and 1980s (Plate 7.5), which had devastating social and environmental consequences (Mortimore and Adams, 2001; Tarhule and Lamb, 2003). The time series of drought for West Africa and streamflow for the Niger River, which is the largest river in the region, are shown in Figure 7.14. The period of the 1970s to 1980s is outstanding in the record, with peaks of drought extent (65 per cent) in 1983. The drought is also identified in the record for the Congo River to the south, which drains the majority of the central African tropical rain forest. Estimates of the loss of life are up to 600,000 in 1972–1975 and again in 1984–1985. The decreasing precipitation trends over the region have been well documented (Hulme, 1992; L'Hôte et al, 2002), but the forcing mechanisms have been the subject of debate. Initially, overgrazing and deforestation were thought to be the main drivers, but more recent studies have shown that a combination of land–atmosphere interactions (Nicholson, 2000; Los et al, 2006), ocean temperatures (Giannini et al, 2003; Lau et al, 2006) and anthropogenic forcing (Held et al, 2005; Biasutti and Giannini, 2006) are likely causes. In particular, research has identified linkages with temperatures in the tropical Atlantic, Indian and Pacific oceans. Assessments of climate model simulations of the 20th century indicate that models generally cannot replicate observed precipitation of the West African monsoon. There is some disagreement over the future projections for the 21st century, and whether the region will become drier, wetter or neither (Cook, 2008).

Severity–area–duration analysis

There has been considerable study of drought conditions in the Sahel (e.g. Hulme, 1992; L'Hôte et al, 2002) because of the devastating environmental impacts, and it is no surprise that the SAD curves are dominated by events that primarily affected this region during the 1980s (Plate 7.2). The 1984–1985 Sahel drought was the most severe for most spatial extents, with the 1987–1988 drought forming the envelope curve for extents greater than about 10 million km^2. Note that the series of droughts in the 1980s could be considered to be the same drought in terms of their spatial and temporal

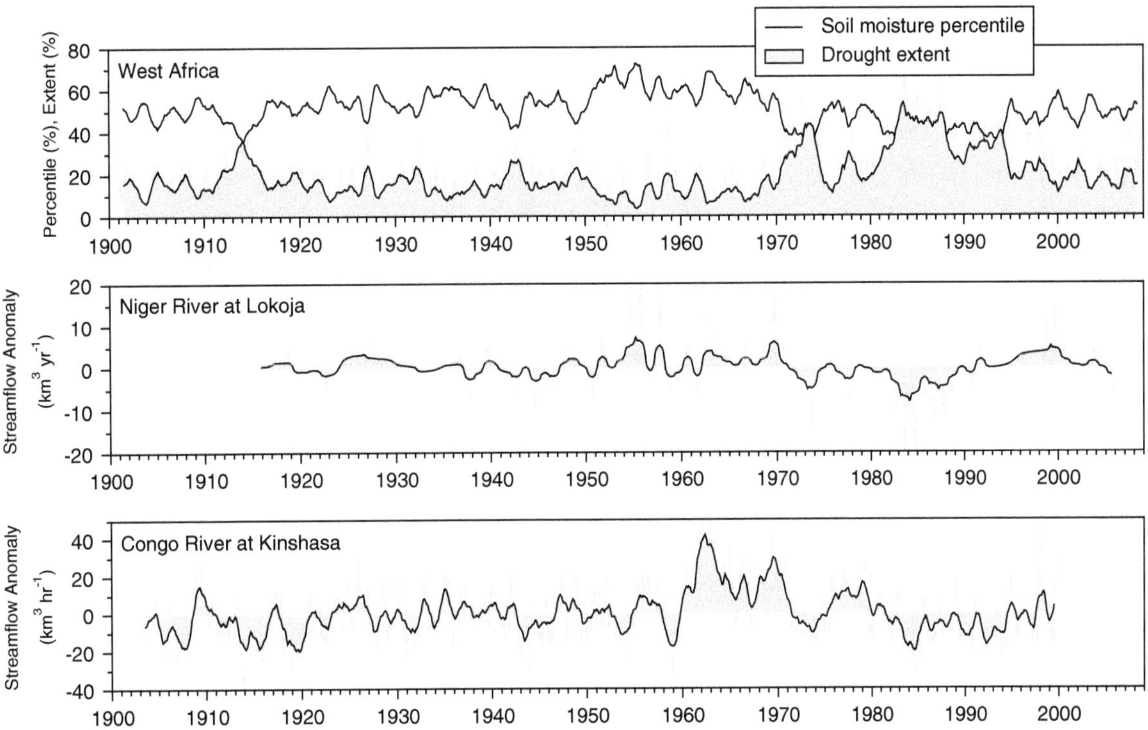

Figure 7.14 *Time series of soil moisture and drought extent for 1901–2008 for West Africa; the lower two panels show streamflow anomalies for the Niger and Congo rivers*

Source: streamflow data from the Global Runoff Data Centre, Koblenz, Germany at http://grdc.bafg.de

proximity, but the chosen thresholds for the cluster algorithm meant that they were treated separately. Elsewhere in Africa, Roualt and Richard (2005) identify the 1994–1995 drought as the most spatially extensive event in southern Africa, in terms of six-month SPI, which is identified here as one of the most severe droughts outside of the Sahel-centred events. The drought in 1933 was also located in southern Africa, but the exact characteristics of these earlier droughts are in some doubt because of the poor data coverage early in the 20th century. Figure 7.15 shows the SAD curves for all drought events at 3-month and 12-month durations, and highlights some of the most severe events. The majority of these occurred in the mid-1980s to early 1990s. In terms of spatial extent, the largest droughts were also in the 1980s and 1990s (Table 7.1). The 1984–1985 drought was the most spatially extensive, reaching more than 8 million km^2 in August 1984. Events in the early 1970s (1970 and 1973–1974) and early 1990s (1990–1991) similarly affected the Sahel region. Subsequent droughts of the early 1990s affected mainly eastern and southern Africa. The longest drought on record extended for 21 months over 1992–1993, beginning across Zambia, Zimbabwe and Mozambique, before migrating south to South Africa, where it stalled for most of 1993. The 1993–1994 drought originated over Ethiopia in late 1993 before drifting towards central East Africa during 1994.

Asia

The Asian continent is vast and experiences a huge range of climates and climate variability, from the Arctic tundra in northern Eurasia, the deserts of the Middle East, to the humid regions of Southeast Asia, which are dominated by the ENSO. Droughts in Asia have been studied by many authors, often on a national basis (Meshcherskaya and Blazhevich, 1997; Gruza et al, 1999; Zhou et al, 2005; Jiang et al, 2006; Patel et al, 2007), but also at regional scales (Barlow et al, 2002;

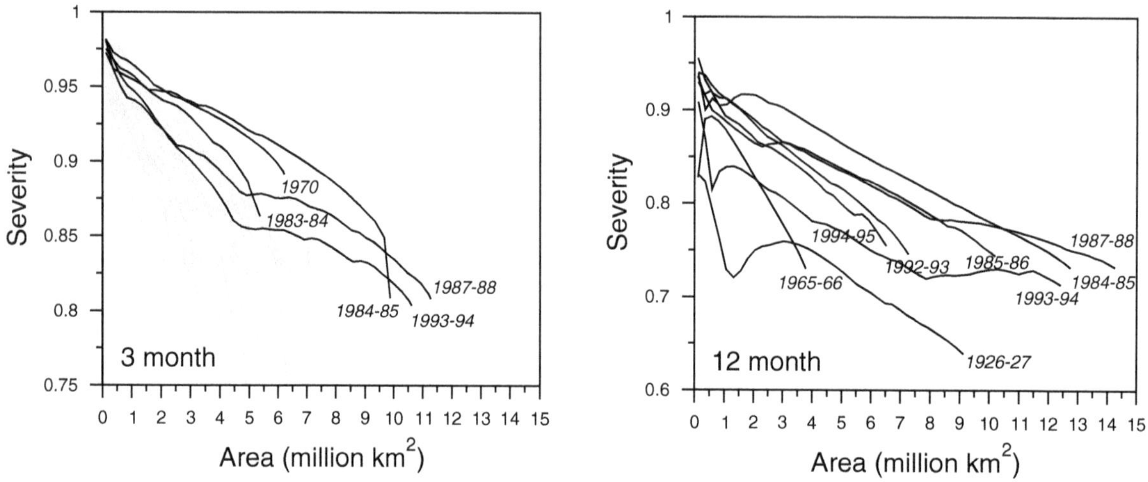

Figure 7.15 *Drought severity–area–duration curves for Africa at 3-month and 12-month durations*

Min et al, 2003). In particular, the Asian monsoon, which impacts from East Africa, through the Indian subcontinent, China, Southeast Asia, and northern Australia, exerts considerable influence on the peoples of these regions via the occurrence of extremes of flooding and drought (Cook et al, 2010), which impacts agriculture and water availability. In this section, we describe 20th-century drought over Asia by region, given the variation in climate, but note that these descriptions have some overlap across regions, which we highlight where possible.

Russia

We begin with Russia and the former Soviet Union (USSR), which have been especially vulnerable to drought because of the susceptibility of agriculture to climate variability, as well as the various socio-political factors that played a role in exacerbating climate impacts. Thus, although droughts over the 20th century have not been severe and widespread they have been sufficient to cause crop yield decrease and failure, which combined with political instability and poor policy decisions have led to famines in the early part of the century and a huge increase in grain imports in the latter part of the century (Golubev and Dronin, 2003), especially after shortages in 1972. The agricultural regions in the east and southeast, stretching from the former Soviet satellite states in eastern Europe, through the Ukraine, northern Caucasus, Volga and Urals to Kazhakstan are subject to high-pressure systems that form from dry Arctic air masses. When combined with high-pressure systems moving across Europe from the west, these give rise to widespread drought conditions. Agriculture is particularly susceptible to conditions in late spring and early summer, which is dependent not only on the prevailing climate conditions but also on the preceding soil moisture conditions and the harshness of cold weather for winter-planted crops in the south. Overall, it is estimated that the Russian agricultural regions have suffered 27 major droughts in the 20th century (Golubev and Dronin, 2003).

As well as affecting agriculture, droughts in Russia are also a factor in forest fires, although human activity is usually the instigator. Like the tropical rain forests, the great swathe of boreal forest that crosses the entire country and the extensive peatlands are a large carbon sink but can also be a source of carbon during fire, burning 12,000–38,000 fires across 0.3 million ha to 3 million ha annually. In 2003, when Europe suffered its worst heat wave on record, 22 million ha of Siberian peatlands were affected by fire, with more than 150 fires across 0.4 million ha detected from satellites in a single day (Malevski-Malevich et al, 2005). Fires in general have been increasing across Asia over the past decades, which is a result of not only warming temperatures but also land-use practices.

The drought in 1921, which affected Europe as well, compounded the social effects of World War I and the Russian Civil War, leading to famine over the next two years. In the Ukraine and Volga region, it is estimated that up to 5 million people may have died. The drought began in the summer of 1920 in central Russia and the Volga River basin, and famine spread to the Ukraine and other important agricultural regions. A drought in 1931 that stretched from the Ukraine, Volga and Urals to Kazakhstan is also associated with the famine that affected millions of people, peaking in the summer of 1932 and 1933. Subsequent severe crop failures in the USSR happened in 1946, 1948, 1957, 1959, 1963, 1972, 1975, 1979 and 1981, which are linked with drought, but are likely to be highly related to socio-political factors. The famine in 1947 was probably the last great famine to hit the Soviet Union. It was instigated by drought that had spread from Europe in the same and preceding years, and affected much of the main agricultural regions. The number of deaths is estimated to be between 1 and 1.5 million, mostly peasants who were denied access to the surplus stocks of grain held by the state (Ellman, 2000). This led to plans being put forward for major water resource engineering to ensure resilience against future natural hazards, including highly ambitious and environmentally drastic plans for the reversal of Siberian rivers to feed the agricultural lands to the south.

Globally, 1972 was a particularly difficult year with food shortages worldwide, including that in Russia, the near collapse of Peruvian fisheries, droughts in the Sahel and frosts in southern Brazil destroying coffee plantations (Glantz, 1978). The strong El Niño of 1972–1973 was partly the cause of this. Plate 7.6 shows the extent of drought in the latter part of 1972, highlighting dry conditions in the Sahel, eastern Russia, eastern Siberia and down through eastern Asia to India and Southeast Asia. The drought in Russia was partly the cause of the drop in global agricultural production, but led to the import of grain from the US, facilitated by political will in the USSR to continue meat production and in the US to reduce grain stocks at cheap prices. Extremely high temperatures (2–4°C above the long-term mean) were endured during the summer of 1972, which combined with low preceding snow levels, caused a 13 per cent drop in production for the Soviet Union as a whole. This initiated the ascent of agricultural prices, including an enormous price spike in wheat prices from US$60 to US$208 per metric tonne (Battisti and Naylor, 2009). The drought was centred over the Ukraine and southwest Russia, which were the breadbaskets of the former Soviet Union. Although this period was dry, the crop declines were mainly driven by high temperatures above 30°C during July and August, which are key months for grain ripening.

Middle East

The Middle East is one of the most water-stressed regions of the world, with the majority of its area classified as arid. This region extends over southwest Asia and is loosely defined as including Egypt, the Arabian peninsular, Turkey, Iraq and Iran, and the smaller countries of the eastern Mediterranean. Although the spatial area is about 10 per cent of global land, the Middle East contains less than 1 per cent of the world's freshwater resources, with 80 per cent of this going to agriculture (AFED, 2009). Total available renewable per capita water resources are less than $1000m^3\ yr^{-1}$ with slightly more in Iraq, Lebanon and Syria (FAO, 2003), which is below levels for economic development and human well-being. Although, there is a shortage of water overall, only 50 per cent of available renewable resources is used in the region. Droughts are difficult to quantify in arid regions, because of the permanently dry conditions and the sporadic and highly variable rainfall. Turkey has experienced major national droughts in 1953–1959, 1972–1979, 1989–1994 and 1999–2001 (Sönmez et al, 2005), driven by high precipitation variability, especially in the central and southeast of the country, where agriculture is dominant. The drought in the early 1990s led to the development of rain-enhancement projects (Komuscu et al, 1998). Dry conditions have dominated in recent years, with the most widespread years being 1973, 1984, 1989 and 1990, especially for winter rainfall (Türke, 1996). In Saudi Arabia, despite being categorized as arid (rainfall averages between 50 and $375mm\ yr^{-1}$, with the highest amounts in the southwestern mountains), the climate still supports extensive livestock grazing, potentially across about half the region (Tag El-Din, 1989). This has decreased dramatically over the past 50 years, especially since long-term drought conditions in 1958–1966, when between 50 per cent and 90 per cent of livestock were lost (Al-Taher, 1994). In response to growing water needs, desalination has increased from about 7 million m^3 in 1970 to over 500 million m^3 in 1990, and is expected to exceed 1 billion m^3 in this decade.

Central Asia

West and Central Asia, which covers mainly Iran, Afghanistan, Pakistan and Kazakhstan, is semi-arid steppe where agriculture is highly dependent on the limited rainfall. Rain mainly occurs in the winter and early spring, brought eastwards from the Atlantic and Mediterranean. Rainfall has generally decreased in recent years, affecting water availability, agriculture and grazing (Lioubimtseva and Henebry, 2009). Wang et al (2010a) analysed trends in moisture across Central Asia over the past 100 years and found a drying trend over the west of Uzbekistan and Turkmenistan, central Kazakhstan and southern Xinjiang in China and central Mongolia. Recent drought since 1998 (Barlow et al, 2002) has affected Iran, Afghanistan and Pakistan, with rainfall down to 50 per cent of normal. Drought in this region is usually linked to La Niña conditions, but was probably enhanced during this period by warm SSTs in the western Pacific. The region is also suffering from the impacts of melting glaciers, which normally account for more than 10 per cent of freshwater supply across these drier regions (Cruz et al, 2007).

Kazakhstan has also been suffering from drought recently, but has previously been subjected to one of the most extreme efforts to make the predominantly dry region fertile. During the 1950s–1960s over 13 million ha of steppe were transformed into irrigated agricultural land, supplying a third of the Soviet wheat crop (Kaser, 1997). This was sharply reversed in the early 1990s with the collapse of the Soviet Union and the decline in agricultural land use during that decade. Since then (during the 2000s), this region has been showing extremely fast declines in growing-season vegetation, as derived from satellite data (de Beurs et al, 2009), which have been related to drying conditions, especially in 2004 – which affected much of Kazakhstan and other Central Asian countries.

Over the past 50 years there has been a general decrease in precipitation across the region, yet increases have been noted around expanding irrigated lands, which may be linked to feedbacks with the land (Pielke et al, 2007). There have been extreme changes for the Aral Sea, which has virtually disappeared because of water withdrawals for cotton production in Kazakhstan from the Syr Darya and Amu Darya rivers that once fed it (Glantz, 2005). This has led to changes in the hydrological cycle feedbacks to the atmosphere across western parts of Kazakhstan, Uzbekistan and Turkmenistan (Varushchenko et al, 2000).

Mongolia

The climate of Mongolia is characterized by low precipitation, with long cold winters and dry hot summers (Batima et al, 2005). It particularly suffers from heat waves, which have increased in frequency by 8 to 18 days per year across the region over past decades, reaching 70 days in the mountains and 70 days in the Gobi Desert at their peak in 1998. The temperature increase across arid Central Asia is about 0.18°C per decade for the last 100 years (Chen et al, 2009) and this is twice the rate of increase compared to the northern hemisphere. A major multiyear drought occurred in the late 1920s that is coincident with drought conditions in northern China (Davi et al, 2006). Although PDSI values for the region indicate non-significant changes in moisture availability over the past 50 years, there has been significant drying in recent years in the eastern part and drying over much of the rest of the region in the late 1990s and early 2000s, with some recovery during the mid-2000s (Li et al, 2009). It is thought that the intensity and duration of droughts has increased in recent years, driven by decreasing spring precipitation and rising temperatures. This has also led to increases in the frequency and area of fires in the forests and steppe regions. The long-term drought from 1999 to 2002 had huge impacts on the livestock-dominated economy, reducing water supply and forage resources. It is estimated that between 50 and 70 per cent of Mongolia was affected during these years, with huge numbers of water sources (lakes and rivers) drying up (Batima et al, 2005). Up to 12 million livestock are estimated to have died.

China

China is influenced by the East Asian monsoon, which is the most significant harbinger of natural disasters, inducing floods and droughts affecting agriculture and livelihoods (Qian and Zhu, 2001). Like India and

Russia, China has endured famine brought on in part by droughts, including those in 1907, 1928–1930 in northwest China, 1936–1935 and 1941–1942 in Sichuan Province, in which millions of people may have perished. Qian and Zhu (2001) recorded 12 major drought events during 1949–1995, mostly in the periods 1959–1961, 1978–1982 and 1986–1994, with 6 events occurring in 1986–1994. In the northwest, from the late 1970s to the early 1990s, there was drought but not on the scale of the 1940s (Qian and Zhu, 2001). In the northeast, drying has been occurring since 1970, and particularly in 1982 and 1989. In the north and northwest, droughts also peaked in the 1940s. In east China, drought occurred mostly around 1960, 1978 and 1988. In the south, droughts arose mainly in the 1960s and 1980s (Qian and Zhu, 2001). Ding and Sikka (2006) identify 1957, 1965, 1968, 1972 and 1986 as drought years in northern China, which lasted for between five and seven months, and 1965 as one of the most extensive, affecting eight provinces. This drought was also accompanied by a heat wave with summer air temperature 1–2℃ higher than normal. There were between 5 and 18 more extreme hot days over 35°C than normal (Ding and Sikka, 2006), with similar numbers of extreme hot days for droughts in 1971 and 2001.

Wang et al (2010b) identified 73 major drought events across China since 1950 that lasted for three months and were larger than 150,000 km^2, based on soil moisture data from multiple models. The longest drought was from 1975 to 1981 over southwest China and the most spatially extensive was during 1997–2003, which reached a peak in October 1997, covering 43 per cent (4.2 million km^2) of the land. This latter drought had rainfall amounts that were between 20 and 50 per cent lower than normal, with 2001 having between 40 and 70 per cent less rainfall than normal, making it the worst drought year since 1949 (Ding and Sikka, 2006). Other prominent droughts occurred in 1964–1970 and 2004–2006, although the latter may have continued beyond the end of the Wang et al (2010b) study.

Drying since the 1980s in the north and northwest is starting to mirror the severe conditions of the early 20th century (Wang and Zhai, 2003; Ma and Fu, 2003; Wang et al, 2010a). Figure 7.16 shows soil moisture averaged over East Asia (which encompasses Korea and Japan as well) and indicates that drying is widespread across the region. Zhou et al (2005) identify the increases in dry areas in northern China as unprecedented during the past 50 years, with drought conditions persisting for up to five years in some parts from 1997. Wang et al (2010b) found that the area of China with significant drying trends was 37 per cent and for wetting trends was 22 per cent. Drying was identified over spatially contiguous areas of the northeast and north central China. As these regions are relatively dry to start with and rely on irrigation for much of their agricultural water supply, they are vulnerable to continued drying. Reliance on groundwater has overcome many of these problems, yet this is likely to be unsustainable in the long term.

As with other countries, these more recent droughts do not have the same negative effects on livelihoods as the early 20th-century events, yet still cause severe economic impacts and hardship. In particular, severe droughts have affected parts of northern China in 1997, and during 1999 to 2002, mostly affecting agriculture, but also causing water shortages and dust storms. In 2000, over 40 million ha of agricultural land was impacted. The drought in 2001 caused cessation of drinking water for 33 million rural people and 22 million livestock, and crop damage costs of US$6.4 billion (Shen et al, 2007). The occurrence of dust

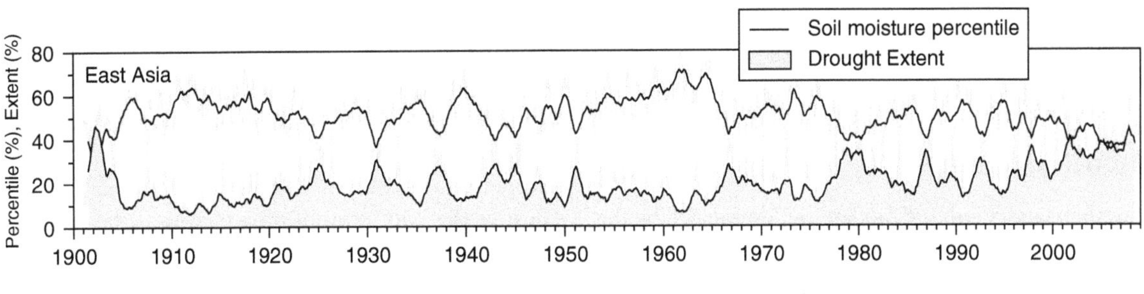

Figure 7.16 *Soil moisture and drought extent time series for East Asia*

storms over northern China has decreased between the 1950s and mid-1990s, but has increased since about 1997 (Zou and Zhai, 2004) in response to increased drought occurrence.

The general drying over much of northern China in recent decades hints at the possible role of climate change, including the direct effect of warming temperatures. There is also the direct human influence in the form of overexploitation of water resources (Wang et al, 2010a), which has led to a drying-up of lakes and rivers (Liu and Chen, 2006). For example, Zhou et al (2005) analyse the increase in the occurrence of days with zero streamflow in the Yellow River since the 1990s, and show seasonally earlier and longer dry periods. The drought in northern China in 1997 resulted in a record duration of 226 days of zero flow.

However, climate variability and drought across China are dominated by the strength of the East Asian monsoon, which was strong in the 1960s (Qian et al, 2003), but has weakened over the past 30 years in terms of land–ocean pressure gradients (Yu et al, 2004). The reasons for this are the subject of current research but are probably related to changes in land–ocean thermal contrasts, and connections to ENSO activity (Lau et al, 2000; Li et al, 2010) and northern Eurasian snow cover (Wu et al, 2009). El Niño is thought to be the main culprit for weakening the East Asian monsoon (Li et al, 2010), but the location of the western Pacific subtropical high is also associated with drought in this region (Shen et al, 2007; Zhou et al, 2009). Whether these changes have been forced by global warming or are part of the longer-term decadal variability seen over the past 100–150 years is unclear. Nevertheless, climate models project that a warmer and moister atmosphere in the future will actually lead to an enhancement of the circulation strength and precipitation of the summer monsoon over most of China (e.g. Sun and Ding, 2009) that will offset enhanced drying due to increased atmospheric evaporative demand in a warmer world (Sheffield and Wood, 2008). Chapter 8 discusses in more detail the possible changes to the Asian monsoon under future climate change.

Southeast Asia

Southern and Southeast Asia are also dominated by the monsoons, which give rise to a dry season in which dry cool air is drawn from the continental interior, and a distinct wet season that brings moist humid air from the oceans onto the land and large amounts of rainfall. The wet season, coupled with high solar radiation, bring ideal conditions for agrarian communities. Herein lies the problem: agriculture is dependent on this seasonal cycle, but highly sensitive to slight changes in not only the amount of monsoon rainfall, but perhaps more importantly to the length of the monsoon and its date of onset and departure (Webster et al, 1998). In Southeast Asia (Myanmar, Laos, Vietnam, Philippines, Indonesia), drought is usually linked to ENSO activity, with El Niño causing drying across the region. Figure 7.17 shows the time series of soil moisture averaged over Southeast Asia, and highlights the influence of El Niño on drought, and especially in the last 40 years. The correlation between the ENSO and soil moisture is very high, as shown in the central panel of the figure. This was particularly borne out in the strong El Niño of 1997–1998, which brought about devastating crop failures and forest/peatland fires in the Philippines, Indonesia and Laos, affecting over 9 million ha in Indonesia alone (Plate 7.6). Accompanying this were carbon emissions into the atmosphere estimated to be up to 2.57Pg (Page et al, 2002). Figure 7.17 also shows streamflow anomalies for the Mekong River, which rises in the Tibetan Plateau and traverses continental Southeast Asia. Again, the increased intensity of droughts is evident in the last 30–40 years and coincides with El Niño events.

India

India has experienced a multitude of severe droughts over the last century, which are generally forced by interannual variability in the Indian monsoon during June to September (Krishnamurthy and Shukla, 2000), which accounts for at least 75 per cent of annual rainfall. Agriculture (as well as hydropower and industry) is heavily dependent on the monsoon rains, but late onset or reduced monsoonal rains can lead to drought and reduced yields. Conversely, an extended monsoon period and heavy rains can bring about devastating flooding. Prior to the mid-20th century, droughts have led to widespread famine, including during 1770, 1876–1877 and 1899, when millions of people are estimated to have died. The drought at the turn of the 20th century prompted the establishment of the Indian Meteorological Service and the work of Gilbert Walker, who first identified low-frequency variability in global climate and the link between the

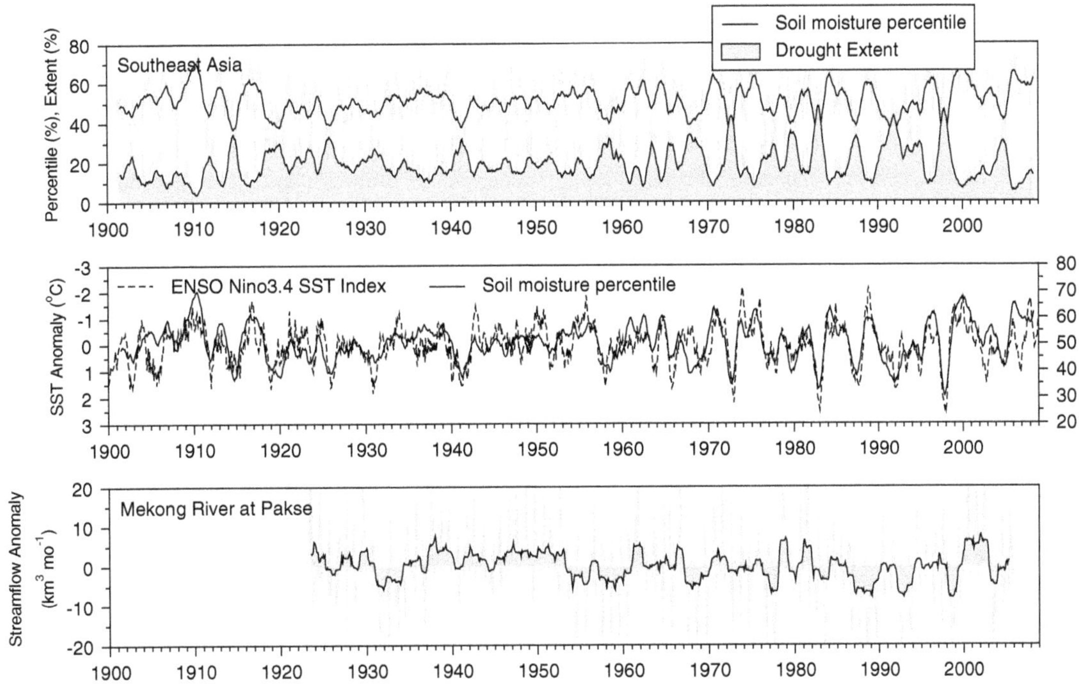

Figure 7.17 *Time series of soil moisture and drought extent for Southeast Asia and a comparison of ENSO (characterized by Niño3.4 SST anomalies) and soil moisture; also shown are streamflow anomalies for the Mekong River*

Note: The y-axis of the SST data is flipped.

Source: streamflow data from the Global Runoff Data Centre, Koblenz, Germany at http://grdc.bafg.de

Indian monsoon and the ENSO. Parathsarathy et al (1987) indicate that the periods of 1899–1920 and 1965–1982 were generally drought-prone, with the years of most widespread drought in 1899 and 1918, covering about 70 per cent of India in terms of rainfall deficits. Other drought years in the early 20th century include 1902, 1905 and 1920 (Ding and Sikka, 2006). The drought in 1918–1919 (Parthasarathy et al, 1994) was forced by a severe El Niño (Giese et al, 2010) and is one of the most severe on record. The impacts that included an estimated 18 million deaths were compounded by the global influenza pandemic. The 1918 event may only have been matched by the 1972 event. Other droughts of the first half of the 20th century include the event associated with the Bengal famine and disease in 1943, which is estimated to have led to the deaths of 4.5 million people.

In past decades, 1965, 1966, 1972, 1974, 1979, 1987 and 2002 were severe drought years, with the 1965–1966, 1972, 1987 and 2002 droughts being the most widespread and damaging (OFDA/CRED, 2006). The droughts of 1965–1966 and 1979 affected humid regions of India, whereas the 1972, 1987 and 2002 droughts impacted mostly in drier regions of the northwest and Pakistan (see Plate 7.6). The 1965–1966 event was particularly devastating because of two consecutive years of drought, and is estimated to have been linked to the deaths of more than 1.5 million people. The link between drought and famine has thankfully been broken in the past 30 years because of the Green Revolution in agriculture, which has introduced efficient technologies and irrigation practices, although at the expense of groundwater reserves. Similarly, recession would normally follow major drought events with negative GDP, although since the 1990s this has abated and GDP reductions have been limited, mainly as a result of the shift of economic income away from agriculture (PACS, 2004). Nevertheless, recent droughts have had tremendous impacts. The 1987 drought was one of the worst in the 20th century, with an overall deficit of about 20 per cent in rainfall, affecting about two-thirds of agricultural areas and 285 million people. The drought of 2002 brought about half the average rainfall in July, the core month of the monsoon, which was the lowest recorded amount in more than 130 years (Mujumdar et al, 2007).

Many of these drought events are associated with the ENSO (Ding and Sikka, 2006). The strength of the monsoon is governed in part by the ENSO (Krishna Kumar et al, 2006), with El Niño events bringing low pressure over the southern Indian Ocean. This causes dry air to be drawn from the Asian land mass and brings drought, although this may be tempered by a warmer Indian Ocean. This relationship is not perfect and there are other mitigating and forcing factors that play a role, such as interdecadal variations, relationships with land-surface processes, such as Asian snow cover extent variations (Turner and Slingo, 2010) and so on. For example, the causes of the 2002 drought are linked with intraseasonal oscillations (Kripalani et al, 2004) and convective activity in the northwest Pacific (Mujumdar et al, 2007). Nevertheless, the association between the ENSO and drought is well documented (Pant and Parthasarathy, 1981; Rasmusson and Carpenter, 1983; Shukla and Paolino, 1983), although this relationship has weakened in recent years (Kripalani and Kulkarni, 1997; Krishna Kumar et al, 1999; Ashok et al, 2001).

Severity–area–duration analysis

Plate 7.2 summarizes the characteristics of the most severe drought in Asia. The high-latitude droughts (e.g. 1982–1984, 1986–1990) are not well documented in the literature, given the sparse population and low economic activity in these regions. However, indirect evidence from, for example, remotely sensed fire activity across the Boreal forest (e.g. Balzter et al, 2007) is consistent with large-scale dry conditions during the 1980s. The drought events in southwest and Central Asia (1944–1945 and 1950–1952) are very severe, especially at 3-month and 12-month durations (Plate 7.2). The 2007–2008 event began in northeast China, but eventually linked up with dry conditions in western Asia, forming a band of drought from Kazakhstan through northwest China and Mongolia to northeast China. This drought forms much of the 12-month SAD envelope curve for extents greater than 10 million km^2 and is close to the 1944–1945 drought in its 3-month duration (Figure 7.18). This and other similar droughts in East Asia in the late 1990s and 2000s are typical of the reported increase in drought in northern China and over the past two decades. The 1997–1998 drought (from eastern China to Central Asia) is also reported by Zhou et al (2005), who analysed the PDSI over China from 1951 to 2003. The early 1950s drought (1950–1952) was the most severe and extensive, when it centred on Kazakhstan in April 1951. This drought forms most of the 24-month SAD envelope curve and the lower extent portion of the 12-month curve. It is comparable to other major events at 3-month duration (Figure 7.18).

The longest drought identified (1986–1990; Table 7.1) originated over eastern Siberia before migrating slowly westwards and persisting over the Ob River basin, east of the Ural Mountains during 1989–1990. The 1932–1935 drought was similar in duration and location, although

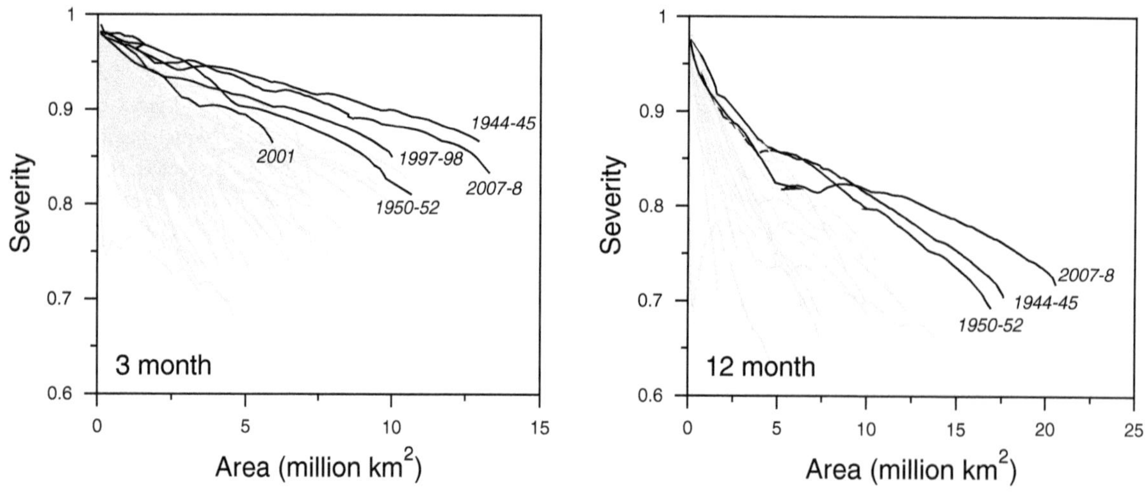

Figure 7.18 *Drought severity–area–duration curves for Asia at 3-month and 12-month durations*

the sparsity of precipitation gauges in this period reduces confidence in the characteristics of this event. Other long-duration droughts are the 1950–1952 (Iran–Turkmenistan–Kazakhstan region), 1982–1984 (central Siberia) and 1992–1994 (eastern Siberia) events. Similar to the high-latitude droughts in North America, the Siberian events are prolonged by freezing temperatures over the winter. The 1944–1945 drought is the most spatially extensive, closely followed by droughts in the 1980 and the 1990s, which mainly impacted on East Asia.

Oceania

The Australian climate ensures that drought is not an applicable term in the normal sense of the word. Over the 20th century Australia has endured either wet or dry spells, with little in between. Thus when Australia is dry, it is really dry, often for many years to decades. This variability is shown Figure 7.19, which illustrates the 20th-century time series of precipitation and drought. The persistence of dry or wet conditions for multiple years and then a switch to the opposite state is clear. Australia has thus experienced multiple long-term drought events over the 20th century, which have shaped the country and its inhabitants. These dry periods are generally forced by variability at interannual and interdecadal scales associated with the El Niño (Chiew et al, 1998), the Southern Annual Mode (SAM) and northwestern SSTs (Verdon-Kidd and Kiem, 2009), but modified by Pacific interdecadal variability (Power et al, 1999; White et al, 2004) and the influence of the Indian Ocean (Ummenhofer et al, 2009).

Australia has suffered from several major drought events in living memory, often referred to as '100-year droughts', based on the expected probability that they might occur once a century. The first was the Federation drought of 1895–1902, which only just overlaps with the start of our 20th century analysis time period, but warrants mention here because of the profound impacts and partial causes due to the plague of rabbits that infested New South Wales in the preceding years. This drought is associated with the ENSO and Interdecadal Pacific Oscillation (IPO) activity (Verdon-Kidd and Kiem, 2009), which is a broader-scale variation of SSTs across the Pacific that is associated with the PDO and ENSO. The drought saw arid conditions across eastern Australia during 1901, drying up rivers in western Queensland, lowering the Murray River to near empty in New South Wales, and reducing the water supply to Sydney.

Plate 7.7 shows the extent of drought by the autumn of 1902. Livestock was reduced by almost half, from 91 million sheep to 54 million and 12 million cattle to 7 million. This was followed by drought in 1914, which devastated the wheat crop in that year. The drought began in some parts of the south in 1913 but quickly covered a large extent from July to October 1914, which was a strong El Niño year. The Murray River dropped to its lowest recorded level in some places. The Australian wheat crop was very low, with complete failure in some parts of the southwest (BOM, 2010). Between 1937 and 1945, dry conditions known as the 'World War II' drought covered much of eastern Australia, beginning in New South Wales and Victoria and then spreading to the southwest grain regions, where yields in 1938 dropped to their lowest level since 1914. In Victoria, extremely dry conditions led to the January 1939 'Black Friday' bushfires, which claimed the lives of 71 people. Some respite was to be had at the end of 1939 but this was immediately followed by the El Niño of 1940, which spread drought across more than 50 per cent of the country (Plate 7.7). 1940 turned out to be the driest year of the century over South and Western Australia, causing municipal water restrictions. Drought continued in many places through to 1945, despite some intermittent relief in 1942, leading to the drying up of many major rivers, such as the Murray, Hunter and Hawkesbury, and major dust storms. The drought finally ended in the winter of 1945, but continued in parts of southern Queensland and northern New South Wales through to 1946 and 1947. It is thought that a combination of the ENSO and SAM are associated with this prolonged event, as well as persistently cool SSTs off northwestern Australia (Verdon-Kidd and Kiem, 2009).

The next dry period came between 1957 and 1968 over much of central Australia. The worst of the drought came in the later years from 1965 onwards when it affected the east, where most of the population and agriculture existed. The event was most severe in the beginning of 1965. The soil moisture data indicate that this was the peak drought extent of the 20th century, covering over 75 per cent of the country in February 1965 (Plate 7.7). Some relief came in 1966 but the drought peaked again in 1967 in the south, where vast areas of Tasmania were burnt in just one day, killing 62 people around the capital Hobart. The drought was very

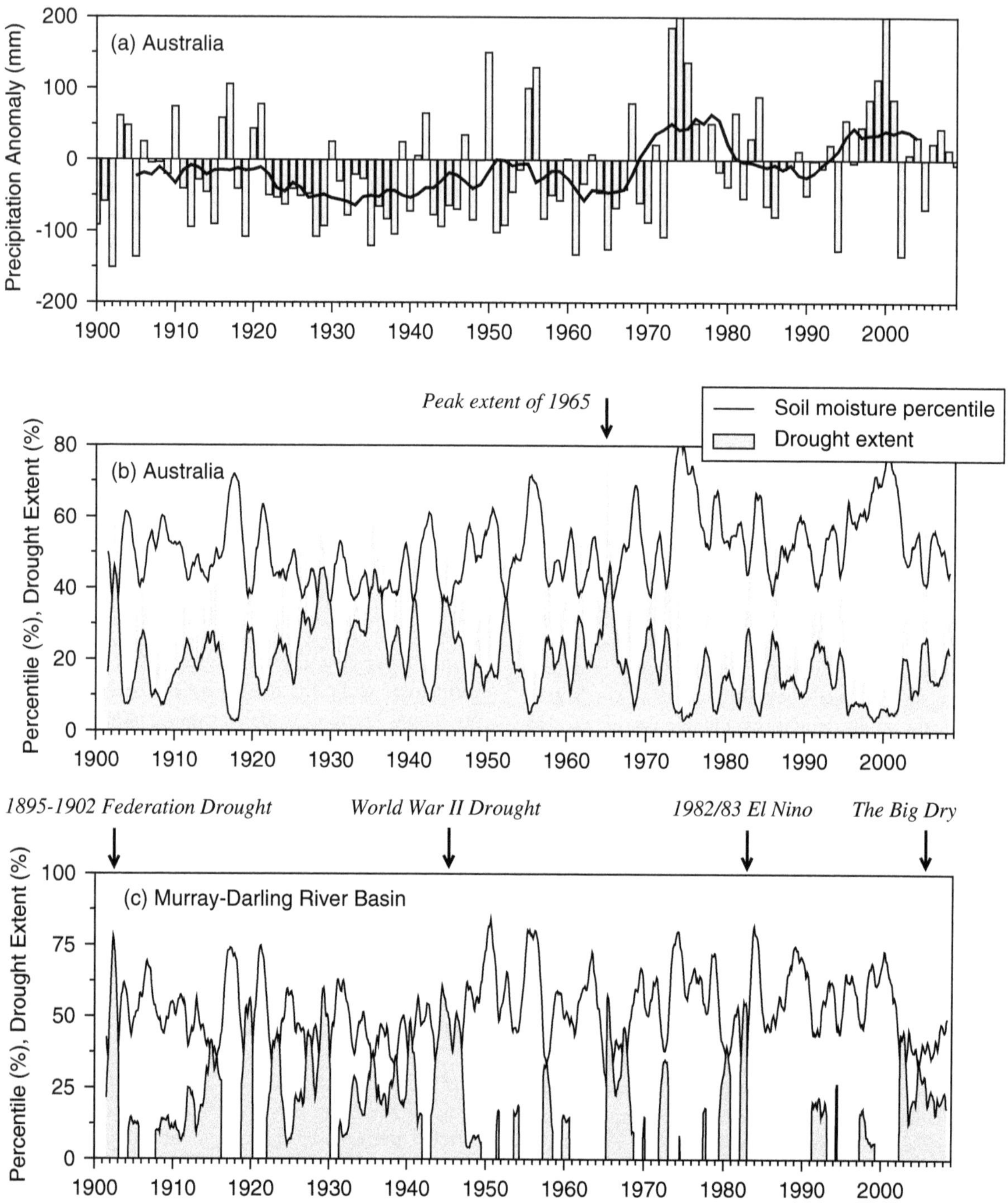

Figure 7.19 *(a) Annual time series of Australian precipitation for 1901–2008. The black lines are the 11-year running means. Data are taken from the Australian Bureau of Meteorology (BOM) (available at: http://reg.bom.gov.au/silo/products/cli_chg/index.shtml). (b) Time series of soil moisture and drought extent for Australia, 1901–2008. (c) Time series of soil moisture and drought extent for the Murray-Darling River basin in southeast Australia. The soil moisture and drought extent data have been smoothed*

intense across southeastern Australia, with dust storms occurring in South Australia and crops failing in Victoria. Melbourne experienced the driest year in 140 years of records. The exports of Australian wool dropped by AUS$27 million (US$27 million) (Butler, 1978).

1982–1983 Southeast drought

The drought in 1982–1983 was driven by the extreme El Niño event of the time, but was also probably associated with the drought-inducing phases of the IOD and SAM (Verdon-Kidd and Kiem, 2009). This was probably the first drought associated with El Niño by the public. It mainly affected large parts of central and eastern Australia, and in particular the southeast. Total costs have been estimated to be about AUS$7 billion (US$3 billion), which ranks it as one of the worst droughts to hit Australia. This was because the drought was short but intense, with rainfall totals in April to December at record or near-record levels in many parts of the southeast (BOM, 2010). Agriculture was severely affected, resulting in high unemployment in rural areas. At the end of 1982, water restrictions were in place in Melbourne, and the upper Murrumbidgee River ran dry, with some reservoirs so low that historic towns that had been submerged when the dams were built reappeared from the waters. The drought peaked in February 1983 (Plate 7.7), with record lows of rainfall across the southeast and Tasmania. Dust storms and bushfires were to take the lives of 75 in the Ash Wednesday fires in Victoria and South Australia, when dry forests experienced temperatures of over 40°C. The drought abruptly ended in March, with large rainfall causing flooding across much of the southeast.

Recent droughts and the 'Big Dry'

Another El Niño caused drought in 1991–1995 over Queensland and northern New South Wales. This was unusual because the El Niño lasted for several years, whereas it usually lasted for a year at the most, such as the 1982–1983 event. This had only happened previously in 1911–1915 and 1939–1941. The drought was initiated by dry conditions in the north in 1991, during the normally wet winter season. It then merged into the effects of El Niño during 1992 to 1993, which brought about drought in Queensland and northern New South Wales, and eventually to most of the country in the summer of 1993–1994. The drought in this year was estimated to have cut agricultural production by almost 10 per cent, leading to mass unemployment in the agricultural industry. The El Niño conditions abated in 1995, leading to the end of the drought.

Drought has resurged in the early 2000s in southeast Australia, probably persisting since before 2000 in some regions. This drought, known as the 'Big Dry', has severely impacted on agriculture in the Murray-Darling River basin, which is responsible for providing much of the nation's agricultural produce via irrigation drawn from the river system. This drought is probably mainly associated with SAM variability, compounded by El Niño during 2002–2004 and the complete lack of precipitation-bringing La Niñas, which also tend to be blocked by the SAM when in positive mode (Kiem and Verdon-Kidd, 2009). Record high temperatures have also occurred, implying that drought is linked to the emergence of global warming (Cai et al, 2009), although this may partly be a feedback from the drought itself (Lockart et al, 2009). River flows have been at near-record lows, making the recent event more severe than the Federation drought (1895–1903) or the World War II drought (1937–1945).

New Zealand

New Zealand has suffered from a number of severe droughts during the 20th century, especially in the drier east, which have led to economic impacts and environmental damage. Some of the more severe events occurred in 1945–1946, 1972–1973, 1977–1978, 1982–1983 and 1997–1998, indicating a link with ENSO activity. The 1997–1998 drought resulted in a loss of about US$620 million to the economy. During an El Niño there are generally stronger westerly winds in the summer, which tend to bring about more rain in the west and drought in the east. In the winter, the more southerly winds tend to bring cooler conditions. The amount of variation explained by the ENSO is small (about 25 per cent), since the relationships between the ENSO and drought are irregular and regionally changeable. The risk of drought is complicated because of the large regional variations in climate across the islands, shaped by the mountain range that bisects the two main islands and drives large orographic effects on precipitation. Drought impacts mainly on New Zealand's major industry of pastoral agriculture, especially when drought occurs during the autumn or spring, when grass development is most crucial (Keen and Rotorua, 1996).

Fowler and Adams (2004) looked at the region around the city of Auckland and identified droughts in 1913–1914, 1983–1984 and 1993–1994. The 1993–1994 drought led to a water supply crisis for the city, driven by the lack of recharge to already depleted reservoirs in the winter of 1993 and a dry summer in 1994, which led to record decreases in storage.

Severity–area–duration analysis

In Oceania, the nine-month 1982–1983 drought is considered to be one of the most severe and damaging events of the second half of the 20th century, mainly because it was intense (short and severe), and coincided with major agricultural and populated regions (Nicholls, 2004; Mpelasoka et al, 2007). It does not appear on the three-month SAD envelope curve (Plate 7.2), but ranks sixth on the list of most spatially extensive and lengthy events. Its severity is nevertheless comparable to the more extensive events at three-month durations, especially for relatively small spatial extents (Figure 7.20). The 1964–1965 drought dominates the SAD envelope curves for durations of 3 and 12 months, but is only slightly more severe than other events at low spatial extents. Beyond about 1 million km^2 it is clearly the worst 3-month duration event. The Australian Bureau of Meteorology also highlights the 1965 drought as particularly severe (BOM, 2010), with an estimated AUS$300–500 million (US$290–485 million) in agricultural losses (Mpelasoka et al, 2007). It affected the majority of Australia at its peak (Plate 7.7), sparing only the extreme north and west, while the earlier 1961–1962 drought originated in the interior before migrating westwards. The most recent drought of 2008 that is part of the long-term 'Big Dry' is the most severe short-term drought below 1 million km^2. The drought of 1928–1930 is the longest event on record (Table 7.1), and is the only event that registers on the 24-month SAD envelope curve. This drought affected mainly the dry interior of Australia (White et al, 2004) and so had relatively little impact. Some of the longest droughts were among the most spatially extensive events (1951–1953, 1961–1962, 1964–1965; see Table 7.1) (Plummer et al, 1999; White et al, 2004). However, some droughts were more regional, such as the 1951–1953 drought, which persisted for 15 months over northern Australia and the 9-month 1982–1983 drought in the east and southeast. Figure 7.20 presents all the events and highlights the severity of the 1982–1983 drought and the Federation drought, which is captured here for 1901–1902. Interestingly, two of the most recent droughts in 2005 and 2008 have a short-term severity that is comparable to any drought events which have occurred over the course of the 20th century. The 12-month SAD curves are all from the 1960s and before, with the majority clustered around the 1930s to the early 1950s, reflecting the high incidence of large-scale drought events in this period.

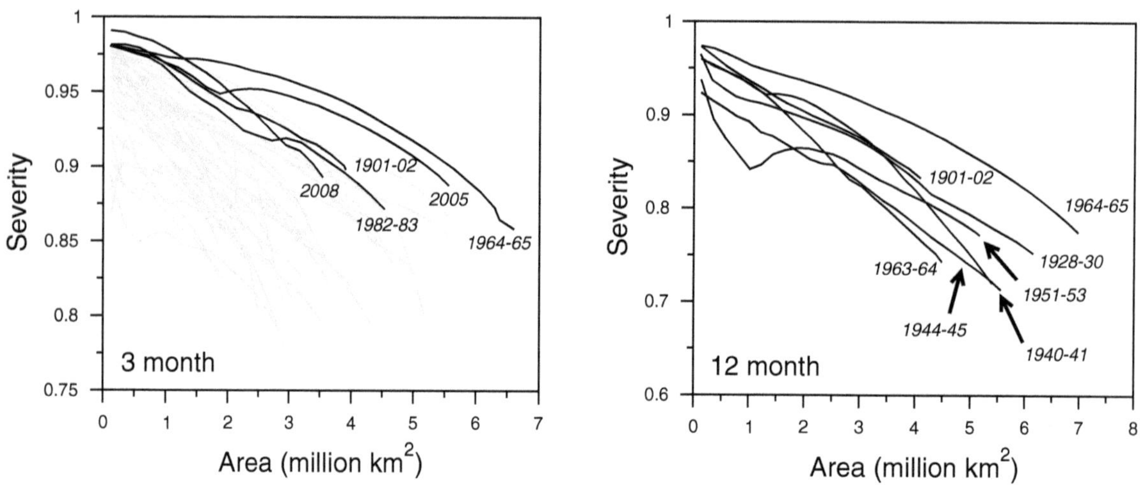

Figure 7.20 *Drought severity–area–duration curves for Oceania at 3-month and 12-month durations*

The Most Severe Droughts of the 20th Century

Which were the most severe droughts of the 20th century? To answer this question, we use the SAD analysis as an objective method to quantify the exact start and end dates of drought events, and their spatial extent and severity during their lifetime. This does not take into account the impacts of the droughts, as these are difficult to quantify and compare, especially given other non-climatic contributing factors, such as population vulnerability and the prevailing economic state. Nevertheless, this provides a reasonable method to rank droughts based on their physical characteristics.

The bottom panel in Plate 7.2 merges the SAD curves for each continent to form SAD envelope curves for the world. The droughts are listed in the figure legend and ranked according to the number of points that they contribute to the SAD envelope curves. The curves are dominated by events in Asia and North America. The 1975–1977 Asian drought that covered much of Kazakhstan and western Russia (and joins with the European 1975–1977 drought) contributes to most of the 12-month and 24-month curves. The remainder of the 12-month and 24-month curves for larger areas are derived from the 1944–1945 and 1950–1952 Central Asian droughts, respectively. The 24-month curve is actually made up of data from six different events, with contributions from the 1954–1957 North American and 1928–1930 Australian droughts at smaller areas. The three-month curve is derived from data for the 1963–1964 South American drought, which generally affected the Amazon region, the 1944–1945 Central Asian drought and the 1976–1977 winter North American drought. Only one drought from each of Europe (1975–1977) and South America (1963–1964) contribute to the curves, but together they make up a substantial part, and three droughts come from Oceania. The recent Australian drought in 2008 is the severest drought for short durations and small areas. Interestingly, the African droughts of the 1980s do not contribute at all to the curves yet dominate the African envelope curves. Table 7.2 lists the longest and most spatially extensive droughts across the globe over 1901–2008.

Table 7.2 *List of top ten ranked drought events for the world in terms of duration and maximum spatial extent*

Event	Location	Duration (months)	Spatial extent (million km^2)
		Longest duration	
1986–1990	Siberia	42	3.3
1954–1957	North America	41	7.5
1950–1952	Central Asia	26	7
1975–1977	Europe	25	4
1928–1930	Australia	24	4
1920–1922	Europe	23	5
1982–1984	Eastern Siberia	23	5.9
1992–1993	Africa	21	7.3
1992–1994	Eastern Siberia	21	5.7
1963–1964	North America	20	5.5
		Largest Extent	
1963–1964	South America	18	10.7
1944–1945	Central Asia	15	10.0
1982–1983	Africa	6	9.9
1987–1988	Central/Eastern Siberia	10	9.8
1968–1969	Siberia	14	9.7
1995	South America	3	8.7
1997–1998	Asia	12	8.7
1982–1983	Asia	9	8.5
1967–1968	South America	16	8.5
2002–2003	Asia	9	8.3

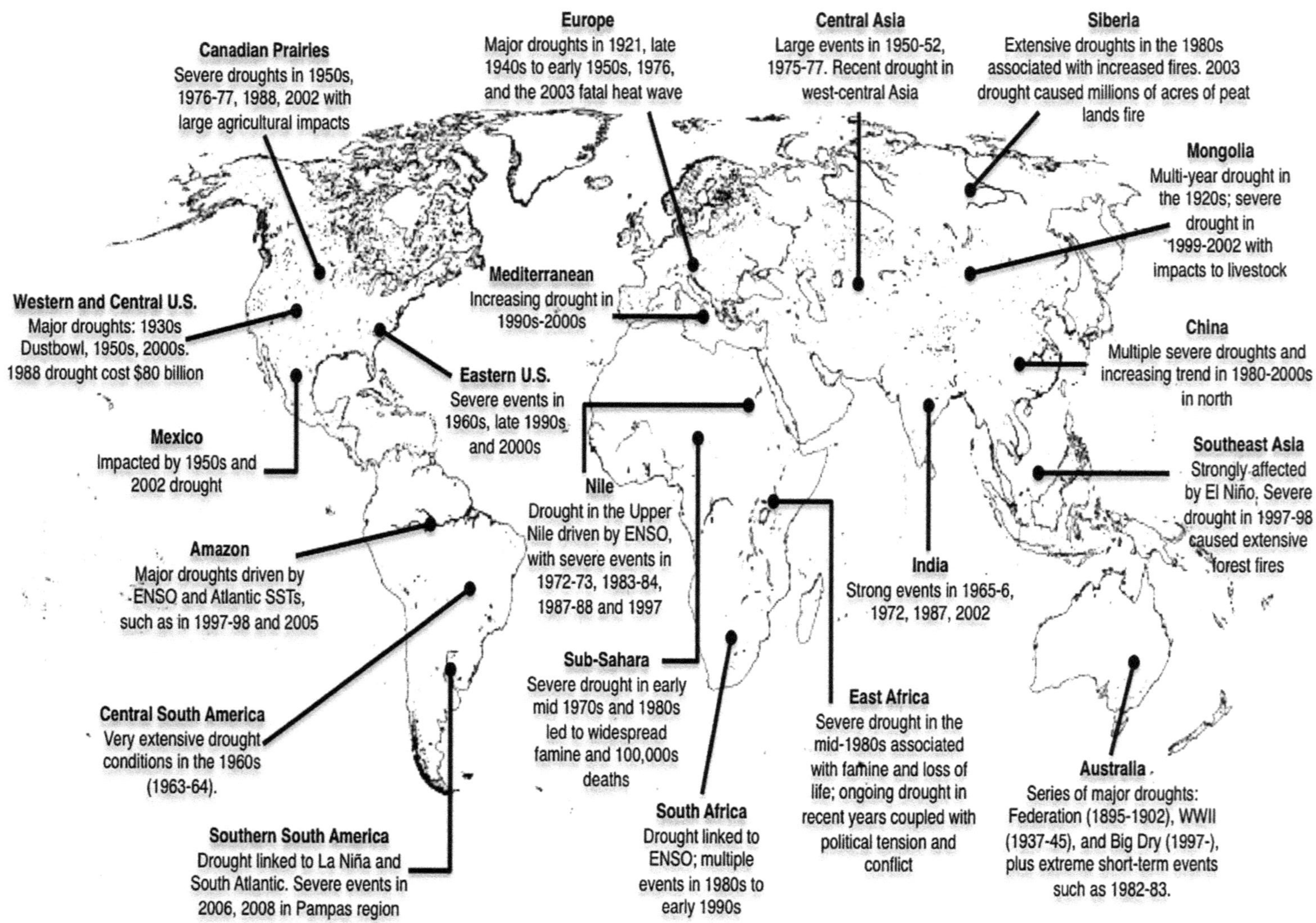

Figure 7.21 *Major droughts of the 20th century*

We finish by summarizing the occurrence of major droughts in the map in Figure 7.21. This presents some of most severe and extensive (in terms of duration and area) events of the 20th century, as gleaned from the modelled soil moisture and ancillary data, which includes other hydrological measurements and reported impacts. In general, the identified droughts in this chapter are well known and consistent with those reported in the scientific and popular literature. Notable exceptions are the less well-known high-latitude droughts in Siberia and Canada/Alaska, whose long durations (up to four years) are, nevertheless, likely to have impacted on the local populations (albeit sparse) and fragile ecosystems (Hinzman et al, 2005). Other droughts are ranked highly in terms of severity and spatial extent, yet are not well documented or analysed, such as the 1965 Australian and 1963–1964 South American droughts. In some regions, for example, the Mediterranean, East Asia and Africa, increases in drought in recent decades hint at an influence of climate change, although it is clear from this and the previous chapter that there is large variability in climate and drought on decadal timescales. The next chapter discusses these issues and the projected changes in drought occurrence and severity over the 21st century.

References

Abtew, W., Melesse, A. M. and Dessalegne, T. (2009) 'El Nino Southern Oscillation link to the Blue Nile River Basin hydrology', in *Hydrological Processes*, vol 23, pp3653–3660

AFED (2009) Tolba, M. K. and Saab, N. W. (eds), 'Arab environment: Climate change impact of climate change on Arab countries', Report of the Arab Forum for Environment and Development

Al-Taher, A. A. (1994) 'Drought and human adjustment in Saudi Arabia', in *GeoJournal*, vol 33.4, pp411–422

Andreadis, K. M., Clark, E. A., Wood, A. W., Hamlet, A. F. and Lettenmaier, D. P. (2005) 'Twentieth-century drought in the conterminous United States', in *Journal of Hydrometeorology*, vol 6, pp985–1001

Ashok K., Guan Z. and Yamagata T. (2001) 'Impact of the Indian Ocean Dipole on the relationship between the Indian Monsoon rainfall and ENSO', in *Geophysical Research Letter*, vol 28, pp4499–4502

Balzter, H., Gerard, F., Weedon, G., Grey, W., Combal, B., Bartholome, E., Bartalev, S. and Los, S. (2007) 'Coupling of vegetation growing season anomalies with hemispheric and regional scale climate patterns in Central and East Siberia', in *Journal of Climate*, vol 20, no 15, pp3713–3729, doi:10.1175/JCLI4226

Barlow, M., Nigam, S. and Berbery, E. H. (2001) 'ENSO, Pacific decadal variability, and US summertime precipitation, drought, and streamflow', in *Journal of Climate*, vol 14, pp2105–2128

Barlow, M., Cullen, H. and Lyon, B. (2002) 'Drought in central and southwest Asia: La Niña, the warm pool, and Indian Ocean precipitation', in *Journal of Climate*, vol 15, pp697–700

Barrucand, M. G., Vargas, W. M. and Rusticucci, M. M. (2007) 'Dry conditions over Argentina and the related monthly circulation patterns', in *Meteorology Atmospheric Physics*, vol 98, nos 1–2, pp99–114

Batima, P., Natsagdorj, L., Gombluudev, P. and Erdenetsetseg, B. (2005) 'Observed climate change in Mongolia' (Working Paper), *Assessments of Impacts and Adaptation to Climate Change*, vol 13

Battisti, D. S. and Naylor, R. L. (2009) 'Historical warnings of future food insecurity with unprecedented seasonal heat', in *Science*, vol 323, no 5911, pp 240–244, doi:10.1126/science.1164363

Berri, G. S., Ghietto, M. A. and García, N. O. (2002) 'The influence of ENSO in the flows of the Upper Paraná River of South America over the past 100 years', in *Journal of Hydrometeorology*, vol 3, pp57–65

Biasutti, M. and Giannini, A. (2006) 'Robust Sahel drying in response to late 20th century forcings', in *Geophysical Research Letters*, vol 33, L11706, p4

Bischoff, S. A. et al (2000) 'Climate variability and Uruguay River flows', in *Water International*, vol 25, pp446–456

Bonacina, L. C. W. (1923) 'The European drought of 1921', in *Nature*, vol 112, no 2813, pp488–489

Boulanger, J. P., Leloup, J., Penalba, O., Rusticucci, M., Lafon, F. and Vargas, W. (2005) 'Observed precipitation in the Parana-Plata hydrological basin: Long-term trends, extreme conditions and ENSO teleconnections', in *Climate Dynamics*, vol 24, no 4, pp393–413

Bradford, R. B. (2000) 'Drought events in Europe', in J. V. Vogt and F. Somma (eds) *Drought and Drought Mitigation in Europe: Advances in Natural and Technological Hazards Research*, vol 14, pp7–20, Kluwer Academic Publishers, Boston

Brázdil, R., Trnka, M., Dobrovolný, P., Chromá, K., Hlavinka, P. and Žalud, Z. (2009) 'Variability of droughts in the Czech Republic, 1881–2006', in *Theoretical and Applied Climatology*, vol 97, nos 3–4, pp297–315, doi:10.1007/s00704-008-0065-x

Briffa, K. R., Jones, P. D. and Hulme, M. (1994) 'Summer moisture variability across Europe, 1892–1991: An analysis based on the Palmer Drought Severity Index', in *International Journal of Climatology*, vol 14, no 5, pp475–506

Briffa, K. R., van der Schrier, G. and Jones, P. D. (2009) 'Wet and dry summers in Europe since 1750: Evidence of increasing drought', in *International Journal of Climatology*, vol 29, no 13, pp1894–1905

Brooks, C. E. P. and Glasspoole, A. I. C. (1992) 'The drought of 1921', in *Quarterly Journal of the Royal Meteorological Society*, vol 48, no 1992, pp139–150

BOM (2010) Australian Bureau of Meteorology, www.bom.gov.au/lam/climate/levelthree/c20thc/drought.htm

Büntgen, U., Brázdil, R., Fran, D. and Esper, J. (2009) 'Three centuries of Slovakian drought dynamics', in *Climate Dynamics*, vol 35, nos 2–3, pp315–329, doi:10.1007/s00382-009-0563-2

Burton, M. L. and Hicks, M. J. (2005) 'Hurricane Katrina: Preliminary estimates of commercial and public sector damages', www.marshall.edu/cber/research/katrina/Katrina-Estimates.pdf

Butler, J. E. (1978) *Natural Disasters*, Heinemann Educational Publishers, London

Cai, W., Cowan, T., Briggs, P. and Raupach, M. (2009) 'Rising temperature depletes soil moisture and exacerbates severe drought conditions across southeast Australia', in *Geophysical Research Letters*, vol 36, L21709, p5, doi:10.1029/2009GL040334

Calanca, P. (2007) 'Climate change and drought occurrence in the Alpine region: How severe are becoming the extremes?' *Global and Planetary Change*, vol 57, pp151–160

Camilloni, I. A. and Barros, V. R. (2000) 'The Paraná River response to El Niño 1982–83 and 1997–98 events', in *Journal of Hydrometeorology*, vol 1, pp412–430

Chase, T. N., Wolter, K., Pielke Sr., R. A. and Rasool, I. (2006). 'Was the 2003 European summer heat wave unusual in a global context?', in *Geophysical Research Letters*, vol 33, L23709, p5, doi:10.1029/2006GL027470

Chen, F., Wang, J., Jin, L., Zhang, Q., Li, J. and Chen, J. (2009) 'Rapid warming in mid-latitude Central Asia for the past 100 years', in *Frontiers of Earth Science in China*, vol 3, no 1, pp42–50, doi:10.1007/s11707-009-0013-9

Chiew, F. H. S., Piechota, T. C., Dracup, J. A. and McMahon, T. A. (1998) 'El Niño/Southern Oscillation and Australian rainfall, streamflow and drought: Links and potential for forecasting', in *Journal of Hydrology*, vol 204, pp138–149

Ciais, P., et al (2005) 'Europe-wide reduction in primary productivity caused by the heat and drought in 2003', in *Nature*, vol 437, pp529–533

Cook, B. I., Miller, R. L. and Seager, R. (2009) 'Amplification of the North American "Dust Bowl" drought through human-induced land degradation', in *Proceedings of the National Academy of Sciences*, vol 106, pp4997–5001

Cook, E. R., Meko, D. M., Stahle, D. W. and Cleaveland, M. K. (1999) 'Drought reconstructions for the continental United States', in *Journal of Climate*, vol 12, no 4, pp1145–1162

Cook, E. R., Woodhouse, C. A., Eakin, C. M., Meko, D. M. and Stahle, D. W. (2004) 'Long-term aridity changes in the Western United States', in *Science*, vol 306, pp1015–1018

Cook, E. R., Anchukaitis, K. J., Buckley, B. M., D'Arrigo, R. D., Jacoby, G. C. and Wright, W. E. (2010) 'Asian Monsoon failure and megadrought during the last millennium', in *Science*, vol 328, no 5977, pp486–489, doi:10.1126/science.1185188

Cook, K. H. (2008) 'The mysteries of Sahel droughts', in *Nature Geoscience*, vol 1, pp647–648

Cox, P. M., Harris, P. P., Huntingford, C., Betts, R. A., Collins, M., Jones, C. D., Jupp, T. E., Marengo, J. A. and Nobre, C. A. (2008) 'Increasing risk of Amazonian drought due to decreasing aerosol pollution', in *Nature*, no 453, pp212–215, doi:10.1038/nature06960

Cruz, R. V., Harasawa, H., Lal, M., Wu, S., Anokhin, Y., Punsalmaa, B., Honda, Y., Jafari, M., Li, C. and Huu Ninh, N. (2007) 'Asia', in M. L. Parry, O. F. Canziani, J. P. Palutikof, P. J. van der Linden and C. E. Hanson (eds) *Climate Change 2007: Impacts, Adaptation and Vulnerability. Contribution of Working Group II to the Fourth Assessment Report of the Intergovernmental Panel on Climate Change*, Cambridge University Press, Cambridge, pp469–506

Dai, A., Trenberth, K. E. and Qian, T. (2004) 'A global data set of Palmer Drought Severity Index for 1870–2002: Relationship with soil moisture and effects of surface warming', in *Journal of Hydrometeorology*, vol 5, pp1117–1130

Davi, N. K., Jacoby, G. C., Curtis, A. E., Bachin, B. (2006) 'Extension of drought records for central Asia using tree rings: West central Mongolia', in *Journal of Climate*, vol 19, pp288–299

de Beurs, K. M., Wright, C. K. and Henebry. G. M. (2009) 'Dual scale trend analysis for evaluating climatic and anthropogenic effects on the vegetated land surface in Russia and Kazakhstan', *Environmental Research Letters*, 4 045012

De Bono, A., Peduzzi, P., Giuliani, G. and Kluser, S. (2004) 'Impacts of summer 2003 heat wave in Europe', in *Early Warning on Emerging Environmental Threats 2*, United Nations Environment Programme, Nairobi

Demuth, S. and Kulls, C. (1997) 'Probability analysis and regional aspects of droughts in southern Germany', in *Sustainability of Water Resources under Increasing Uncertainty*, Proceedings of Rabat Symposium SI, April 1997, International Association of Hydrological Sciences, no 240, pp97–104

Diaz, H. (1983) 'Drought in the United States: Some aspects of major dry and wet periods in the contiguous United States, 1895–1981', in *Journal of Climate Applied Meteorology*, no 22, pp3–16

Ding, Y. and Sikka, D. (2006) 'Synoptic systems and weather', in B. Wang (ed) *Asian Monsoon*, Springer, Berlin, doi:10.1007/3-540-37722-0_4, pp131–201

EC (European Commission) (2007) 'Communication addressing the challenge of water scarcity and droughts in the European Union, (COM 2007 414)', Commission of the European Communities, Brussels, Belgium

Ellman, M. (2000) 'The 1947 Soviet famine and the entitlement approach to famines', in *Cambridge Journal of Economics*, vol 24, pp603–630

FAO (Food and Agriculture Organization) (2003) 'Review of world water resources by country', Water Report, FAO, Rome, no 23

Feyen, L. and Dankers, R. (2009) 'Impact of global warming on streamflow drought in Europe', in *Journal of Geophysical Research-Atmosphere*, vol 114, D17116, p17, doi:17110.11029/12008JD011438

Fink, A. H., Brücher, T., Krüger, A., Leckebusch, G. C., Pinto, J. G. and Ulbrich, U. (2004) 'The 2003 European summer heat waves and drought – synoptic diagnosis and impacts', in *Weather*, vol 59, no 8, pp209–216

Fowler, A. and Adams, K. (2004) 'Twentieth century droughts and wet periods in Auckland (New Zealand) and their relationship to ENSO', in *International Journal of Climatology*, vol 24, pp1947–1961

Garrido, A. and Gómez-Ramos, A. (2000) 'Socio-economic aspects of droughts', in J. V. Vogt and F. Somma (eds) *Drought and Drought Mitigation in Europe*, Advances in Natural and Technological Hazards Research, Kluwer Academic Publishers, Boston, vol 14, pp197–207

Giese, B. S., Campo, G. P., Slowey, N. C., Sardeshmukh, P. D. and Carton, J. A. (2010) 'The 1918/19 El Niño', in *Bulletin of the American Meteorological Society*, vol 91, pp177–183

Giannini, A., Saravanan, R. and Chang, P. (2003) 'Oceanic forcing of Sahel rainfall on interannual to interdecadal time scales', in *Science*, vol 302, pp1027–1030

Glantz, M. H. (1978) 'Render unto weather', in *Climatic Change*, vol 1, 305–306

Glantz, M. H. (2005) 'Water, climate, and development issues in the Amu Darya basin', in *Mitigation and Adaptation Strategies for Global Change*, vol 10, no 1, pp23–50

Golubev, G. and Dronin, N. (2003) 'Geography of droughts and food problems in Russia of the twentieth century', in *Research Monograph of the Centre for Environmental Systems Research*, University of Kassel and Moscow State University

Gruza, G., Rankova, E., Razuvaev, V. and Bulygina, O. (1999) 'Indicators of climate change for the Russian Federation', in *Climatic Change*, vol 42, no 1, pp219–242

Held, I. M., Delworth, T. L., Lu, J., Findell, K. L. and Knutson, T. R. (2005) 'Simulation of Sahel drought in the 20th and 21st centuries', *Proceedings of the National Academy of Sciences of the United States of America*, no 102, pp17,891–17,896

Hinzman, L. D. et al (2005) 'Evidence and implications of recent climate change in northern Alaska and other arctic regions', in *Climate Change*, vol 72, pp251–298, doi:10.1007/s10584-005-5352-2

Hulme, M. (1992) 'Rainfall changes in Africa: 1931–1960 to 1961–1990', in *International Journal of Climatology*, vol 12, pp685–699

Janowiak, J. E. (1988) 'The global climate for March-May 1988: The end of the 1986-87 Pacific warm episode and the onset of wide-spread drought in the United States', in *Journal of Climate*, vol 1, pp1019–1040

Jiang, T., Qiang, Z., Deming, Z. and Yijin, W. (2006) 'Yangtze floods and droughts (China) and teleconnections with ENSO activities (1470–2003)', in *Quaternary International*, vol 144, pp29–37

Karl, T. R. and Quayle, R. G. (1981) 'The 1980 summer heat wave and drought in historical perspective', in *Monthly Weather Review*, no 109, pp2055–2073

Kaser, M. (1997) 'The economics of Kazakhstan and Uzbekistan', Royal Institute of International Affairs, London

Keen, H., and Rotorua, L. (1996) 'Changes in drought policies in New Zealand', Drought Network website, www.drought.unl.edu/pubs/dnn/arch28.pdf

Keyantash, J. A. and Dracup, J. A. (2004) 'An aggregate drought index: Assessing drought severity based on fluctuations in the hydrological cycle and surface water storage', in *Water Resources Research*, vol 40, no 9, p14

Kiem, A. S. and Verdon-Kidd, D. C. (2009) 'Towards understanding hydro-climatic change in Victoria, Australia – Why was the last decade so dry?', in *Hydrology and Earth System Sciences*, vol 6, pp6181–6206

Komuscu, A. U., Erkan, A. and Oz, S. (1998) 'Possible impacts of climate change on soil moisture variability in the Southeast Anatolian development project (GAP) region: An analysis from agricultural drought perspective', in *Climate Change*, vol 40, pp519–545

Kripalani, R. H. and Kulkarni, A. (1997) 'Climatic impact of El Niño/La Niña on the Indian monsoon: A new perspective', in *Weather*, no 52, pp39–46

Kripalani, R. H., Kulkarni, A., Sabades, S., Ravadekar, J. V., Patwardhan, S. K. and Kulkarni, J. R. (2004) 'Intraseasonal oscillations during monsoon 2002 and 2003', in *Current Science*, vol 87, no 3, pp325–331

Krishna Kumar, K., Rajagopalan, B. and Cane, M. A. (1999) 'On the weakening relationship between Indian Monsoon and ENSO', in *Science*, vol 284, pp2156–2159

Krishna Kumar, K., Rajagopalan, B., Hoerling, M., Bates, B. and Cane, M. (2006) 'Unraveling the mystery of Indian monsoon failure during El Niño', in *Science*, vol 314, pp115–119

Krishnamurthy, V. and Shukla, J. (2000) 'Intraseasonal and interannual variability of rainfall over India', in *Journal of Climate*, vol 13, no 24, pp4366–4377

Lau, K. M., Kim, K. M. and Yang, S. (2000) 'Dynamical and boundary forcing characteristics of regional components of the Asian summer monsoon', in *Journal of Climate*, vol 13, no 14, pp2461–2482

Lau, K. M., Shen, S. S. P., Kim, K. M. and Wang, H. (2006) 'A multimodel study of the twentieth-century simulations of Sahel drought from the 1970s to 1990s', in *Journal of Geophysical Research-Atmospheres*, vol 111, D07111, p9

Lawrimore, J. and Stephens, S. (2003) 'Climate of 2002 Annual Review', NOAA National Climatic Data Center website, http://lwf.ncdc.noaa.gov/oa/climate/research/2002/ann/drought-summary.html

L'Hôte, Y., Mahé, G., Somé, B. and Triboulet, J. P. (2002) 'Analysis of a Sahelian annual rainfall index from 1896 to 2000: The drought continues', in *Hydrological Sciences Journal–Journal Des Sciences Hydrologiques*, vol 47, pp563–572

Li, J., Cook, E. R., D'arrigo, R., Chen, F. and Gou, X. (2009) 'Moisture variability across China and Mongolia: 1951–2005', in *Climate Dynamics*, vol 32, pp1173–1186, doi: 10.1007/s00382-008-0436-0

Li, H. M., Dai, A. G., Zhou, T. J. et al (2010) 'Responses of East Asian summer monsoon to historical SST and atmospheric forcing during 1950-2000', in *Climate Dynamics*, vol 34, no 4, pp501–514

Liebmann, B. et al (2004) 'An observed trend in Central South American precipitation', in *Journal of Climate*, vol 22, pp4357–4367

Lioubimtseva, E. and Henebry, G. M. (2009) 'Climate and environmental change in arid Central Asia: Impacts, vulnerability, and adaptations', in *Journal of Arid Environments*, no 73, vol 11, pp963–977

Liu, Y. B. and Chen, Y. N. (2006) 'Impact of population growth and land-use change on water resources and ecosystems of the arid Tarim River Basin in western China', in *International Journal of Sustainable Development and World Ecology*, vol 13, pp295–305

Llano, M. P. and Penalba, O. C. (in press) 'A climatic analysis of dry sequences in Argentina', in *International Journal of Climatology*, doi:10.1002/joc.2092

Lloyd-Hughes, B. and Saunders, M. A. (2002) 'A drought climatology for Europe', in *International Journal of Climatology*, vol 22, pp1571–1592

Lockart, N., Kavetski, D. and Franks, S. W. (2009) 'On the recent warming in the Murray-Darling Basin: Land surface interactions misunderstood', in *Geophysical Research Letters*, vol 36, L24405, p6, doi:10.1029/2009GL040598

Los, S. O., Weedon, G. P., North, P. R. J., Kaduk, J. D., Taylor, C. M. and Cox, P. M. (2006) 'An observation-based estimate of the strength of rainfall-vegetation interactions in the Sahel', in *Geophysical Research Letters*, vol 33, L16402, p5

Lyon, B. and Dole, R. (1995) 'A diagnostic comparison of the 1980 and 1988 US summer heat wave-droughts', in *Journal of Climate*, vol 8, pp1658–1676

Ma, Z. G. and Fu, C. B. (2003) 'Interannual characteristics of the surface hydrological variables over the arid and semi-arid areas of northern China', in *Global Planet. Change*, vol 37, pp189–200

Magrin, G., Gay García, C., Cruz Choque, D., Giménez, J. C., Moreno, A. R., Nagy, G. J., Nobre, C. and Villamizar, A. (2007) 'Latin America', in M. L. Parry, O. F. Canziani, J. P. Palutikof, P. J. van der Linden and C. E. Hanson (eds) *Climate Change 2007: Impacts, Adaptation and Vulnerability. Contribution of Working Group II to the Fourth Assessment Report of the Intergovernmental Panel on Climate Change*, Cambridge University Press, Cambridge, pp581–615

Malevski-Malevich, S. P., Molketin, E. K., Nadezdina, E. D. and Skliarevich, O. B. (2005) 'Assessment of forest fire scales in the Russia under the projected climate change by 2100', in *Meteorology and Hydrology*, vol 3, pp36–44

Marengo, J. A. (1995) 'Variations and change in South American streamflow', in *Climatic Change*, vol 31, pp99–117

Marengo J. A., Tomasella, J. and Uvo, C. (1998) 'Trends in streamflow and rainfall in tropical South America: Amazonia, eastern Brazil, and northwestern Peru', in *Journal of Geophysical Research*, vol 103, pp1775–1784

Marengo, J. A., Nobre, C. A., Tomasella, J., Oyama, M. D., De Oliveira, G. S., De Oliveira, R., Camargo, H., Alves, L. M. and Brown, I. F. (2008) 'The drought of Amazonia in 2005', in *Journal of Climate*, vol 21, pp495–516

Marengo, J. (2009) 'Long-term trends and cycles in the hydrometeorology of the Amazon basin since the late 1920s', in *Hydrological Processes*, vol 23, pp3236–3244, doi:10.1002/hyp.7396

Meggers, B. (1994) 'Archeological evidence for the impact of mega-Niño events on Amazonia during the past two millennia', in *Climatic Change*, vol 28, pp321–338

Meshcherskaya, A. V. and Blazhevich, V. G. (1997) 'The drought and excessive moisture indices in a historical perspective in the principal grain-producing regions of the Former Soviet Union', in *Journal of Climate*, vol 10, pp2670–2682

Min, S. K., Kwon, W. T., Park, E. H. and Choi, Y. (2003) 'Spatial and temporal comparisons of droughts over Korea with East Asia', in *International Journal of Climatology*, vol 23, pp223–233, doi:10.1002/joc.872

Mo, K., Zimmerman, J., Kalnay, E. and Kanamitsu, M. (1991) 'AGCM study of the 1988 United States drought', in *Monthly Weather Review*, vol 119, pp1512–1532

Mortimore, M. J. and Adams, W. M. (2001) 'Farmer adaptation, change and crisis in the Sahel', in *Global Environmental Change*, vol 11, pp49–57

Mpelasoka, F., Hennessy, K., Jones, R. and Bates, B. (2007) 'Comparison of suitable drought indices for climate change impacts assessment over Australia towards resource management', in *International Journal of Climatology*, vol 28, pp1283–1292, doi:10.1002/joc.1649

Mujumda, M., Kumar, V. and Krishnan, R. (2007) 'The Indian summer monsoon drought of 2002 and its linkage with tropical convective activity over northwest Pacific', in *Climate Dynamics*, vol 28, pp743–758, doi:10.1007/s00382-006-0208-7

Namias, J. (1966) 'Nature and possible causes of the northeastern United States drought during 1962–65', in *Monthly Weather Review*, vol 94, pp543–554

Namias, J. (1967) 'Further studies of drought over northeastern United States', in *Monthly Weather Review*, vol 95, pp497–508

Namias, J. (1982) 'Anatomy of Great Plains protracted heat waves (especially the 1980 US summer drought)', in *Monthly Weather Review*, vol 110, pp824–838

Namias, J. (1991) 'Spring and summer 1988 drought over the contiguous United States: Causes and prediction', in *Journal of Climate*, vol 4, pp54–65

Nicholls, N. (2004) 'The changing nature of Australian droughts', in *Climatic Change*, vol 63, pp323–336

Nicholson, S. (2000) 'Land surface processes and Sahel climate', in *Reviews of Geophysics*, vol 38, pp117–139

Oba, G., Post, E. and Stenseth, N. C. (2001) 'Sub-saharan desertification and productivity are linked to hemispheric climate variability', in *Global Change Biology*, vol 7, no 3, pp241–246

OFDA/CRED (2006) EM-DAT: The OFDA/CRED International Disaster Database (available online at the website www.em-dat.net), Université Catholique de Louvain, Brussels, Belgium

PACS (Poorest Areas Civil Society) (2004) 'Drought in India: Challenges & initiatives', New Dehli (India) PACS Programs, www.empowerpoor.org/backgrounder.asp?report=192

Page, S. E., Slegert, F., Riely, J. O., Boehm, H. V., Jaya, A. and Limin, S. (2002) 'The amount of carbon released from peat and forest fires in Indonesia during 1997', in *Nature*, vol 420, pp61–65

Palmer, W. (1965) 'Meteorological drought' (Research Paper 45), US Department of Commerce, Washington, DC

Pankhurst R. (1966). 'The Great Ethiopian famine of 1888–1892: A new assessment part one', in *Journal of the History of Medicine and Allied Sciences*, vol 21, pp95–124

Pant, G. B. and Parthasarathy, S. B. (1981) 'Some aspects of an association between the southern oscillation and Indian summer monsoon', in *Meteorology and Atmospheric Physics*, vol 29, no 3, pp245–252, doi:10.1007/BF02263246

Parthasarathy, B., Sontakke, N. A., Monot, A. A. and Kothawale, D. R. (1987) 'Drought/floods in the summer monsoon season over different meteorological subdividions of India for the period 1871–1984', in *International Journal of Climatology*, vol 7, pp57–70

Parthasarathy, B., Munot, A. A. and Kothawale, D. R. (1994) 'All-India monthly and seasonal rainfall series: 1871–1993', in *Theoretical Applied Climatology*, vol 49, pp217–224

Patel, N. R., Chopra, P. and Dadhwal, V. K. (2007) 'Analyzing spatial patterns of meteorological drought using standardized precipitation index', in *Meteorological Applications*, vol 14, no 4, pp329–336

Pekarova, P., Miklanek, P. and Pekar, J. (2006) 'Long-term trends and runoff fluctuations of European rivers', in *Climate Variability and Change – Hydrological Impacts* (Proceedings of the Fifth FRIEND World Conference held at Havana, Cuba, November 2006), International Association of Hydrological Sciences Publications, vol 308, p520

Phillips, O. L., Aragão, L. E. O. C., Lewis, S. L., Fisher, J. B., Lloyd, J., Lopez-Gonzalez, G., Malhi, Y., Monteagudo, A., Peacock, J., Quesada, C. A. et al (2009) 'Drought Sensitivity of the Amazon Rainforest', in *Science*, vol 323, pp1344–1347

Pielke, R. A. Sr., Adegoke, J., Beltran-Przekurat, A., Hiemstra, C. A., Lin, J., Nair, U. S., Niyogi, D. and Nobis, T. E. (2007) 'An overview of regional land use and land cover impacts on rainfall', in *Tellus B*, vol 59, pp587–601

Plummer, N., Salinger, M. J., Nicholls, N., Suppiah, R., Hennessy, K. J., Leighton, R. M., Trewin, B., Page, C. M. and Lough, J. M. (1999) 'Changes in climate extremes over the Australian region and New Zealand during the twentieth century', in *Climatic Change*, vol 42, no 1, pp183–202

Poumadère, M., Mays, C., Le Mer, S. and Blong, R. (2005) 'The 2003 heat wave in France: Dangerous climate change here and now', in *Risk Analysis*, vol 5, no 6, pp1483–1494, doi:10.1111/j.1539-6924.2005.00694.x

Power, S., Casey, T., Folland, C., Colman, A. and Mehta, V. (1999) 'Inter-decadal modulation of the impact of ENSO on Australia', in *Climate Dynamics*, vol 15, pp319–324

Qian, W. and Zhu, Y. (2001) 'Climate change in China from 1880 to 1998 and its impact on the environmental condition', in *Climatic Change*, vol 50, pp419–444

Qian, W., Hu, Q., Zh, Y. and Lee, D.-K. (2003) 'Centennial-scale dry-wet variations in East Asia', in *Climate Dynamics*, vol 21, pp77–89, doi:10.1007/s00382-003-0319-3

Quintana, J. (2000) 'The Drought in Chile and La Niña', in *Drought Network News*, vol 12, no 2, p4

Raffa, K. F., Aukema, B. H., Bentz, B. J., Carroll, A. L., Hicke, J. A., Turner, M. G. and Romme, W. H. (2008) 'Cross-scale drivers of natural disturbances prone to anthropogenic amplification: The dynamics of bark beetle eruptions', in *Bioscience* 58, pp501–517

Rasmusson, E. M. and Carpenter, T. H. (1983) 'The relationship between the eastern Pacific sea surface temperature and rainfall over India and Sri Lanka', in *Monthly Weather Review*, vol 111, pp354–384

Richard, Y., Fauchereau, N., Poccard, I., Rouault, M. and Trzaska, S. (2001) '20th century droughts in Southern Africa: Spatial and temporal variability, teleconnections with oceanic and atmospheric conditions', in *International Journal of Climatology*, no 21, pp873–895

Rouault, M. and Richard, Y. (2005) 'Intensity and spatial extent of droughts in southern Africa', in *Geophysical Research Letters*, vol 32, no 15, p4

Ropelewski, C. F. (1988) 'The global climate for June-August 1988: A swing to the positive phase of the southern oscillation, drought in the United States, and abundant rain in monsoon areas', in *Journal of Climate*, vol 1, pp1153–1174

Rosenberg, N. J. (1993) 'A methodology called mink for study of climate-change impacts and responses on the regional-scale: An introductory editorial', in *Climatic Change*, vol 24, pp1–6

Saleska, S. R., Didan, K., Huete, A. R. and da Rocha, H. R. (2007) 'Amazon forests green-up during 2005 drought', in *Science*, vol 318, no 5850, p612, doi:10.1126/science.1146663

Samanta, A., Ganguly, S., Hashimoto, H., Devadiga, S., Vermote, E., Knyazikhin, Y., Nemani, R. R. and Myneni, R. B. (2010) 'Amazon forests did not green-up during the 2005 drought', in *Geophysical Research Letters*, vol 37, L05401, p5, doi:10.1029/2009GL042154

Santos, M. J., Veríssimo, R., Fernandes, S., Orlando, M. and Rodrigues, R. (2000) 'Overview of regional meteorological drought analysis on Western Europe', ARIDE Techical Report no 10, National Institute for Water, Lisbon, Portugal

Schär, C. and Jendritzky, G. (2004) 'Climate change: Hot news from summer 2003', in *Nature*, vol 432, pp559–560, doi:10.1038/432559a

Schubert, S., Suarez, M., Pegion, P., Koster, R. and Bacmeister, J. (2004a) 'On the cause of the 1930s Dust Bowl', in *Science*, vol 303, no 5665, pp1855–1859

Schubert, S. D., Suarez, M. J., Pegion, P. J., Koster, R. D. and Bacmeister, J. T. (2004b) 'Causes of long-term drought in the US Great Plains', in *Journal of Climate*, vol 17, no 3, pp485–503

Sheffield, J. and Wood, E. F. (2007) 'Characteristics of global and regional drought, 1950–2000: Analysis of soil moisture data from off-line simulation of the terrestrial hydrological cycle', in *Journal of Geophysical Research*, vol 112, D17115, p21, doi:10.1029/2006JD008288

Sheffield, J. and Wood, E. F. (2008) 'Projected changes in drought occurrence under future global warming from multi-model, multi-scenario, IPCC AR4 simulations', in *Climate Dynamics*, vol 13 no 1, pp79–105, doi:10.1007/s00382-007-0340-z

Shen, C., Wang, W. C., Hao, Z. and Gong, W. (2007) 'Exceptional drought events over eastern China during the last five centuries', in *Climatic Change*, vol 85, pp453–471

Shukla, J. and Paolina, D. A. (1983) 'The southern oscillation and long range forecasting of the summer monsoon rainfall over India', in *Monthly Weather Review*, no 111, pp1830–1837

Sönmez, F. K., Kömüscü, A. Ü., Erkan, A. and Turgu, E. (2005) 'An analysis of spatial and temporal dimension of drought vulnerability in Turkey using the standardized precipitation index', in *Natural Hazards*, vol 35, no 2, pp243–264, doi:10.1007/s11069-004-5704-7

Stahle, D. W. and Cleaveland, M. K. (1988) 'Texas drought history reconstructed and analyzed from 1698 to 1980', in *Journal of Climate*, vol 1, pp69–74

Stott, P. A., Stone, D. A. and Allen, M. R. (2004) 'Human contribution to the European heat wave of 2003', in *Nature*, vol 432, pp610–614

Sud, Y. C., Mocko, D. M., Lau, K.-M. and Atlas, R. (2003) 'Simulating the Midwestern US drought of 1988 with a GCM', in *Journal of Climate*, vol 16, pp3946–3965

Sun, Y. and Ding, Y.-H. (2009) 'A projection of future changes in summer precipitation and monsoon in East Asia', in *Science in China Series D: Earth Sciences*, vol 53, pp284–300

Tag El-Din, S. S. (1989) 'Range development in the north of the Kingdom of Saudi Arabia', Workshop on Desert Studies in the Kingdom of Saudi Arabia, King Saud University, Center for Desert Studies

Tarhule, A. and Lamb, P. J. (2003) 'Climate research and seasonal forecasting for West Africans: Perceptions, dissemination, and use?' in *Bulletin of the American Meteorological Society*, vol 84, pp1741–1759

Tran, L., Knight, C. G. and Wesner, V. (2002) 'Drought in Bulgaria and atmospheric synoptic conditions over Europe', in *GeoJournal*, vol 57, pp149–157

Trenberth, K. and Branstator, G. (1992) 'Issues in establishing causes of the 1988 drought over N. America', in *Climate*, vol 5, pp159–172

Trnka, M., Hlavinka, P., Semerádová, D., Dubrovský, M., Žalud, Z. and Možný, M. (2007) 'Agricultural drought and spring barley yields in the Czech Republic', in *Plant-Soil Environment*, vol 53, no 7, pp306–316

Türke, M. (1996) 'Meteorological drought in Turkey: A historical perspective, 1930–03', in *Drought Network News*, October 1996

Turner, G. and Slingo, J. M. (2010) 'Using idealized snow forcing to test teleconnections with the Indian summer monsoon in the Hadley Centre GCM', in *Climate Dynamics*, doi:10.1007/s00382-010-0805-3

Ummenhofer, C. C., England, M. H., McIntosh, P. C., Meyers, G. A., Pook, M. J., Risbey, J. S., Sen Gupta, A. and Taschetto, A. S. (2009) 'What causes Southeast Australia's worst droughts?' in *Geophysical Research Letters*, vol 36, L04706, p5, doi:10.1029/2008GL036801

UNEP (2006) *GEO Year Book*, Earthscan, London

van der Schrier, G., Briffa, K. R., Jones, P. D. and Osborn, T. J. (2006) 'Summer moisture variability across Europe', in *Journal of Climate*, vol 19, pp2818–2834

Varushchenko, A. N., Lukyanova, S. A., Solovieva, G. D., Kosarev, A. N. and Kuraev, A. V. (2000) 'Evolution of the Gulf of Kara-Bogaz-Gol in the past century', in K.P. Lulla and L.V. Dessinov (eds) *Dynamic Earth Environments: Remote Sensing Observations from Shuttle-Mir Missions*, John Wiley and Sons, New York, pp201–210

Verdon-Kidd, D. C. and Kiem, A. S. (2009) 'Nature and causes of protracted droughts in southeast Australia: Comparison between the Federation, WWII, and Big Dry droughts', in *Geophysical Research Letters*, vol 36, L22707, p6, doi:10.1029/2009GL041067.12

Vicente-Serrano, S. M. (2006) 'Spatial and temporal analysis of droughts in the Iberian Peninsula (1910–2000)', in *Hydrological Sciences–Journal–des Sciences Hydrologiques*, vol 51, no 1, pp83–97

Vuille, M. and Milana, J.-P. (2007) 'High-latitude forcing of regional aridification along the subtropical west coast of South America', in *Geophysical Research Letters*, vol 34, L23703, p6, doi:10.1029/2007GL031899

Wang, Z. W. and Zhai, P. M. (2003) 'Climate change in drought over northern China during 1950–2000', in *Acta Geographica Sinica*, vol 58, pp61–68

Wang, J., Chen, F., Jin, L. and Bai, H. (2010a) 'Characteristics of the dry/wet trend over arid central Asia over the past 100 years', in *Climatic Research*, vol 41, pp51–59

Wang, A., Lettenmaier, D. P. and Sheffield, J. (2010b) 'Soil moisture drought in China, 1950–2006', in *Journal of Climate* (in review)

Warrick, R. A. (1980) 'Drought in the Great Plains: A case study of Research on Climate and Society in the USA', in J. Ausubel and A. K. Biswas (eds) *Climatic Constraints and Human Activities*, IIASA Proceedings Series, Pergamon Press, New York, vol 10, pp93–123

Webster, P. J., Magafia, V. O., Palmer, T. N., Shukla, J., Tomas, R. A., Yanai, M. and Yasunari. T. (1998) 'Monsoons: Processes, predictability, and the prospects for prediction', in *Journal of Geophysical Research*, vol 103, C7, pp14,451–14,510

Webster, P. J., Moore, A. M., Loschnigg, J. P. and Leben, R. R. (1999) 'Coupled ocean-atmosphere dynamics in the Indian Ocean during 1997–98', in *Nature*, vol 401, pp356–360

White, W. B., Gershunov, A., Annis, J. L., McKeon, G. and Syktus, J. (2004) 'Forecasting Australian drought using southern hemisphere modes of sea-surface temperature variability', in *International Journal of Climatology*, vol 24, pp1911–1927

Williams, E. et al (2005) 'The drought of the century in the Amazon Basin: an analysis of the regional variation of rainfall in South America in 1926', in *Acta Amazonia*, vol 35, no 2, pp231–238

Williamson, G. B., Laurance, W. F., Oliveira, A. A., Delamônica, P., Gascon, C., Lovejoy, T. E. and Pohl, L. (2000) 'Amazonian tree mortality during the 1997 El Niño drought', in *Conservation Biology*, vol 14, no 5, pp1538–1542

Worster, D. (1979) *Dust Bowl: The Southern Plains in the 1930s*, Oxford University Press, New York

Wu, B., Yang, K. and Zhang, R. (2009) 'Eurasian snow cover variability and its association with summer rainfall in China', in *Advances in Atmospheric Sciences*, vol 26, no 1, pp31–44

Xoplaki, E., González-Rouco, J. F., Luterbacher, J. and Wanner, H. (2004) 'Wet season Mediterranean precipitation variability: Influence of large-scale dynamics and trends', in *Climate Dynamics*, vol 23, no 1, pp63–78, doi:10.1007/s00382-004-0422-0

Yu, R., Wang, B. and Zhou, T. (2004) 'Tropospheric cooling and summer monsoon weakening trend over East Asia', in *Geophysical Research Letters*, vol 31, L22212, p4, doi:10.1029/2004GL021270

Zaidman, M. D., Rees, H. G. and Young, A. R. (2001) 'Spatio-temporal development of streamflow droughts in north-west Europe', in *Hydrology and Earth System Sciences*, vol 5, no 4, pp733–751

Zeng, N., Yoon, J.-H., Marengo, J. A., Subramaniam, A., Nobre, C. A., Mariotti, A. and Neelin, J. D. (2008) 'Causes and impact of the 2005 Amazon drought', in *Environmental Research Letters*, vol 3, 014002, p9

Zhou, X., Zhai, P. and Zhang Q. (2005) 'Variations in droughts over China: 1951–2003', in *Geophysical Research Letters*, vol 32, L04707, p4, doi:10.1029/2004GL021853

Zhou, T., Yu, R., Zhang, J., Drange, H., Cassou, C., Deser, C., Hodson, D. L. R., Sanchez-Gomez, E., Li, J., Keenlyside, N., Xin, X. and Okumura, Y. (2009) 'Why the Western Pacific subtropical high has extended westward since the late 1970s', in *Journal of Climate*, vol 22, pp2199–2215

Zou, X. K. and Zhai, P. M. (2004) 'Relationship between vegetation coverage and spring dust storms over northern China', in *Journal of Geophysical Research*, vol 109, D03104, p9, doi:10.1029/2003JD003913

8

Drought in the 21st Century

Introduction

Climate change is now a household phrase that holds meaning for many, and implications for everyone on the planet. It is no longer restricted to the realm of scientific research but is a major concern for politicians and businesses that see the expensive social, economic and environmental ramifications (and opportunities) of change. At the level of individuals, there is a growing awareness of the potential impacts, although there may be a detachment between the broader picture and how it may affect each of us. This is in part due to the uncertainty surrounding how climate change will manifest itself at the time and space scales of our day-to-day lives. Scientifically, there are key concerns about how global warming will propagate through the climate and hydrological systems, what the feedbacks may be and how best to mitigate or adapt to climate change. Unfortunately, the consequences of global warming on the climate and terrestrial systems, and their extremes, are at best uncertain and at worst exceptionally worrying.

Climate change is often measured by changes in primary climate variables, such as global surface air temperature and precipitation. These variables are first-order drivers of climate impacts, inducing changes in weather extremes, sea-ice thinning and glacier retreat, and thus are appropriate for studying the broader issues in climate change. They are also the best-observed variables over large scales with relatively long historical records. Recent changes in climate may be large enough to be detectable now, although this will depend on the climate variable and our level of confidence in detecting change (Hegerl et al, 2006). Analysis of the instrumental record indicates that recent increases in global annual temperature are anomalous and more rapid, when compared to the long-term record (e.g. Jones et al, 1999; Hansen et al, 1999, Brohan et al, 2006) and model results suggest that this cannot be due to natural variability alone (Jansen et al, 2007). However, changes in global, annual variables may bear little relation to regional and seasonal changes, especially for precipitation (Giorgi and Bi, 2005), and therefore changes in hazards such as droughts and floods that have the most serious impacts on human and environmental welfare.

The apparent increase of drought in recent years across the world has hinted at the influence of global warming and climate change. As discussed in previous chapters, there is ample evidence that drought has been prevalent and ongoing in some regions at the time of writing: northern China, southern South America, eastern Africa, and southeast Australia. Some of these droughts may dissipate as the natural cycles of climate return wetter conditions to these regions. On the other hand, it is a worrying possibility that climate change may force these droughts to persist for longer, with the ultimate danger of a switch to permanently dry conditions. The general consensus from the latest IPCC report is that droughts have become more frequent since the 1970s, and that this is being driven partly by increasing global temperatures (IPCC, 2007). However, it was argued in Chapter 6 that the magnitude of the increase in drought is likely to be overestimated because of simplifications in the underlying data and to some extent the uncertainties in trends in precipitation. Figure 8.1 shows the global time series of soil moisture from the Variable Infiltration Capacity (VIC) hydrological model simulations for 1901–2008 (see Chapter 4), which we argue is the best estimate of large-scale hydrological changes over the 20th and early 21st centuries. These data similarly indicate global drying in recent years, with an accompanying increase in the area that is experiencing drought. In the first order, this is driven by the decrease in precipitation (especially relative to the high levels of precipitation during the strong El Niño of 1997–1998), but may also be related to increasing temperatures and rising evapotranspiration. It appears that the increase in evapotranspiration has slowed in recent years also, despite of the continuation of global warming, which is probably a feedback from the drying soil moisture (Jung et al, 2010). Nevertheless, the drying and increase in drought extent is not particularly outstanding at global scales when compared to the previous 100 years. Plate 8.1 shows that this drying is restricted mainly to the regions mentioned above and is being forced by decreased precipitation. Again this highlights the disparity between measures of change at global and regional scales. It now remains to be seen whether these trends will continue or whether the decadal variability in precipitation that is evident in Figure 8.1 will bring wetter conditions in the next decade.

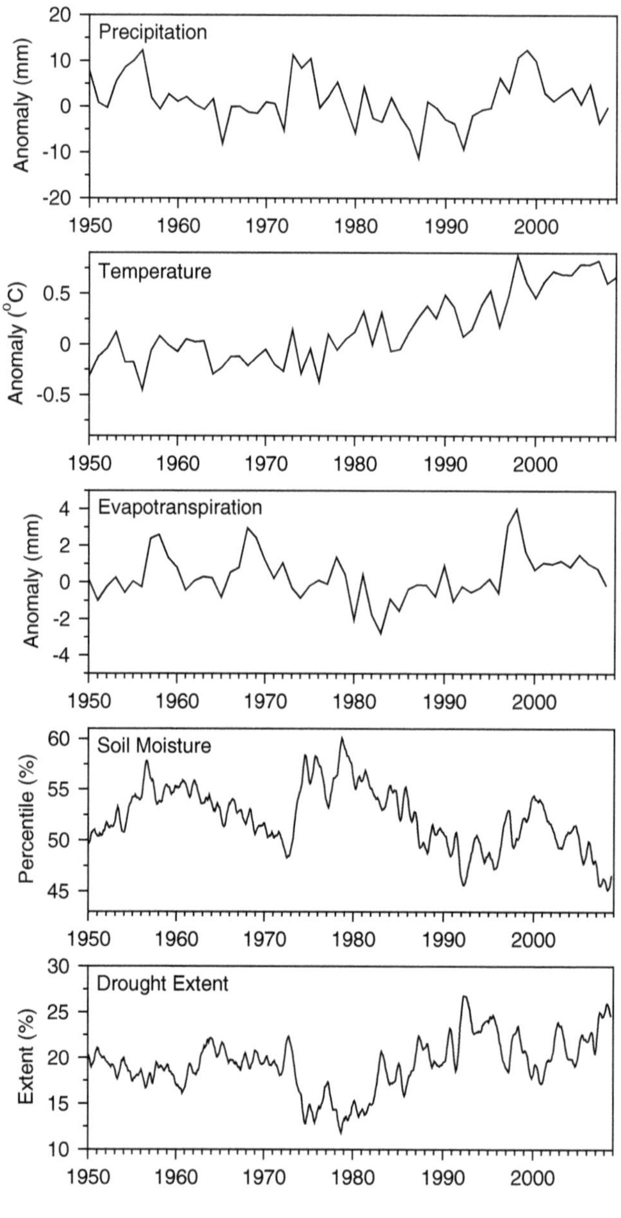

Figure 8.1 *Time series of global average precipitation, surface air temperature, evapotranspiration, soil moisture and drought extent for 1950–2008*

While there appears to be an emerging trend of increasing occurrence and severity of drought, there remain large uncertainties derived from the underlying data (e.g. the sign of trends in precipitation) and modelling, as well as uncertainties over the attribution of change to either natural variability or man-made global warming and climate change. If these trends continue, it may yet be several years to decades before we are able to definitely discern them from natural variability, and attribute global warming as the main cause (Sheffield and Wood, 2008a). What are certain are the escalating impacts of drought in human, environmental and economic terms. This is, in part, due to increasing exposure of economic assets and vulnerability of people to drought, as well as to increasing population and water demand.

Given that global warming has taken place over the 20th century and that this is likely to have been caused in part by human emissions of greenhouse gases (IPCC, 2007), it is of interest to know what the impact of this warming has been on drought. Answering this question may also help to understand the potential for future changes, given that warming is expected to continue. Figure 8.2 shows the difference in trends in regionally averaged soil moisture between the VIC hydrological model simulation and a second identical simulation, but driven by climatological temperature (i.e. without the long-term rising temperature trend) (Sheffield and Wood, 2008b). This shows where the changes are attributable to trends in temperature. For most regions, the differences are less than 1 per cent per year. However, outstanding are regions in northern high latitudes with differences of up to 6 per cent per year (Alaska, Northeast Canada) and 2–3 per cent per year (northern Europe, northern Asia). It is useful to note that these are differences in trends over 11-year windows and generally much higher than the trend over the full 50-year period of the experiment. Of particular interest are the large differences in North American regions during the last 20 years of the record, which indicate that decreasing soil moisture is exaggerated by the increasing trend in air temperature. The temperature trend is particularly pronounced during 1990–2000 (not shown). Even at the global scale a difference is noticeable in the last 20 years. Despite some long-term variability over the full period in most regions, the evidence points towards a temperature effect in recent years that tends to exaggerate or force decreasing trends in soil moisture.

A key question is whether such human-induced global warming and climate change will impact on drought in the long term, and thus affect levels of socio-economic and ecosystem well-being? Although we will never know the full answer to this question until after the fact, reasonable estimates can be made based on our knowledge of how the climate and hydrological systems work. This is currently the topic of intense scientific research, with the goal of understanding and quantifying potential changes in the future and the likelihood of these happening. The main tool in answering these questions is climate modelling. Around the world, centres run models for the 21st century and beyond, driven by plausible scenarios of greenhouse gas emissions to show how climate may change in the future, and the impacts it may have on the land. These models show that drying will be a feature of the 21st century, with some regions experiencing changes that may stretch levels of human adaptation. Although there is great uncertainty in these projections, they suggest that drought will increase in frequency, areal extent and severity, and this will mostly happen in semi-arid regions and those that are dependent on glaciers and snow accumulation.

There is also the influence of natural variability to consider. It has been shown that large-scale and long-term drought events, such as the 1930s Dustbowl drought, are possible due to natural variability alone, and even these pale into insignificance against the naturally occurring events from the palaeoclimate record of the Holocene. When coupled with potential climate change, which may impact regionally and exaggerate the influence of natural variability, the extremes of climate may become more pronounced (Easterling et al, 2000; Palmer and Räisänen, 2002). Of considerable interest is the change in variability and extremes under recent and future global warming, and the potential acceleration of the water cycle, which may alter the occurrence and severity of drought. Climate model studies have shown that variability is likely to increase under plausible future climate scenarios (Wetherald and Manabe, 2002), dependent upon climate sensitivity, with large regional changes in the water cycle. The potential for more droughts of greater severity is a worrisome possibility (Gregory et al, 1997; Wetherald and Manabe, 1999; Wang 2005), compounded by positive feedbacks, such as increases in the frequency of heat waves (Seneviratne et al, 2006b)

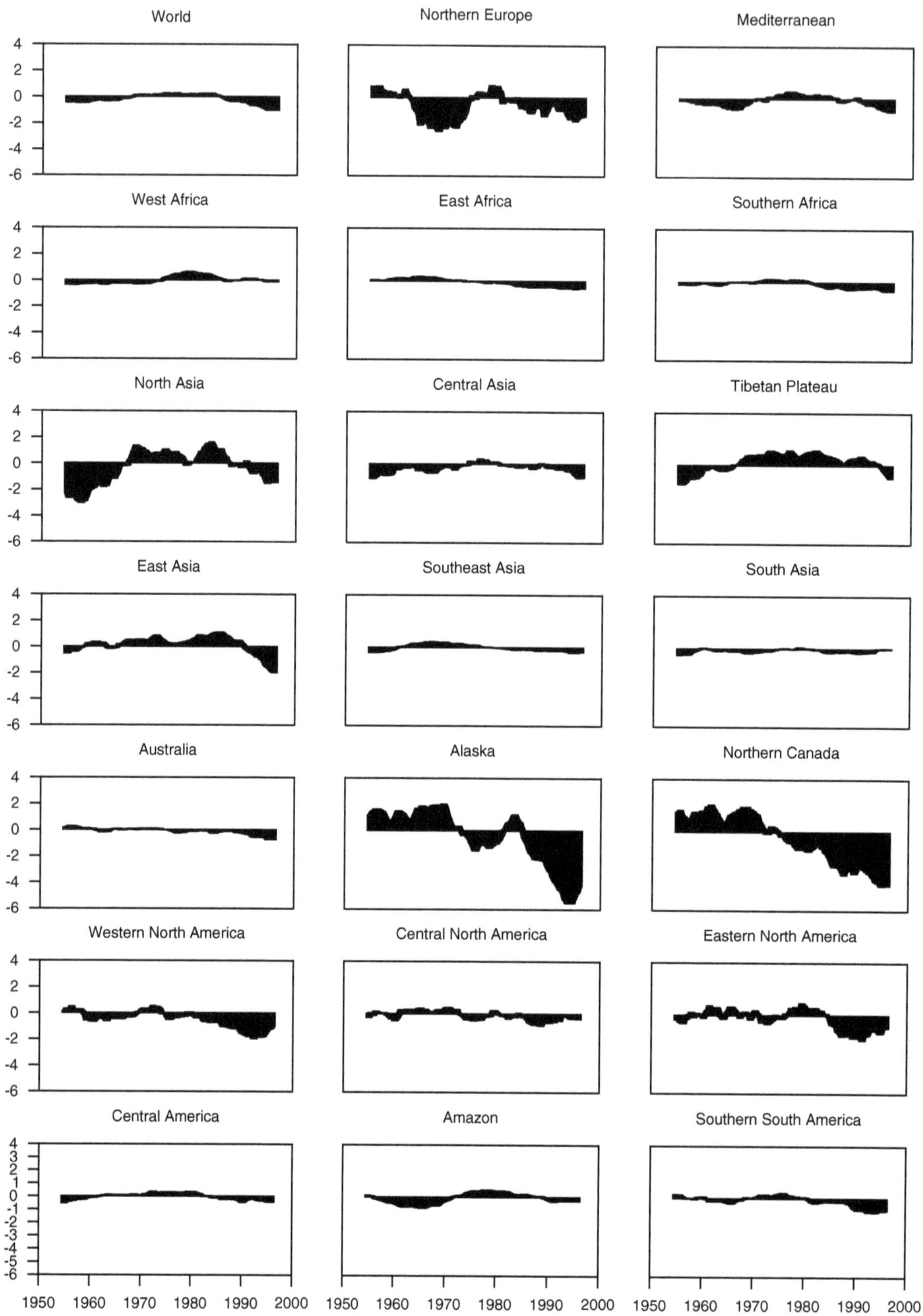

Figure 8.2 *Difference in 11-year moving window trends in soil moisture percentile between the original and climatological temperature-driven VIC hydrological model simulations*

Source: adapted from Sheffield and Wood (2008b)

and decreased carbon uptake by vegetation (Ciais et al, 2005).

This chapter examines the latest climate model projections for the 21st century and explores how and why drought is expected to change. We begin by providing some background to our current understanding of the physical mechanisms of why drought may increase in the future. This includes an overview of some of the recent changes in hydroclimate and drought that have occurred regionally and globally, which appear to be a result of these mechanisms. We analyse potential changes in drought under future global warming, as characterized by persistence in severe soil moisture deficits, from multiple climate models and scenarios. This builds on the analysis of drought in the 20th century based on observation-forced hydrological modelling presented in earlier chapters. Previous studies have assessed predicted changes in mean climate and specifically soil moisture (Wetherald and Manabe, 2002; Wang 2005) that will probably (but not necessarily) induce changes in drought. Here, we take into consideration changes in the full distribution of pertinent variables and not just the mean or some other tendency measure. Furthermore, we directly analyse model output, as opposed to derived products such as the Palmer Drought Severity Index (PDSI), which may suffer from inadequacies that will enhance uncertainty in the results. In terms of models and scenarios analysed, previous studies have focused on single models and/or single scenarios (e.g. Wetherald and Manabe, 2002; Wang 2005; Burke et al, 2006; Kim and Byun, 2009; Feyen and Dankers, 2009). To instil confidence in the robustness of assessment of future change, these must take into account the uncertainty in future climates due to model differences, as well as the diversity of possible emission pathways, as represented by the different scenarios. Nevertheless, uncertainties are inherent in this study, such as biases induced by the specific models and methods of analysis used, and we highlight these where relevant.

After presenting the climate model datasets and methods for assessing future drought, we briefly evaluate how well the models represent drought during the 20th century against observation-based estimates. This is critical to instilling confidence in the models to project future change, as there are many uncertainties in climate modelling, among which is their depiction of current climate. We then show how drought is expected to change over the 21st century for three future climate scenarios, globally and regionally, and where these changes are statistically significant relative to contemporary climate variability, as derived from 20th-century simulations. We also explore the driving mechanisms of these changes and conclude by discussing the potential ramifications of these changes to human and ecological systems.

Mechanisms of Changes in Drought

In theory, global warming should induce more land and ocean evaporation that leads to more precipitation, because of a more humid atmosphere. On the other hand, more land evaporation will dry out soil moisture. This is expected to result in more precipitation in terms of monthly and annual amounts but also to exacerbate year-to-year variability and extreme weather events. There are also many feedbacks and regional variations in these mechanisms that complicate how they manifest across the world. On top of this, changes in atmospheric circulation are possible due to increasing temperatures impacting, for example, on the intensity and extent of tropical convection, monsoonal systems and mid-latitude storm tracks. In cases where precipitation decreases and evapotranspiration, or more precisely evaporative demand, increases there will be increased drought occurrence and severity, as well as possible aridity (see Chapter 3 for an overview of the drivers of drought). However, this may happen through a variety of mechanisms. There can also be indirect impacts, such as higher temperatures causing melting of snowpacks and earlier spring melt, leading to drier conditions earlier in the summer. In general, the different mechanisms can be classified as local thermodynamic or large-scale circulation changes.

Thermodynamic mechanisms

The main local thermodynamic mechanism is the increase in evaporative demand. For this to happen there are a number of different but related factors that can change. As temperatures rise, the capacity of the atmosphere to hold moisture increases, as governed by the Clausius-Clapeyron equation (Held and Soden, 2000), with potential for rising levels of evaporation and/or precipitation (Trenberth, 1999), although these may be limited by other factors such as available energy and aerosol concentration. There may also be a drying of the atmosphere (caused by changes in atmospheric circulation, as discussed later), which can lead to more demand. Increased solar radiation can also increase demand, which may happen for a number of reasons (Wild et al, 2004): increases in natural solar output, decreases in cloud cover and changes in aerosols. Aerosols can reduce solar radiation reaching the ground, but can also reduce precipitation because of the less efficient condensation of water vapour. Feedback mechanisms may also play a role through shifts in the amount of radiation reflected back to the atmosphere because of changes in the albedo of the land surface, for example, through the melting of snow and ice (which are more reflective than bare ground or vegetation), or changes in land cover. Radiation will also increase because of higher long-wave emissions from the atmosphere, due to higher concentrations of greenhouse gases, such as CO_2 and water vapour (the greenhouse effect), although this may be offset by the emissions from the land surface back to the atmosphere. Finally, higher wind speeds may cause increased transport of moist air away from a location and increase the potential for evaporation. Again, this is driven by changes in atmospheric circulation. If increased evaporative demand is not offset by increased precipitation, then drying will take place.

When available moisture is not a restriction, the increase in atmospheric water-holding capacity and evaporative demand will lead to faster rates of evapotranspiration and precipitation. This is termed the acceleration or intensification of the hydrological cycle (Trenberth, 1999; Held and Soden, 2000; Huntington, 2006) and is an expected outcome of global warming. As well as faster cycling of water, this is likely also to result in more extremes, with more intense but less frequent precipitation events. This will increase the frequency of flash flooding and run-off, but decrease soil moisture and increase drought risk as less water infiltrates into the soil. Given the warming that has occurred over the 20th century, it would be expected that intensification of the hydrological cycle would be discernable in the observational record. Huntington (2006) reviews the observational evidence so far for water cycle intensification to date, and concludes that despite some contradictions the overall picture points towards intensification. Detection of intensification is, however, problematic because of the short length of observational records, natural variability at decadal and longer timescales and modification of hydrological systems by human activity.

Observational evidence shows that global temperatures have increased at a significant rate over the 20th and early 21st centuries (Figure 8.1). This is virtually irrefutable, given corroborative evidence from temperature-sensitive systems, such as glaciers and ice sheets, and from independent sources such as satellite measurements of upper tropospheric temperatures and from climate models driven by known greenhouse gas concentrations and other forcings (IPCC, 2007). Current evidence shows that precipitation increased over the 20th century, although there were large decadal variations and there is some uncertainty because of the differences among datasets. However, this general increase in precipitation is also consistent with a rise in the intensity of the hydrological cycle as a result of global warming, which would lead to wetter conditions (Trenberth and Shea, 2005). In the past decade, precipitation has been decreasing in some regions, causing long-term drought (see Figure 8.1), and this is being enhanced by increased demand from the atmosphere, driving higher evapotranspiration. In some regions, such as Australia, southeast South America, the western US and eastern Africa, evapotranspiration and thus the intensity of the hydrological cycle actually appears to be declining because of declining moisture availability (Jung et al, 2010).

Increases in snow water equivalent (the amount of water in the snowpack) have been reported across northern high latitudes (e.g. Ye et al, 1998; Zhang et al, 2001; Karl et al, 1993; Brown, 2000). The most recent updates to 2007 for northern Eurasia indicate that snow amount has increased, but the length of the snow season has decreased over much of the region due to a later onset and earlier melting (Bulygina et al, 2009). This is generally consistent with warming temperatures and increased moisture and precipitation. This is also confirmed by increases in evapotranspiration (Zhang et al, 2009), derived from satellite measurements of vegetation and radiation, and in measures of precipitation recycling derived from reanalysis data over the northern hemisphere (Dirmeyer and Brubaker, 2006).

With increasing precipitation, it is likely that streamflow has increased, although this is not guaranteed. Increases have been reported in the past 50 or more years in the US (Groisman et al, 2001), into the Arctic Ocean (Peterson et al, 2002) and from major global rivers (Labat et al, 2004). However, flows from northern Canada have decreased (Déry and Wood, 2005), driven by falling levels of precipitation, but may also be related to temperature-induced impacts on spring melt and timing, as exemplified by the climatological temperature experiment (Figure 8.2). Changes in streamflow can be induced by other factors, such as human interference in the form of land-use change, reservoir building, irrigation, water withdrawals and inter-basin transfers. Furthermore, glacier retreat causes decreased streamflow (Khromova et al, 2003), whereas the melting of glaciers and permafrost causes increased streamflow (Zhang et al, 2005). For some regions, such as the Mississippi River basin in the US, the increase in precipitation over the past half century has not been matched by the same increase in streamflow, implying an increase in evapotranspiration (Milly and Dunne, 2001) and an acceleration of the hydrological cycle, although this may partly be due to reservoir building and land abandonment. Water vapour has also been reported to increase, confirming theoretical predictions from the Clausius-Clapeyron equation. Willett et al (2008) found changes of 0.11g kg^{-1} $10yr^{-1}$ in humidity over global land from quality-controlled station observations, which is confirmed by measurements from weather balloons (radiosondes) (Ross and Elliot, 2001) and satellite retrievals (Trenberth et al, 2005). Willett et al (2008) also showed a slight decrease of –0.03 per cent $10yr^{-1}$ in relative humidity, with the most recent updates to 2007 (Simmons et al, 2010) indicating relative humidity has decreased even further in some regions not only because of drying, but also because of larger increases in land temperatures relative to ocean temperatures.

Solar radiation at the Earth's surface decreased up to about the 1980s, during a period known as global dimming. This is related to increases in cloud cover (Dai et al, 1999) and aerosols because of increasing industrial activity. Since the 1980s, increased solar radiation (global brightening) has been measured at surface stations (Wild et al, 2005) and from satellites (Pinker et al, 2005), associated with a large-scale reduction of industry since the collapse of the former Soviet Union. The diurnal temperature range (difference between daytime and night-time temperatures) is inversely related to cloud cover, and decreases have been observed in concert with increased cloudiness (Easterling et al, 1997). Global brightening appears to have slowed in the 21st century in many locations (Wild, 2009).

Evapotranspiration is a key variable in the intensification of the hydrological cycle, providing the link between the land and atmospheric branches of the cycle. Measurements of evaporation from pans, which are closely aligned with potential evaporation where water is not a limitation, indicate decreasing trends over the past 50 years (e.g. Peterson et al, 1995). This 'evaporation paradox' is at odds with the idea of hydrological intensification, which would predict more evaporation in a warmer world. Yet this has been explained in terms of a decreasing level of solar radiation linked to a decreasing diurnal temperature range and increasing cloud cover, and may also be connected to the intensification of the hydrological cycle because more actual evaporation increases atmospheric moisture and therefore hinders potential evaporation (Brutsaert and Parlange, 1998; Roderick and Farquhar, 2002).

Remote thermodynamic mechanisms

Thermodynamic mechanisms may also increase the land–sea thermal gradient (which is partly the reason for the existence of the monsoons). This can lead to reduced relative humidity and precipitation over continental interiors (Rind et al, 1990; Manabe et al, 1992; Rowell and Jones, 2006) and thus increased propensity for drought. Evidence for these changes and links to global warming in the observed record are not clear for the Indian monsoon (e.g. Kripalani et al, 2003), either as a direct impact, indirectly via northern hemisphere snow cover and temperatures, or through tropical connectivity. However, it appears that northern hemisphere warming may be playing a more dominant role in modulating monsoon activity than the ENSO (Krishna Kumar et al, 1999) and decreasing land–sea thermal contrasts may be affecting rainfall in the high Asia region of the Himalayas, Nepal, Bangladesh and northern India (Duan et al, 2006). As discussed in Chapter 7, the East Asian monsoon appears to have weakened in the past three decades for various reasons, including changes in land–sea thermal gradients. However, the role played by global warming in effecting these changes remains unclear. In general, a warmer future will probably increase the amount of precipitation associated with the monsoons (Sun and Ding, 2009).

There is also the potential for changes in drought via connections with local and remote soil moisture. Drying of soil moisture in the spring can also influence seasonal precipitation and drying in the summer (Gregory et al, 1997; Wetherald and Manabe, 1995, 1999). Reduced spring soil moisture can be instigated by earlier snowmelt in cold regions, or by increased evaporation driven by a lengthening growing season. With less soil moisture feeding evaporation, there will be less moisture recycled that can be made available for precipitation in the following weeks and into the summer. This mechanism may reduce precipitation locally or remotely because of advection of air masses out of a region due to prevailing winds. As discussed in Chapter 3, the state of the land surface can influence the persistence of droughts by impacting on seasonal precipitation. Observation-based evidence for shifts in the recycling of precipitation is, however, limited by the quality of data required to calculate it. For example, Dirmeyer and Brubaker (2006) used atmospheric reanalysis data, constrained in part by available observations, and found positive trends for much of northern Europe and North America through the expansion of recycling into the springtime, but no change in Asia where the observational density is much lower.

Large-scale circulation changes

At larger scales, global warming can force changes in circulation patterns and even the extent of relatively stable climate regions. There are well-understood thermodynamic arguments (Allen and Ingram, 2002; Held and Soden, 2006) that predict the Hadley circulation will expand, causing storm tracks to move polewards and reducing precipitation over the sub-tropics (Frierson et al, 2007). Model simulations of future climate scenarios have reinforced this prediction, which we will discuss later. In fact, there are multiple lines of evidence that show the tropics have expanded in width since the late 1970s. This evidence includes: changes in the difference between tropical and northern hemispheric ozone concentrations of about 2.5° latitude, as measured from satellites (Hudson et al, 2006). Fu et al (2006) argue that atmospheric temperature measurements from satellite microwave sensors indicate a widening of the tropical belt of about 2° latitude. Data from radiosondes and reanalysis data show that the height of the tropopause (the boundary between the top of the troposphere and the bottom of the stratosphere), which is an indicator of the location of the tropics, has expanded by 5° to 8° since 1979 (Seidel and Randel, 2007). Hu and Fu (2007) examined wind data from model reanalysis for the location of net zero north–south wind and satellite observations of outgoing long-wave, which depends on cloud cover and water vapour and is thus an indicator of subsidence regions at the edges of Hadley cells. Both these measures indicate an expansion of the tropics of 2° to 4.5° latitude. As a result, studies of reanalysis data suggest that northern hemisphere storm tracks have shifted polewards in the last decades (IPCC, 2007), although the underlying data are somewhat uncertain in their temporal fidelity. There is also consistency between atmospheric blocking frequency, a major instigator of drought in northern Europe, and NAO variability (see Chapter 3).

Future projections from climate models indicate that this expansion will continue up to 2° latitude over the 21st century, with a general drying of the sub-tropics (Frierson et al, 2007). This is, however, less than what has been observed in recent decades and there is a large degree of uncertainty between models (Seidel et al, 2007). The predicted changes include the warming of the troposphere and cooling of the stratosphere, the latter of which may induce polewards movement of the jet streams. There is an overall projected shift in the edges of the Hadley cell (Lu et al, 2007), a weakening of tropical circulation patterns, migration of mid-latitude storm tracks towards the poles (Kushner et al, 2001; Yin, 2005; Bengtsson et al, 2006), and increased precipitation in the tropics. Finally, there is expected to be a shift of the subtropical dry zones polewards, and increased atmospheric stability at the margins of tropical convection (Neelin et al, 2003, 2006), decreasing precipitation in the subtropics (Seidel et al, 2007). The impacts that the expansion of the tropics may have on the ground are the drying of the sub-tropics, which are already stressed because of limited water availability and growing populations, such as in southwestern North America (Seager et al, 2007), the Mediterranean, southern South America and Australia. On the other hand, the polewards shift of storm tracks and jet streams will increase precipitation in some regions and cause changes in areas impacted by tropical storms and cyclones.

Climate Models, Emission Scenarios and Future Projections

Climate models

It is impossible to know exactly what might happen to climate and drought in the future. But informed estimates of potential changes can be made, based on our understanding of how the climate system works and possible future trajectories of greenhouse gas and aerosol emissions. This can be achieved using climate models (more formally known as general circulation models or GCMs; see Chapter 4) run for a future time period, forced by greenhouse gas emission scenarios. In the rest of this chapter, we present the latest climate model projections of future changes in drought. GCMs simulate the global coupled land–atmosphere–ocean climate system, typically at a spatial resolution of between 2° and 4° latitude and longitude (although this is decreasing rapidly with the expanding power of new computer technologies). They are forced by plausible future anthropogenic changes in atmospheric composition and simulate the response of the Earth system to these forcings. Although model simulations of current climate are by and large reasonable at global scales and are improving all the time, they are generally poor at representing climate processes at regional and smaller scales, at which the impacts are most important. We will discuss this in more detail later.

Future emissions scenarios

To estimate potential future climate change we use data from the Intergovernmental Panel on Climate Change (IPCC) Fourth Assessment Report (AR4) GCM simulations. The range of future climates predicted by GCMs is large and regionally dependent (NRC, 2003; Giorgi and Bi, 2005) and we therefore use data from multiple GCMs to cover this uncertainty. The data are taken from simulations run for three future scenarios: the Special Report on Emissions Scenarios (SRES) A2, A1B and B1 (Nakićenović et al, 2000). Each scenario represents different mixes of changes in population, economic output, land use, and energy and technology use. The scenarios can be characterized by the maximum atmospheric CO_2 concentration resulting from these changes in human activity. Figure 8.3 shows the evolution of global CO_2 emissions for these scenarios, and the corresponding climate model-predicted changes in temperature. SRES B1 represents relatively slow population growth and an emphasis on environmental protection, with CO_2 concentrations stabilized at 550ppm by the end of the century. SRES A1B describes a future of very rapid economic growth, global population that peaks in mid-century and declines thereafter, and the rapid introduction of new and more efficient technologies with a balance between fossil and non-fossil energy sources. This scenario is characterized by maximum concentrations of 720ppm. SRES A2 describes a heterogeneous world with continuously increasing global population and regionally orientated economic development and fragmented technological change. This is generally regarded as a worst-case scenario that

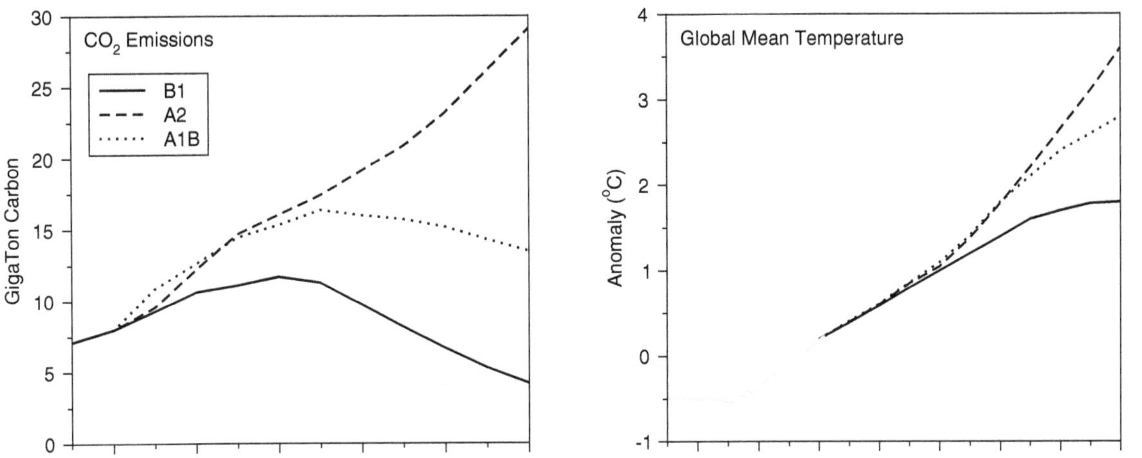

Figure 8.3 *Projected global CO_2 emissions and changes in global surface air temperature for the 21st century for three SRES emissions scenarios*

sees a four- to five-fold increase in CO_2 emissions over 2000–2099, during which CO_2 concentrations increase from about 350 to 850ppm. We additionally use data from the corresponding 20th-century simulations (20C3M), which were also run in a coupled mode, i.e. with a free-running ocean component. The 20C3M simulations are driven by prescribed historical greenhouse gas concentrations, sulphate-aerosol loadings and other forcings since the start of the industrial revolution. We use data for eight models (Table 8.1), which were selected as those that had climate, hydrological and ancillary data available for all these scenarios. Where ensemble simulations for a particular scenario were available we used the first ensemble member. However, it should be noted that single simulations were available more often than not.

Drought Estimation

We characterize future changes in drought in terms of soil moisture deficits, in the same manner as for the 20th-century VIC hydrological model simulations that were forced by observations of precipitation, radiation and other meteorology (see Chapters 6 and 7). Again, we define drought occurrence as an extended period of anomalously low soil moisture and calculate a number of metrics of drought, including the duration, intensity, severity and areal extent (see Chapter 4). Empirical probability distributions are derived for the modelled soil moisture fields for each month from the 20th-century GCM simulations (20C3M), and the current state of drought is characterized by the percentile of the current soil moisture value in relation to the 20C3M period distribution. The data used here are taken from Sheffield and Wood (2008a), which used a soil moisture threshold, $q_0\ (\theta) = 10$ per cent, which is lower than the 20 per cent threshold used previously, and reflects slightly more severe drought conditions that occur only once every 10 years for a particular month on average.

Changes in the occurrence of drought are calculated between the 20th century (20C3M) and future climate (SRES) simulations. The 20C3M simulations are used to represent present-day drought conditions under the contemporary climate. Changes are estimated for the ensemble of climate models to take into account the uncertainty due to model differences. For each model, we calculated the frequency of drought and other statistics in 30-year periods. For the 20C3M simulation this was done for the period 1961–1990, assuming this to be representative of the contemporary climate. For the future climate simulations, we calculated statistics over a 30-year moving window. In order to identify changes in drought, tests were undertaken to see whether future climate drought is statistically different from that of the 20th century. More formally, we test the null hypothesis that the distribution of a drought statistic across all models has

Table 8.1 *Models used in this study, which have monthly soil moisture data available for 20C3M and SRES B1, A1B and A2 scenarios*

Model name	Modelling group	Country
CGCM3.1 (T47)	Canadian Centre for Climate Modelling & Analysis	Canada
GFDL-CM2.1	US Department of Commerce / NOAA / Geophysical Fluid Dynamics Laboratory	USA
GISS-ER	NASA / Goddard Institute for Space Studies	USA
INM-CM3.0	Institute for Numerical Mathematics	Russia
IPSL-CM4	Institut Pierre Simon Laplace	France
MIROC3.2 (medres)	Center for Climate System Research (University of Tokyo), National Institute for Environmental Studies and Frontier Research Center for Global Change (JAMSTEC)	Japan
MRI-CGCM2.3.2	Meteorological Research Institute	Japan
ECHAM5/MPI-OM	Max Planck Institute for Meteorology	Germany

the same mean as the distribution across all models during a future time period. The Student's t-test statistic, which is a measure of the ratio of the difference in means to the combined variance of the two distributions, is calculated at each grid cell and the null hypothesis is rejected for t-test values greater than the critical value at confidence level α (usually taken as 0.05, or 95 per cent). In this case we conclude that a statistically significant change has occurred. Application of this test assumes that the means of the two samples (20th century and future climate) do not deviate substantially from a normal distribution.

Skill of climate models in representing 20th-century drought

Climate models do not provide a perfect representation of the climate system. Even if we knew the exact future trajectory of emissions of greenhouse gases and aerosols, volcanic eruptions and solar variability, the projections of drought would be subject to the biases and simplifications of the climate model. These uncertainties are due to a number of reasons, including the lack of ecosystem and human feedbacks in the modelling, simplified or absent representation of key processes, difficulty in representing extreme events, especially for precipitation, and underestimation of climate variability and persistence. In general, the models are more accurate at global scales, rather than regional scales.

Simulations of current climate are usually performed in one of two ways: uncoupled (where the ocean model component is not run and sea surface temperatures (SSTs) are prescribed instead, or AGCM); and coupled (for which the ocean model is free-running, SSTs are predicted and the ocean interacts with the atmosphere, or AOGCM). Chapter 4 gives more details of the different types of coupled and uncoupled climate models. In uncoupled mode, with specified SSTs, GCMs can generally replicate the precipitation declines associated with some of the major droughts of the 20th century (Giannini et al, 2003; Schubert et al, 2004). However in coupled mode, the models have trouble simulating the variations in tropical SSTs and climate (e.g. Lin et al, 2006; Lin, 2007) that are key to predicting drought on seasonal and longer timescales. Analysis of the latest IPCC simulations from the World Climate Research Programme (WCRP)/Coupled Model Intercomparison Project (CMIP3) database indicates that some models are capable of reasonably reproducing the Sahel precipitation trends (Lau et al, 2006). However, many models cannot do so, and even the best models cannot reproduce the timing and duration of the drought. It was determined that strong coupling between Sahel rainfall and SSTs plus robust coupling of soil moisture and precipitation were deemed necessary to replicate drought conditions in the late 20th century (see Chapter 7). Similarly limited skill is shown for the US, and only in the central US do a small number of models match observed interannual precipitation variability and reproduce the ratio of advected and locally recycled moisture (Ruiz-Barradas and Nigam, 2005). Comparison of soil moisture data from the IPCC AR4 20C3M simulations with field measurements have been carried out by Li et al (2007) over small regions in the northern hemisphere, based on the database of Robock et al (2000). They found that the models simulated the seasonal cycles for available data for the Ukraine, Russia and Illinois adequately, but were poor for Mongolia and China. Importantly, all models failed to replicate observed summer drying in Russia and the Ukraine during the latter part of the 20th century.

Here we are interested in the large-scale, long-term statistics of drought as characterized by persistent soil moisture deficits. A comparison of global and regional averaged drought characteristics from the 20C3M simulations with observation-based estimates is shown in Figure 8.4 (the regions are defined in Figure 6.7). The observation-based statistics are derived from simulations of the 20th century using the VIC hydrological model, which were described in Chapter 4 and analysed in the previous two chapters. The lack of large-scale, long-term monitoring of soil moisture, means that observation-forced modelling such as this provides the best possible estimate of historic soil moisture values at continental to global scales (Maurer et al, 2002; Sheffield et al, 2006; Guo and Dirmeyer, 2006), and therefore the best dataset to evaluate the skill of the climate models.

The climate model values in Figure 8.4 encapsulate the observation-based values, although the spread among models is large, depending on the region and statistic (e.g. Central America for 4- to 6-month drought frequency, southern Africa for more than 12-month frequency, Australia for mean drought duration). For mean drought duration and accumulated severity, the agreement is reasonably good and the observation-based

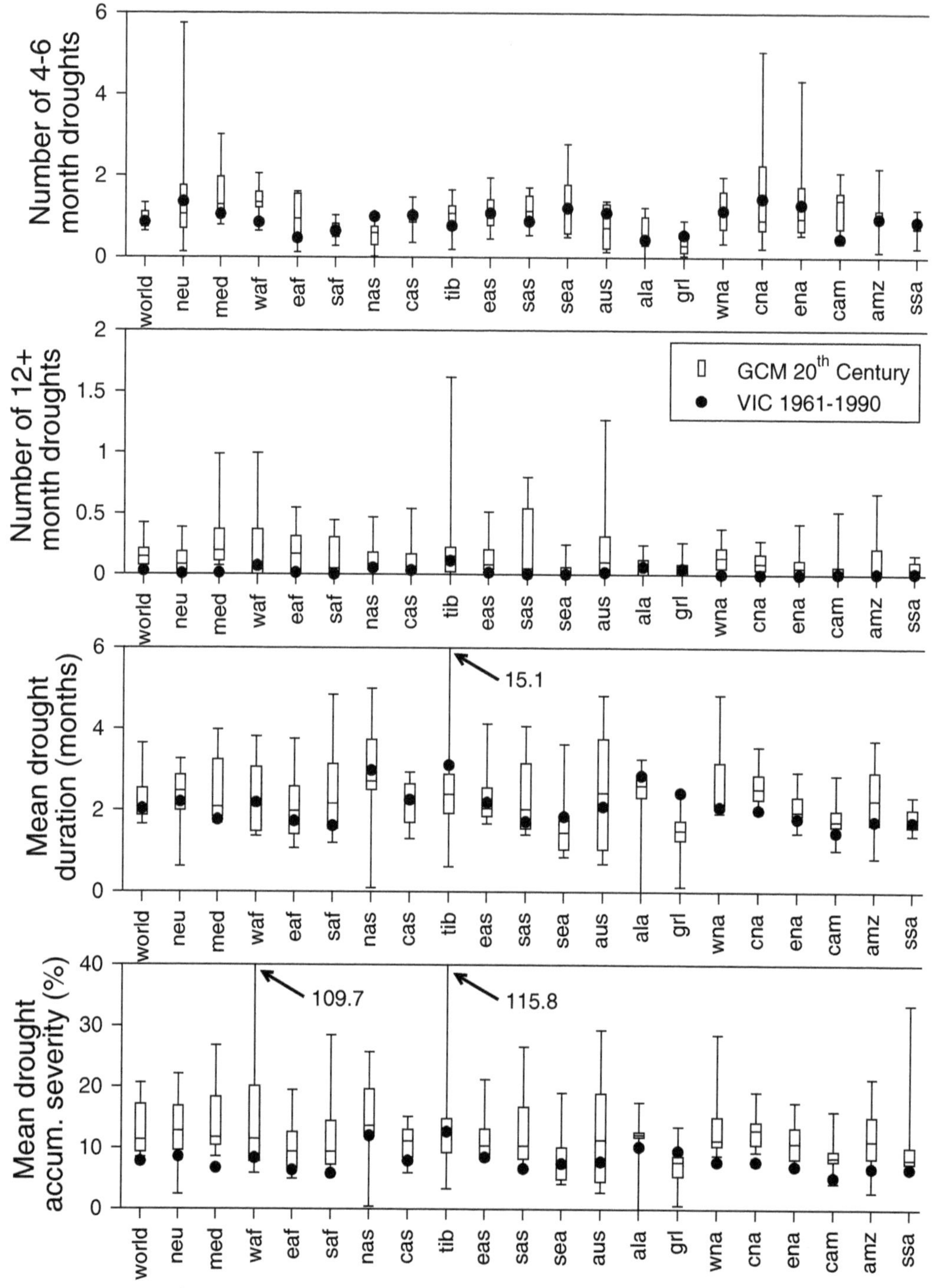

Figure 8.4 *Regionally averaged short- and long-term drought frequencies for 1961–1990 from 20th century (20C3M) IPCC AR4 coupled model simulations (box and whiskers) and the VIC off-line simulation (black dots)*

Note: The box and whiskers represent the interquartile range and the range (maximum and minimum values) of the set of climate models.

Source: reproduced with permission from Sheffield and Wood (2008a)

values fall within the maximum and minimum range for nearly all regions and generally lie within the inter-quartile range (the middle 50 per cent of models) of the model distribution. This is consistent with Seneviratne et al (2006a) who, as part of the Global Land Atmosphere Coupling Experiment (GLACE), showed that GCMs (and particularly AGCMs) were capable of distinguishing between observed regions of high and low soil moisture memory on monthly timescales. For the most part, the climate model values tend to be higher, most notably for more than 12-month frequencies, mean duration and mean accumulated severity, although for the first this is a result, in part, of the zero lower bound in most regions. The general overestimation by the models of long-term drought frequencies and mean drought duration points to an underestimation of climate variability that has been noted previously (e.g. Collins et al, 2002; Hunt, 2006).

Projected Future Changes in Drought

Projections of future global temperature increases are reasonably robust across models (Figure 8.3). As discussed above, the warming has implications for changes in evaporation, water cycle intensification and drought. However, there is much less agreement about changes at regional scales and especially for precipitation, which is difficult to model accurately, and because of uncertainties in the impact of aerosols and feedbacks with the land surface. Projections of precipitation changes are highly dependent on individual scenarios and models (Covey et al, 2003). Meanwhile, regional variability (Giorgi and Bi, 2005) compounds the uncertainty in future changes in drought regime. In fact, for some regions, the sign of the change in precipitation differs among models (Murphy et al, 2004). These changes in terms of precipitation totals and statistics, such as storm intensity and frequency, may affect drought occurrence (Sheffield and Wood, 2008a; Burke and Brown, 2008), as we shall see in the rest of this chapter.

Overview of projected changes in drought

Several earlier studies using climate models have suggested that the interior of the northern hemisphere continents will become drier over the next century, especially in the summer (Rind et al, 1990; Gregory et al, 1997; Wetherald and Manabe, 1995, 1999, 2002). Gregory et al (1997) analysed summer drought over southern Europe and central North America in terms of precipitation and soil moisture from a single integration of the Hadley climate model, forced by 1 per cent per year increasing CO_2 concentrations. They found rises in multivariate drought statistics driven primarily by evaporation through increased temperatures, but also through decreased precipitation in the form of fewer events. Wetherald and Manabe (1999) analysed soil moisture from a single climate model for three scenarios: increasing greenhouse gases, increasing sulphate aerosol and a combination of both scenarios. They similarly found summer dryness and winter wetness in North America and southern Europe, as well as other semi-arid regions, although high latitudes showed increasing wetness. Based on a threshold of one standard deviation, changes in soil moisture did not become detectable for several decades. In summarizing such studies, the third IPCC report (IPCC, 2001) concluded that increased drought risk over these regions was 'likely'.

More recently, Giorgi (2006) analysed a set of IPCC AR4 climate simulations and calculated a climate change index based on changes in precipitation and temperature means and variability. He found major climate change 'hot-spots' in the Mediterranean and northern Europe, followed by high northern latitudes and Central America. Other hot-spots were identified in southern equatorial Africa, the Sahara and eastern North America. Wang (2005) analysed a large set of IPCC AR4 models in terms of consensus changes in precipitation, temperature and soil moisture, and found intermodel consistency in some regions of northern mid- and high-latitudes in predicting summer dryness and winter wetness. In terms of drought specifically, Burke et al (2006) calculated the PDSI from the latest version of the UK Hadley Centre climate model for the SRES A2 scenario and found regionally strong wetting and drying, but a net global drying trend projecting an increase in the area of extreme drought from 1 to 30 per cent by the end of this century. Based on these and other studies, the conclusion of the latest IPCC report (IPCC, 2007) was that 'In a warmer future climate, most Atmosphere-Ocean General Circulation Models project increased summer dryness and winter wetness in most parts of the northern middle and high latitudes. Summer dryness indicates a greater risk of drought.'

Since then a number of studies have looked further at various aspects of future projected drought at global to regional scales, and identified regions of increased and decreased drought risk. For example, Kim and Byun (2009) examined multiple climate model projections of a precipitation-based drought index over Asia for the SRES A1B scenario and found increased drought risk in western Asia and minor increases in parts of the South and East Asian monsoon regions, although decreased drought risk in higher latitudes. For Europe, Lehner et al (2006) and Feyen and Dankers (2009) analysed modelled low flows, driven by downscaled climate model data and high-resolution regional model data, and found streamflow drought to become more severe and persistent over most of Europe by the end of the 21st century, except for northern and northeastern parts of Europe.

The consensus from these and other studies on the hydrological impacts of future warming and the conclusions of the past two IPCC reports point towards a greater risk of drought during the 21st century. Some of the most at-risk regions are in the subtropical regions where the subsiding branch of the Hadley cell directs moisture away to the moist tropics (see Chapter 3). Because of the already dry and variable climate in these regions, and the tenuous ecosystems, the potential for large impacts is high. In the next sections we investigate how drought is expected to change at global and regional scales by analysing soil moisture and drought characteristics (Sheffield and Wood, 2008a) from a suite of the latest climate model simulations carried out under the auspices of the IPCC AR4. We quantify the change in global and regional drought occurrence relative to the present day, as represented by 20th-century simulations, and take into account the uncertainty in regional climate change by using data from multiple models for three future scenarios that represent a range of plausible emission pathways.

Global and regional drought under future climate scenarios

Plate 8.2 shows the global averaged time series of monthly mean soil moisture for the three future scenarios (SRES B1, A1B and A2). Also shown are time series of drought characteristics. The model simulations for these future climates were initialized from the end of the 20C3M simulation and we prepend these data to the future scenario time series for comparison. Globally, soil moisture (*SM*) decreases under all scenarios, with corresponding increases in drought spatial extent (*A*). Changes in the frequency of short-term (from 4 to 6 months in duration, $D_{4\text{-}6}$) and long-term (more than 12 months, D_{12+}) droughts and mean accumulated severity (S_{mean}) are all increasing. In general, the increases in drought statistics are greatest under the higher emissions scenario, A2, and least under the lower emissions scenario, B1. Table 8.2 summarizes these statistics for 30-year periods in the 20th century and at the end of the 21st century. The global spatial extent of drought across all models roughly doubles by 2070–2099 under all scenarios relative to the 20th century. The frequency of $D_{4\text{-}6}$ droughts also doubles and D_{12+} droughts become three times more frequent in relation to the 20th century. However, the spread in model projections, as quantified by the interquartile range is fairly large, even at the beginning of the 21st century, indicating a degree of uncertainty. The range increases by the end of the century in such a way that the lower bound on the projections changes little over the century, although it invariably rises when is does show change.

Figures 8.5–8.7 show the projections on a regional basis, as the multimodel average of $D_{4\text{-}6}$ and D_{12+} frequencies and drought extent (*A*). As for the global results, most regions show increases in drought statistics. There is, however, a large degree of variation between regions and across scenarios, reflecting the variations in climate and differences in the response to climate change. In particular, there are large increases for the Mediterranean, western Africa, Central Asia and Central America regions, most notably for D_{12+} frequencies. The mid-latitude North American regions (western, central and eastern North America) also show increases but with larger variations between scenarios. Changes over high latitudes (Alaska, northeast Canada, northern Europe and northern Asia), eastern mid-latitude Asia (Central Asia and Tibetan Plateau) and regions bordering the Indian Ocean (eastern Africa, southern Asia and Southeast Asia) are relatively small. The frequency of D_{12+} droughts actually decreases in northern Europe. Again, changes under the B1 scenario are the least, and the A1B and A2 results are similar.

Table 8.2. *Summary of multimodel global mean soil moisture and drought characteristics for the 20C3M and future climate SRES B1, A1B and A2 scenarios*

	20C3M (1961–1990)	SRES B1 (2070–2099)	SRES A1B (2070–2099)	SRES A2 (2070–2099)
SM (%)	49.8, 4	45.8, 5.6	44.7, 7.4	44.9, 7.3
A (%)	11.2, 3.1	18.9, 6.2	22.3, 7.2	23, 6.7
$D_{4\text{-}6}$ (30yr^{-1})	1.4, 0.3	2.2, 0.8	2.6, 1	2.8, 1.1
D_{12+} (30yr^{-1})	0.3, 0.2	0.6, 0.6	0.7, 0.7	0.7, 0.6
D_{mean} (months)	3.2, 1.7	4.7, 3.3	5.4, 4.9	5.2, 3.6
S_{mean} (months %)	17.8, 9.3	31.9, 22.2	39.8, 40.7	38.2, 29.3

Note: The first of each pair of values is the multimodel mean and the second is the multimodel interquartile range (IQR).

Source: reproduced with permission from Sheffield and Wood (2008a)

Statistical significance of changes

The statistical significance of changes in drought by the end of the 21st century with respect to the 20th century is shown in Plates 8.3 and 8.4, indicating where the future climate values statistically differ from those of the present climate. In general, $D_{4\text{-}6}$ drought frequency increases globally between the 20th century and the future period (Plate 8.3). Large areas are projected to experience more than five short-term droughts every 30 years (averaged over all models), which are statistically significant at the 95 per cent level. These regions include southern Europe, large parts of the US, the Amazon, southern South America, central West Africa and scattered regions in Asia. Statistically significant changes in D_{12+} drought frequency are less spatially extensive (Plate 8.4), with both increases and decreases scattered across the globe. The small extent of decreases is mostly restricted to high latitudes and generally results in no long-term droughts being recorded. Increases in the southern US, Central America, the Mediterranean and southern Africa, are of the order of two droughts per 30 years. The spatial distribution of changes is fairly robust across scenarios, but with magnitudes greatest under the A2 scenario, and lowest under B1. Figure 8.8 and Table 8.3 summarize the regionally averaged drought characteristics for the 20C3M (1961–1990) simulations and their projected changes for the three future climate scenarios (2070–2099). It is useful to note that rises in $D_{4\text{-}6}$ drought frequency are statistically significant for about half the regions and scenarios, and for most regions in terms of spatial extent. Significant changes in D_{12+} frequencies are restricted to the Mediterranean and central North American regions for all scenarios, and to southern Africa, Central America and southern South America for only the higher carbon emission scenarios. In particular the Mediterranean region shows more than a two, three and three times increase in $D_{4\text{-}6}$ and D_{12+} drought frequency and *A*, respectively, since 1961–1990.

Drivers of changes in drought occurrence

Precipitation and temperature are the primary drivers of drought. Precipitation may change in the future for a number of reasons, including large-scale atmospheric circulation changes, as well as local thermodynamic responses to increasing temperatures and feedbacks with the land surface. Temperature is expected to continue to increase in response to greenhouse gas emissions and may impact drought in direct and indirect ways. Increases in temperature will potentially increase transpiration and direct evaporation from the soil through increased atmospheric demand, whereas increases in precipitation will tend to increase levels of soil moisture. Conversely, decreases in temperature and precipitation will tend to increase and decrease soil moisture respectively. However, the contribution of changes in temperature and precipitation to drought

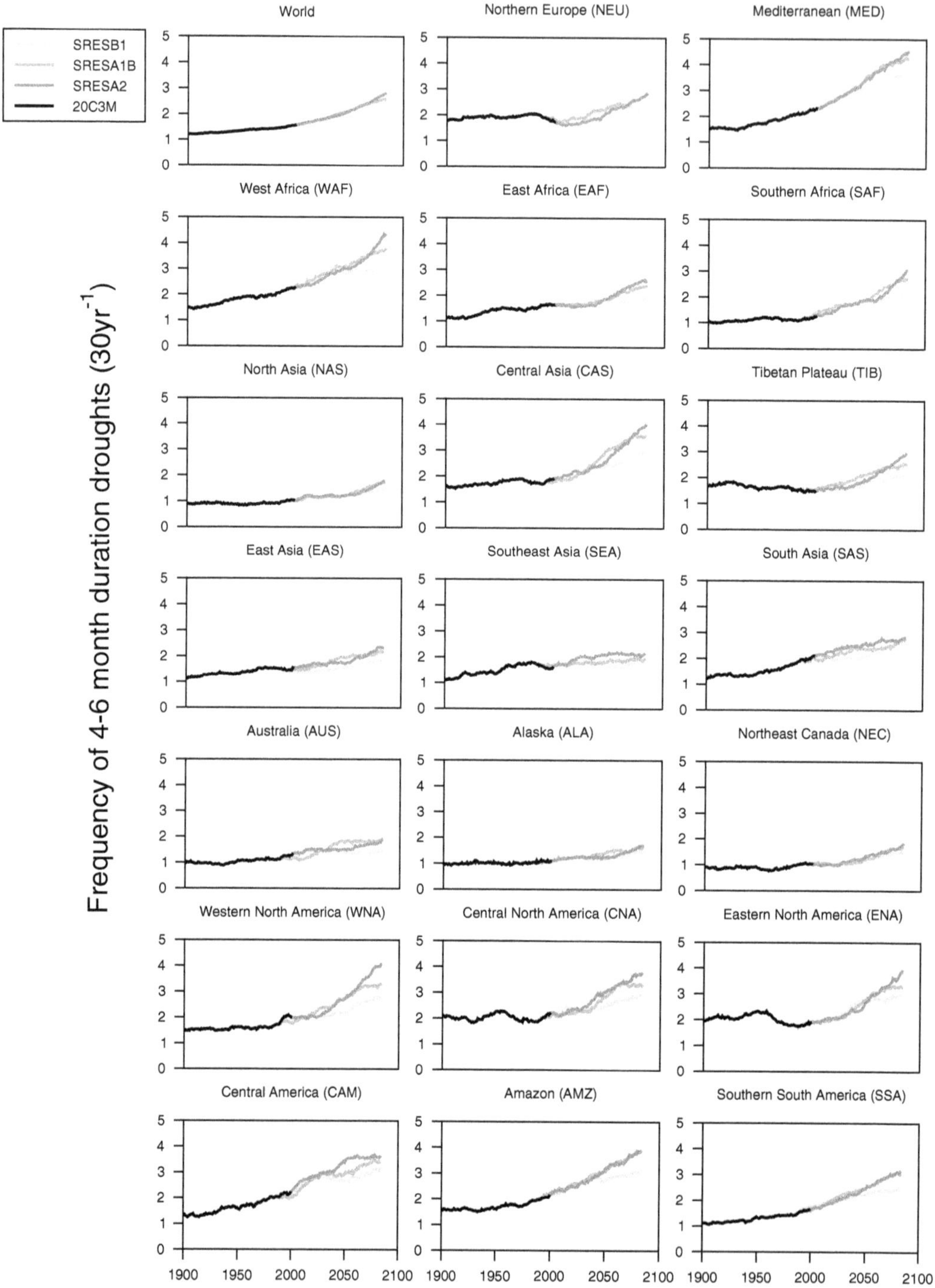

Figure 8.5 *Regional time series of multimodel ensemble mean of frequency of short-term droughts ($D_{4\text{-}6}$, $q_0(\theta)$ = 10 per cent) for the 21st century for the SRES B1, A1B and A2 scenarios*

Note: The values are calculated over a 30-year moving window and then averaged over each region.

Source: reproduced with permission from Sheffield and Wood (2008a)

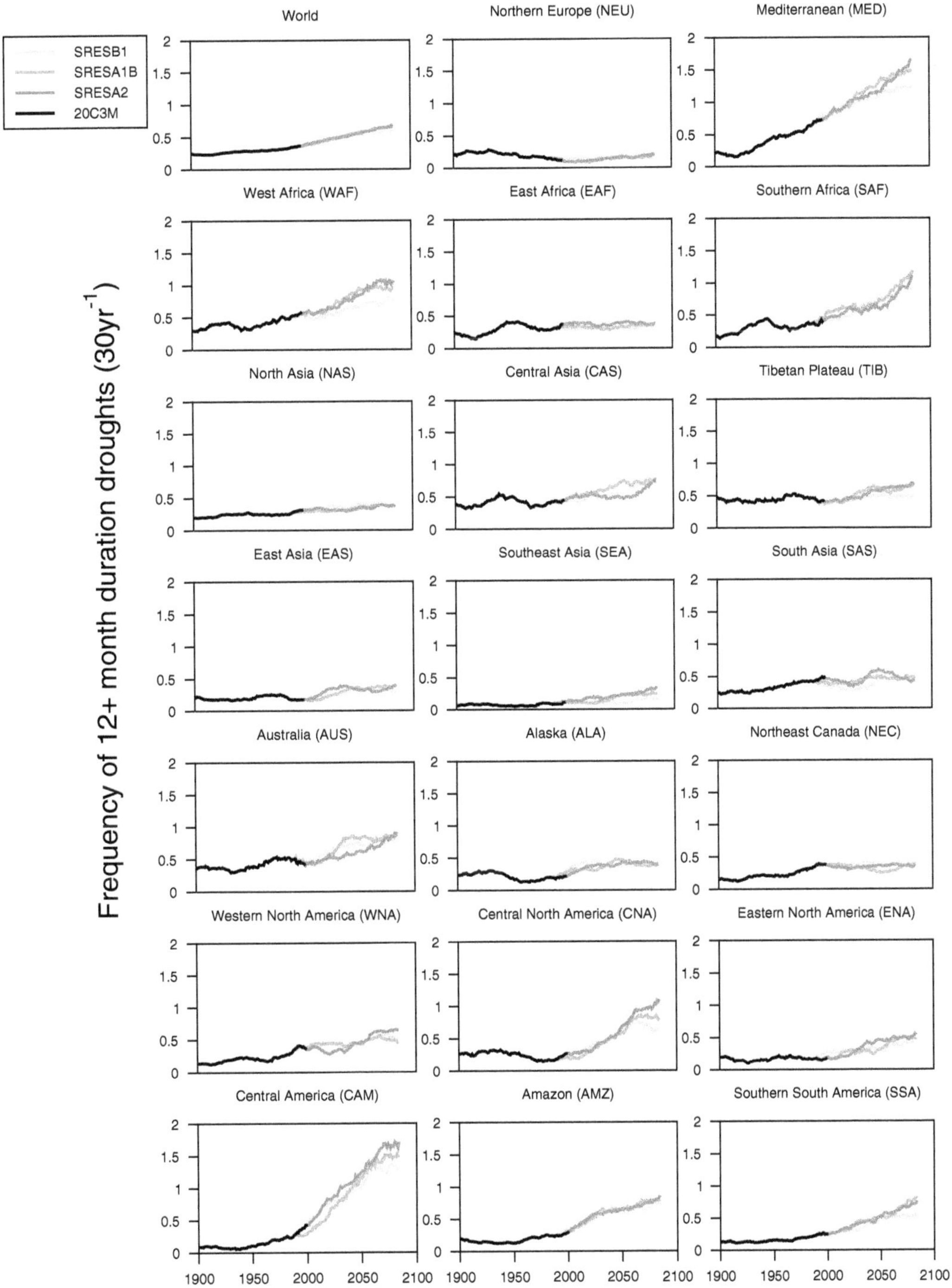

Figure 8.6 *Regional time series of multimodel ensemble mean of frequency of long-term droughts (D_{12+}, $q_0(\theta)$ = 10 per cent) for the 21st century for the SRES B1, A1B and A2 scenarios*

Note: The values are calculated over a 30-year moving window and then averaged over each region.

Source: reproduced with permission from Sheffield and Wood (2008a)

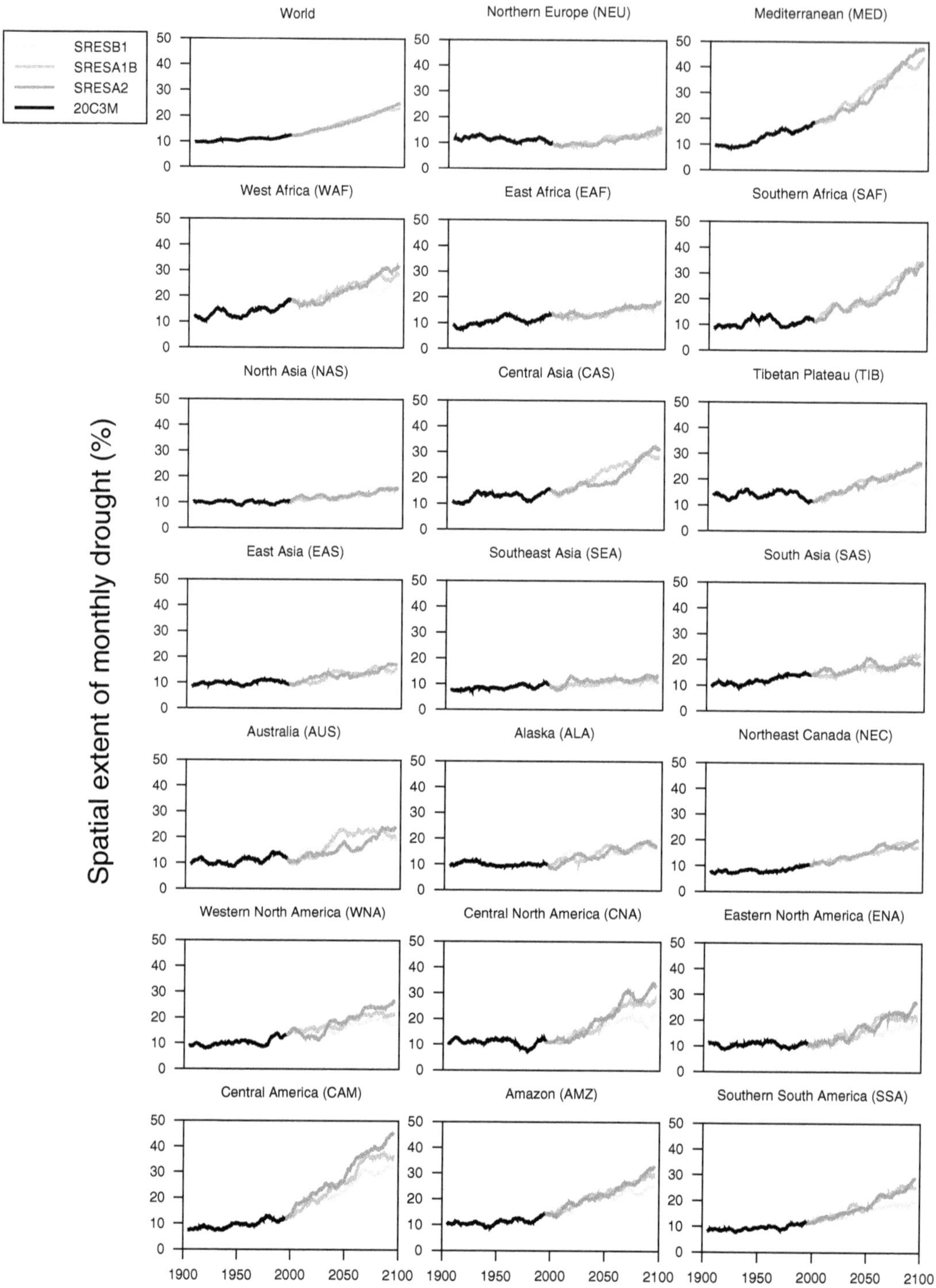

Figure 8.7 *Regional time series of multimodel ensemble mean of spatial extent of monthly drought ($q_0(\theta)$ = 10 per cent) for the 21st century for the SRES B1, A1B and A2 scenarios*

Source: reproduced with permission from Sheffield and Wood (2008a)

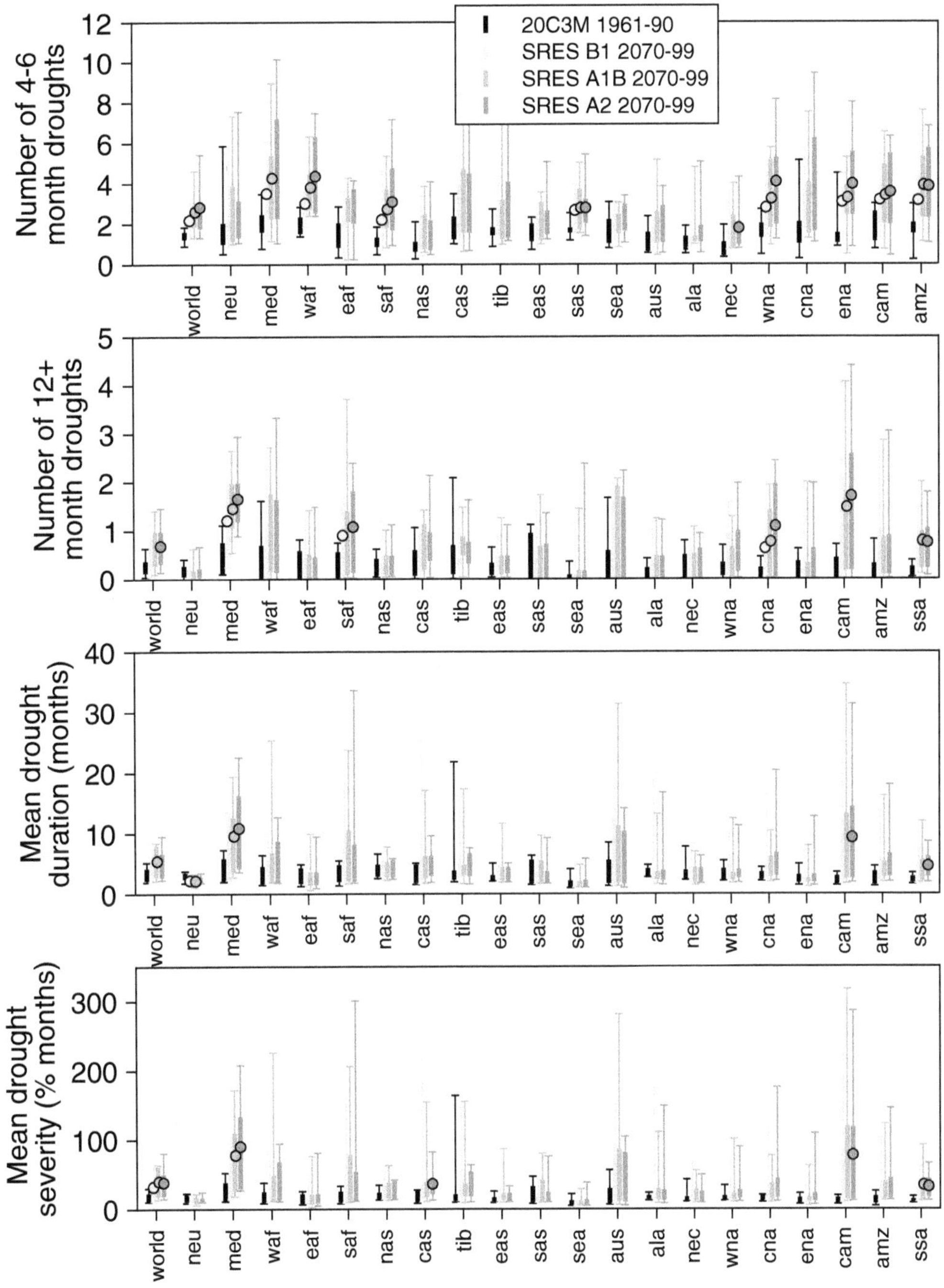

Figure 8.8 *Regionally averaged drought statistics for 1961–1990 (20C3M) and three future climate scenarios for 2070–2099 (SRES B1, A1B, A2)*

Note: The drought statistics are the frequency of four-month to six-month durations, frequency of more than 12-month duration, mean drought duration and mean drought severity. Box and whiskers represent the interquartile range and maximum and minimum values of the set of models. The mean model value is shown as a circle if the future change is statistically significant at the 95 per cent level, relative to 1961–1990 from the 20C3M simulation.

Source: reproduced with permission from Sheffield and Wood (2008a)

Table 8.3. *Regionally averaged, multimodel mean values of short-term ($D_{4\text{-}6}$) and long-term (D_{12+}) drought frequencies and spatial extent of drought (A), for the 20C3M (1961–1990) (20C) and SRES B1, A1B, A2 (2070–2099) scenarios*

	D_{4-6}				D_{12+}				A			
	20C	B1	A1B	A2	20C	B1	A1B	A2	20C	B1	A1B	A2
World	1.4	**2.2**	**2.6**	**2.8**	0.3	0.6	0.7	**0.7**	11.2	**18.9**	**22.3**	**23**
NEU	2	2.3	2.9	2.9	0.2	0.2	0.2	0.2	11.1	12.3	13.3	14.1
MED	2.1	**3.5**	**4.3**	4.6	0.5	**1.2**	**1.5**	**1.7**	15.7	**32.7**	**41.4**	**44.7**
WAF	2	**3**	**3.8**	**4.4**	0.5	0.8	1	1.1	15.1	**23**	**27.9**	**30.2**
EAF	1.5	2	2.4	2.6	0.3	0.3	0.4	0.4	11.2	14.5	16.9	17.1
SAF	1.1	**2.2**	**2.7**	**3.1**	0.3	**0.9**	1.2	**1.1**	10.3	**24.3**	**31.7**	**30.7**
NAS	0.9	1.5	1.7	1.8	0.2	0.4	0.4	0.4	9.9	**13.6**	**15**	**14.9**
CAS	1.8	2.9	3.6	4	0.4	0.7	0.7	0.8	12.8	**25**	**27.9**	**29.2**
TIB	1.7	2.1	2.6	3	0.5	0.5	0.7	0.7	14.8	19.4	24.1	24.4
EAS	1.5	1.8	2.2	2.3	0.2	0.3	0.4	0.4	10.5	13.1	**15.5**	**15.7**
SAS	1.7	**2.7**	**2.8**	**2.8**	0.4	0.5	0.5	0.4	14	17.9	20.4	18.5
SEA	1.7	1.9	2	2.2	0.1	0.2	0.2	0.3	9	10.4	11.6	12.6
AUS	1.2	1.5	1.9	1.8	0.5	0.8	0.9	0.9	12.2	20.2	22.1	22.4
ALA	1	1.4	1.6	1.7	0.1	0.4	0.4	0.4	9.8	15.5	17.4	17.7
NEC	0.9	1.5	1.7	**1.8**	0.3	0.4	0.4	0.4	8.8	**17.3**	**18.3**	**18.2**
WNA	1.6	**2.8**	**3.2**	**4.1**	0.2	0.5	0.4	0.7	10.8	**18.8**	**21.3**	**24.8**
CNA	2	3	3.2	3.8	0.2	**0.6**	**0.8**	**1.1**	9.7	**20.3**	**26.3**	**30.5**
ENA	1.8	**3.1**	**3.3**	**4**	0.2	0.5	0.5	0.5	10.4	**18.2**	21.1	24.1
CAM	1.9	**3.2**	**3.4**	**3.7**	0.2	1.3	**1.5**	**1.7**	10.8	**30.6**	**36**	**41.1**
AMZ	1.7	**3.1**	**3.9**	**3.8**	0.2	0.7	0.8	0.9	11.8	**23.8**	**28**	**29**
SSA	1.4	**2.5**	**3.1**	**3.2**	0.2	0.5	**0.8**	**0.8**	10.1	**19**	**24.8**	**24.9**

Note: Future values in bold are statistically significant, relative to 1961–1990 of the 20C3M simulation, at the 0.05 level with DF=14.

Source: reproduced with permission from Sheffield and Wood (2008a)

occurrence under future climate scenarios is not straightforward because of the complex interactions between these forcing variables and hydrological processes at the Earth's surface. This is complicated further by changes in mean precipitation versus changes in precipitation frequency and intensity, and the seasonality of changes, especially in snow-dominated regions. Apart from precipitation, the key linkage between climate change and changes in soil moisture is evaporation. Although temperature increases will tend to increase potential evaporation, actual evaporation may be enhanced or diminished by precipitation increases and decreases respectively.

Figure 8.9 shows that projected changes in regional drought by the end of the 21st century predominantly occur in the warm season. The figure shows the seasonal cycle of frequency of soil moisture deficits for the 20C3M and the three future climate scenarios. As the threshold for drought is 10 per cent, then by definition the frequencies for the 20th century (20C3M) are 10 per cent for all months. The frequencies for the future SRES scenarios are progressively larger, with 20C3M < B1 < A1B < A2, as noted before. The warm season increases are nearly always greater than those in cooler seasons. For southern hemisphere regions, such as southern Africa, the Amazon, southern South America and Australia, the largest changes are predominantly in the austral spring (September–November). In cooler regions (northern Asia, Alaska, northeast Canada), increased frequencies tend to be concentrated in the warm season (May–August) and increases in the cool season, especially for Alaska and northern Asia, are minimal. This implies that in the warm season either precipitation is decreasing, or evaporation is increasing, or a combination of both. In the cool season, evaporation is generally small outside the tropics, and thus any changes in soil moisture are likely to be driven by changes in precipitation.

Model projections of future climates show increases in temperature globally and for all regions, whereas changes in precipitation are regional and often scenario-dependent. Changes in evaporation are generally projected to increase globally (Meehl et al, 2007), driven by higher temperatures and in some regions by increased levels of precipitation. However, changes in soil moisture do not always follow the sign of precipitation changes, but are also guided by changes in evaporation. This is shown in more detail for three example regions in Figure 8.10, chosen for the diversity they show in terms of future drought scenarios. The Mediterranean region has some of the largest increases in drought frequency, which can be directly attributed to reductions in precipitation totals, especially in the warm season, accompanied by reductions in run-off and soil moisture (Figure 8.10). Changes in evaporation are limited to small decreases in the summer, driven by reduced precipitation that is not offset by increased atmospheric demand from higher temperatures in an already water-limited environment. In the central North American region, drought occurrence is expected to increase, but changes in other variables are mixed. Precipitation is projected to increase in the spring, but decrease in the summer. Coupled with increased temperatures, this leads to increased evaporation in the spring, but little change in the summer as decreased precipitation probably outweighs any increases in potential evaporation.

Changes in drought in the northern Asia region are modest but are focused on the summertime and are representative of other high-latitude regions. The hydrology in these regions is dominated by snow and ice processes which, given increases in air temperature in these regions, can explain the seasonally disproportionate changes in soil moisture deficits. Warmer temperatures tend to reduce the snow cover through melting the existing snowpack and increasing the ratio of rainfall to snowfall. Precipitation is also predicted to increase in the winter (and throughout the rest of the year). Rain falling on the existing snowpack will also increase the rate of melting. This will in turn cause the spring melt to occur earlier in the year. These changes force wetter soil moisture conditions in the winter and spring, although this is offset somewhat by increased evapotranspiration due to higher temperatures. During the summer, soil moisture is reduced, despite increased precipitation, because of earlier spring melt that is compounded by evapotranspiration rises due to higher temperatures. The frequency of soil moisture deficits therefore increases disproportionately in the summertime, in comparison to the wintertime. This mechanism is also seen in the 20th-century model experiment that was driven by climatological air temperature, which implied that increasing temperatures have affected snowmelt and timing, and thus soil moisture in the spring and summer.

Potential Impacts of Future Drought

Despite the uncertainties in the climate model projections, from the models and the emissions scenarios, there are some robust changes in drought projected for the 21st century as a response to already existing and continued increases in greenhouse gas concentrations. In particular, much of the US, Central America, Brazil, Uruguay and northern Argentina, central and southern Africa, the Mediterranean, parts of Southeast Asia and Australia are likely to be affected by more frequent, longer and severe drought. These regions have been identified across several other studies

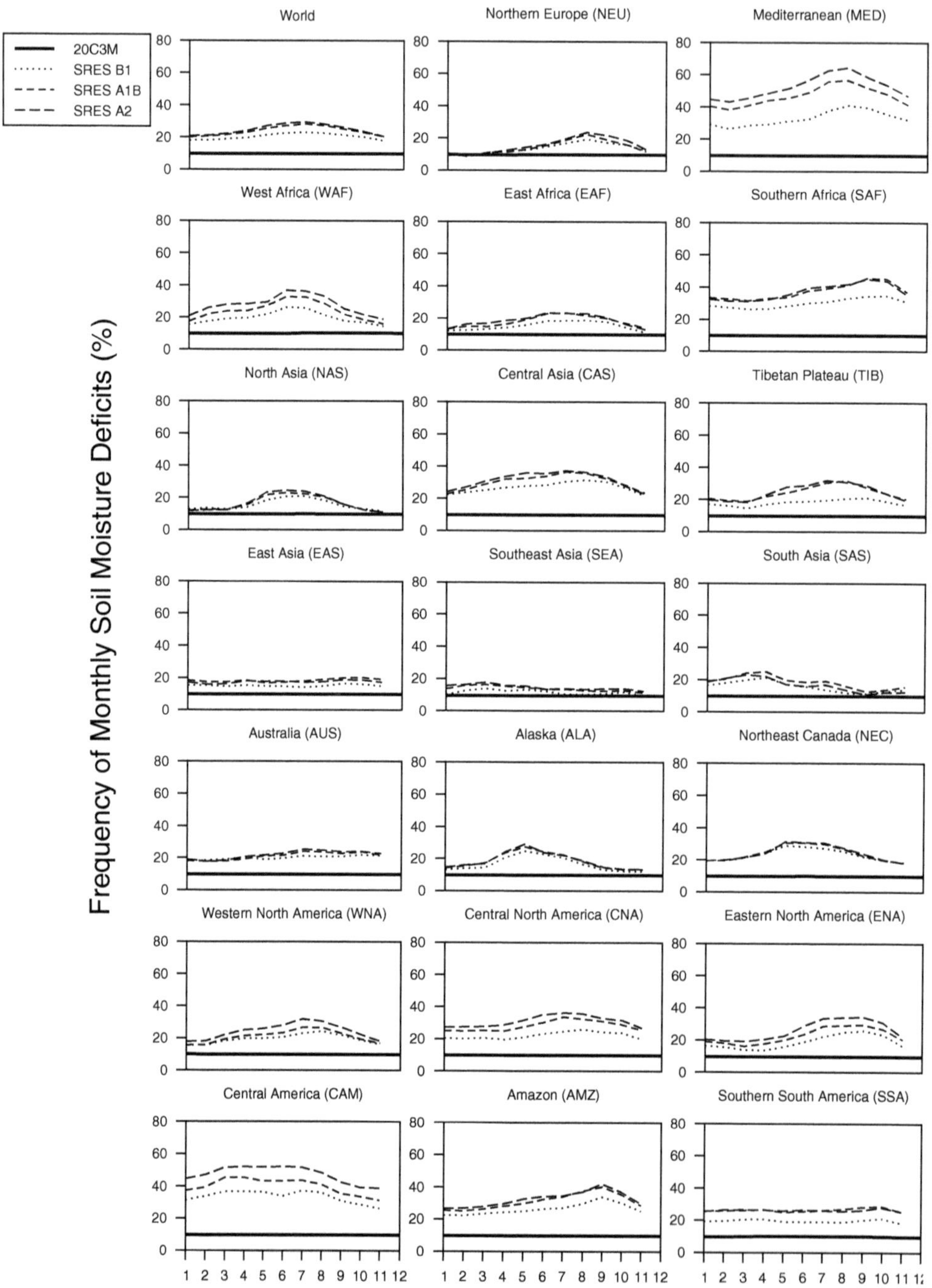

Figure 8.9 *Multimodel regional average of frequency of monthly soil moisture deficits for the 20C3M, SRES B1, A1B and A2 scenarios, relative to the tenth percentile of the 20C3M simulations*

Note: For each model, the number of months with soil moisture deficits below the tenth percentile soil moisture value of the 20C3M simulation are summed up and averaged across all models and the region, and converted into a percentage. The values are calculated separately for each month. The 20C3M data are calculated for 1961–1990 and by definition equal to 10 per cent for all months. The future climate SRES scenarios are calculated for 2070–2099.

Source: reproduced with permission from Sheffield and Wood (2008a)

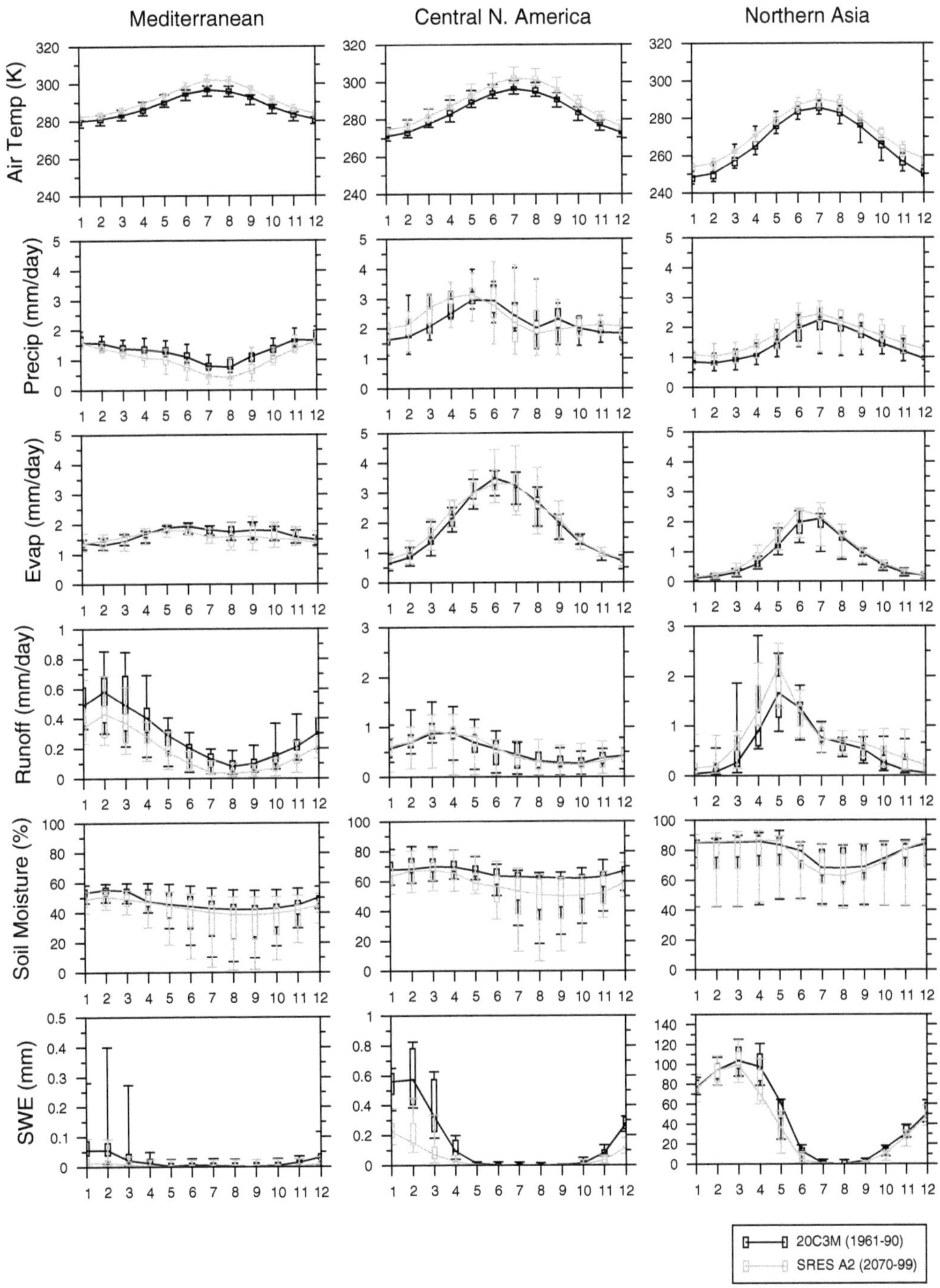

Figure 8.10 *Mean seasonal cycle of air temperature and the main components of the terrestrial hydrological cycle for the Mediterranean, central North American and northern Asia regions for the 20C3M (1961–1990) and SRES A2 (2070–2099) scenarios*

Note: The boxes and whiskers represent the interquartile range across the models.

Source: reproduced with permission from Sheffield and Wood (2008a)

in terms of increased precipitation variability, decreased run-off, reduced soil moisture and increased drought risk (Wang, 2005; Burke et al, 2006; Kundzewicz et al, 2007; Feyen and Dankers, 2009; Kim and Byun, 2009). Most of these regions are currently semi-arid. Whether there will be a transition to a more persistent level of increased drought activity or even aridity depends on what happens to the natural underlying climate variability, and in particular the modes of climate variation on decadal and shorter timescales, such as the ENSO and the PDO. These may act to amplify or counteract the greenhouse gas-induced changes. It also depends on how vegetation responds to the risk of increased droughts that can feed back to the climate and exacerbate drought locally and remotely (Cox et al, 2008). The latest set of climate projections from the IPCC AR4 did not include explicit modelling of vegetation feedbacks, although this will be a key inclusion in the IPCC Fifth Assessment simulations, as part of the move towards Earth System Modelling that includes modelling of vegetation dynamics and biogeochemical processes.

Physically, the projections indicate reduced run-off and streamflow, and more frequent low flows and potentially intermittency of flows (Feyen and Dankers, 2009). Not only are changes expected to the magnitude and extremes of flows, changes in the seasonal distribution of water will present challenges to water managers. The changes in drought are expected to be a result of changes in the full-season hydrological cycle that will manifest themselves in changes in the warm-season low-flow regime. For example, in snow-dominated regions, earlier snowmelt caused by higher temperatures and more rain versus snow will bring earlier high streamflow events, as noted in the US in the Northeast by Hodgkins et al (2003) and in the West by Mote et al (2003). Furthermore, it will create conditions for earlier, longer, more intense low-flow events later in the year. In the summer and autumn, increased evaporation will reduce soil moisture and increase short-term drought, thus extending and amplifying low flows. These changes will be compounded by the direct temperature impact on glaciers. Given that one sixth of the world's population depends upon snowpack and glacier-fed river systems (Barnett et al, 2005), this may play an important role in future drought and water availability, especially in Asia and the rivers fed by the Himalayas (Immerzeel et al, 2010). Sea-level rise may also play a role, through salt-water intrusion into coastal aquifers and estuaries, compounded by streamflow drought, with consequences for freshwater supply. Thus low- and mid-latitude arid and semi-arid regions, coastal areas, snow-fed and glacier-fed regions are likely to be most affected physically by drought.

The human impacts of future drought are likely to be experienced most in regions of large populations, and where population is predicted to increase, leading to higher pressure on water resources. The global population is growing at about 80 million people per year, which is equivalent to an increase in freshwater demand of approximately 64 billion cubic metres a year. Recent droughts in the southeast US have ravaged water supplies, especially in Georgia for Lake Lanier, and this depletion has been exacerbated by a large population expansion. This has resulted in interstate water rights conflicts and ecological impacts on endangered species, such as freshwater mussels that require minimum stream-level habitat. The ongoing drought in the US West, partly driven by global warming impacts on mountain snowpack (Maurer et al, 2007), has reduced reservoir water levels, increased fire frequency and prompted the first state drought emergency in California in decades, while population increases dramatically (e.g. the Colorado basin population rose by 10 per cent from 2000 to 2006). As well as increased demand for water from burgeoning populations, increased living standards also drive per capita water use upwards, including shifts to meat-based diets, non-food agriculture (e.g. fibre and fuel), and direct use for energy generation. Some regions, such as sub-Saharan Africa, southern Asia and the Middle East, which are largely characterized by low per capita income, widespread poverty and rapid urbanization, are already experiencing water stress due to climate and population drivers (WWAP, 2009). In the future it is estimated that of the 3 billion extra people expected by 2050, 90 per cent of them will inhabit developing countries, which are already lacking in basic water amenities. The capacity for populations from developing regions to adapt to climate change impacts on drought and other natural hazards is therefore a critical challenge, and changes in water availability are likely to affect migration patterns.

Agriculture is by far the largest consumer of water globally, using 70 per cent of total water resources, and is thus expected to be impacted by increased drought risk, either directly to rain-fed agriculture or indirectly to irrigated agriculture. Agriculture also contributes to global warming through emissions of methane and nitrous oxide, and through land clearing and burning. It is expected that overall global food production will not be adversely affected, with concentration of cereal production in the northern hemisphere. However, increased droughts and floods may impact locally, especially in subsistence agricultural communities in semi-arid regions of the lower latitudes (WWAP, 2009), which rely on irrigation. Irrigated agriculture will suffer because of reduced flows, as well as increased flooding and rising sea levels, and especially in glacier-fed and snow-fed lowland regions, such as the Ganges and Punjab.

The impact that increased drought will have on water availability, agriculture and riverine ecological function is worrisome (Kabat et al, 2003), and is likely to be compounded by falling water quality as a result of a growing population. These shifts will probably be at the expense of natural ecosystems, and thus trying to meet the challenge of sustainable development, given the potential changes in drought and water resources, is generally not tenable. Competing natural and anthropogenic activities may exacerbate this, with vegetation increasing usage and cities heightening in-stream withdrawals. For example, the Ipswich River, in the US Northeast, which was named one of the country's most endangered rivers by the International Rivers Network, runs dry in the summer due to excessive withdrawals (Zarriello and Ries, 2000). Future projections for this river indicate that flows may drop below minimum thresholds for ecological habitat maintenance for three weeks or longer. This will bring adverse impacts to stream ecology and added pressure on competition for water resources, especially when coupled with rising stream temperatures, as seen in some Northeast rivers (Huntington et al, 2003).

References

Allen, M. R. and Ingram, W. J. (2002) 'Constraints on future changes in climate and the hydrological cycle', in *Nature*, vol 419, pp224–232

Barnett, T. P., Adam, J. C. and Lettenmaier, D. P. (2005) 'Potential impacts of a warming climate on water availability in snow-dominated regions', in *Nature*, vol 438, pp303–309

Bengtsson, L., Hodges, K. I. and Roeckner, E. (2006) 'Storm tracks and climate change', in *Journal of Climate*, vol 19, pp3518–3543

Brohan, P., Kennedy, J. J., Harris, I., Tett, S. F. B. and Jones, P. D. (2006) 'Uncertainty estimates in regional and global observed temperature changes: A new dataset from 1850,' in *Journal of Geophysical Research*, vol 111, D12106, p21, doi:10.1029/2005JD006548

Brown, R. D. (2000) 'Northern Hemisphere snow cover variability and change', in *Journal of Climate*, vol 13, pp2339–2355

Brutsaert, W. and Parlange, M. B. (1998) 'Hydrological cycle explains the evaporation paradox', in *Nature*, vol 396, no 30, pp1327–1332

Bulygina, O. N., Razuvaev, V. N. and Korshunova, N. N. (2009) 'Changes in snow cover over Northern Eurasia in the last few decades', in *Environmental Research Letters*, vol 4, 045026, p6, doi:10.1088/1748-9326/4/4/045026

Burke, E. J. and Brown, S. J. (2008) 'Evaluating uncertainties in the projection of future drought', in *Journal of Hydrometeorology*, vol 9, pp292–299

Burke, E. J., Brown, S. J. and Christidis, N. (2006) 'Modelling the recent evolution of global drought and projections for the twenty-first century with the Hadley Centre climate model', in *Journal of Hydrometeorology*, vol 7, pp1113–1125

Ciais, P., Reichstein, M., Viovy, N., Granier, A., Ogee, J., Allard, V., Aubinet, M., Buchmann, N., Bernhofer, C., Carrara, A., Chevallier, F., De Noblet, N., Friend, A. D., Friedlingstein, P., Grunwald, T., Heinesch, B., Keronen, P., Knohl, A., Krinner, G., Loustau, D., Manca, G., Matteucci, G., Miglietta, F., Ourcival, J. M., Papale, D., Pilegaard, K., Rambal, S., Seufert, G., Soussana, J. F., Sanz, M. J., Schulze, E. D., Vesala, T., Valentini, R. (2005) 'Europe-wide reduction in primary productivity caused by the heat and drought in 2003', in *Nature*, vol 437, pp529–533

Collins, M., Osborn, T. J., Tett, S. F. B., Briffa, K. R. and Schweingruber, F. H. (2002) 'A comparison of the variability of a climate model with paleotemperature estimates from a network of tree-ring densities', in *Journal of Climate*, vol 15, pp1497–1515

Covey, C., AchutaRao, K. M., Cubasch, U., Jones, P., Lambert, S. J., Mann, M. E., Phillips, T. J. and Taylor, K. E. (2003) 'An overview of results from the Coupled Model Intercomparison Project', in *Global Planet Change*, vol 37, pp103–133

Cox, P. M., Harris, P. P., Huntingford, C., Betts, R. A., Collins, M., Jones, C. D., Jupp, T. E., Marengo, J. A. and Nobre, C. A. (2008) 'Increasing risk of Amazonian drought due to decreasing aerosol pollution', in *Nature*, vol 453, pp212–215, doi:10.1038/nature06960

Dai, A., Trenberth, K. E. and Karl, T. R. (1999). 'Effects of clouds, soil moisture, precipitation, and water vapor on diurnal temperature range', in *Journal of Climate*, vol 12, pp2451–2473

Déry, S. J. and Wood, E. F. (2005) 'Decreasing river discharge in northern Canada', in *Geophysical Research Letters*, vol 32, L10401, p4, doi:10.1029/2005GL022845

Dirmeyer, P. A. and Brubaker, K. L. (2006) 'Evidence for trends in the Northern Hemisphere water cycle', in *Geophysical Research Letters*, vol 33, L14712, p4, doi:10.1029/2006GL026359

Duan, K., Yao, T. and Thompson, L. G. (2006) 'Response of monsoon precipitation in the Himalayas to global warming', in *Journal of Geophysical Research*, vol 111, D19110, p8, doi:10.1029/2006JD007084

Easterling, D. R., Horton, B., Jones, P. D., Peterson, T. C., Karl, T. R., Parker, D. E., Salinger, M. J., Razuvayev, V., Plummer, N., Jameson, P. and Folland, C. K. (1997) 'Maximum and minimum temperature trends for the globe', in *Science*, vol 277, pp364–367

Easterling, D. R., Meehl, G. A., Parmesan, C., Changnon, S. A., Karl, T. R. and Mearns, L. O. (2000) 'Climate extremes: Observations, modeling, and impacts', in *Science*, vol 289 no 5487, pp2068–2074, doi:10.1126/science.289.5487.2068

Feyen, L. and Dankers, R. (2009) 'Impact of global warming on streamflow drought in Europe', in *Journal of Geophysical Research-Atmosphere*, vol 114, D17116, p17, doi:17110.11029/12008JD011438

Frierson, D. M. W., Lu, J. and Chen, G. (2007) 'Width of the Hadley cell in simple and comprehensive general circulation models', in *Geophysical Research Letters*, vol 34, L18804, p5, doi:10.1029/2007GL031115

Fu, Q., Johanson, C. M., Wallace, J. M. and Reichler, T. (2006) 'Enhanced mid-latitude tropospheric warming in satellite measurements', in *Science*, vol 312, p1179

Giannini, A., Saravanan, R. and Chang, P. (2003) 'Oceanic forcing of Sahel rainfall on interannual to interdecadal time scales', in *Science*, vol 302, pp1027–1030

Giorgi, F. (2006) 'Climate change hot-spots', in *Geophysical Research Letters*, vol 33, p4

Giorgi, F. and Bi, X. (2005) 'Updated regional precipitation and temperature changes for the 21st century from ensembles of recent AOGCM simulations', in *Geophysical Research Letters*, vol 32, no 21, p4

Gregory, J. M., Mitchell, J. F. B. and Brady, A. J. (1997) 'Summer drought in northern mid-latitudes in a time-dependent CO_2 climate experiment', in *Journal of Climate*, vol 10, pp662–686

Groisman, P. Y., Knight, P. W. and Karl, T. R. (2001) 'Heavy precipitation and high streamflow in the United States: Trends in the 20th century', in *Bulletin of the American Meteorological Society*, vol 82, pp219–246

Guo, Z. C. and P. A. Dirmeyer (2006) 'Evaluation of the Second Global Soil Wetness Project soil moisture simulations: 1. Intermodel comparison', in *Journal of Geophysical Research-Atmospheres*, vol 111, D22S02, p14

Hansen, J., Ruedy, R., Glascoe, J. and Sato, M. (1999) 'GISS analysis of surface temperature change', in *Journal of Geophysical Research*, vol 104, pp30,997–31,022, doi:10.1029/1999JD900835

Hegerl, G. C., Karl, T. R., Allen, M., Bindoff, N. L., Gillett, N., Karoly, D., Zhang, X. B. and Zwiers, F. (2006) 'Climate change detection and attribution: Beyond mean temperature signals', in *Journal of Climate*, vol 19, pp5058–5077

Held, I. M. and Soden, B. J. (2000) 'Water vapor feedback and global warming', *Annual Review of Energy and the Environment*, vol 25, pp441–475

Held, I. M. and Soden, B. J. (2006) 'Robust responses of the hydrological cycle to global warming', in *Journal of Climate*, vol 19, pp5686–5699

Hodgkins, G. A., Dudley, R. W. and Huntington, T. G. (2003) 'Changes in the timing of high river flows in New England over the 20th century', in *Journal of Hydrology*, vol 278, pp244–252

Hu, Y. and Fu, Q. (2007) 'Observed poleward expansion of the Hadley circulation since 1979', in *Atmospheric Chemistry and Physics*, vol 7, pp5229–5236

Hudson R. D., Andrade, M. F., Follette, M. B. and Frolov, A. D. (2006) 'The total ozone field separated into meteorological regimes – Part II: Northern Hemisphere mid-latitude total ozone trends', in *Atmospheric Chemistry and Physics*, vol 6, pp5183–5191

Hunt, B. G. (2006) 'The Medieval Warm Period, the Little Ice Age and simulated climatic variability', in *Climate Dynamics*, vol 27, pp677–694

Huntington, T. G. (2006) 'Evidence for intensification of the global water cycle: Review and synthesis', in *Journal of Hydrology*, vol 319, pp83–95

Huntington, T. G., Hodgkins, G. A. and Dudley, R. W. (2003) 'Historical trend in river ice thickness and coherence in hydroclimatological trends in Maine', in *Climate Change*, vol 61, pp217–236

Immerzeel, W. W., van Beek, L. P. H. and Bierkens, M. F. P. (2010) 'Climate change will affect the Asian water towers', in *Science*, vol 328, no 5984, pp1382–1385, doi:10.1126/science.1183188

IPCC (2001) J. T. Houghton, et al (eds) *Climate Change 2001: The Scientific Basis. Contribution of Working Group I to the Third Assessment Report of the Intergovernmental Panel on Climate Change*, Cambridge University Press, Cambridge

IPCC (2007) S. Solomon, D. Qin, M. Manning, Z. Chen, M. Marquis, K. B. Averyt, M. Tignor and H. L. Miller (eds) *Climate Change 2007: The Physical Science Basis. Contribution of Working Group I to the Fourth Assessment Report of the Intergovernmental Panel on Climate Change*, Cambridge University Press, Cambridge and New York

Jansen, E., Overpeck, J., Briffa, K. R., Duplessy, J. -C., Joos, F., Masson-Delmotte, V., Olago, D., Otto-Bliesner, B., Peltier, W. R., Rahmstorf, S., Ramesh, R., Raynaud, D., Rind, D., Solomina, O., Villalba, R. and Zhang, D. (2007) 'Palaeoclimate', in S. Solomon, D. Qin, M. Manning, Z. Chen, M. Marquis, K. B. Averyt, M. Tignor and H. L. Miller (eds) *Climate Change 2007: The Physical Science Basis. Contribution of Working Group I to the Fourth Assessment Report of the Intergovernmental Panel on Climate Change*, Cambridge University Press, Cambridge and New York

Jones, P. D., New, M., Parker, D. E., Martin, S. and Rigor, I. G. (1999) 'Surface air temperature and its variations over the last 150 years', in *Reviews of Geophysics*, vol 37, pp173–199

Jung, M., Reichstein, M., Ciais, P., Seneviratne, S. I., Sheffield, J., Goulden, M. L., Bonan, G., Cescatti, A., Chen, J., de Jeu, R., Dolman, A. J., Eugster, W., Gerten, D., Gianelle, D., Gobron, N., Heinke, J., Kimball, J., Law, B. E., Montagnani, L., Mu, Q., Mueller, B., Oleson, K., Papale, D., Richardson, A. D., Roupsard, O., Running, S., Tomelleri, E., Viovy, N., Weber, U., Williams, C., Wood, E., Zaehle, S. and Zhang, K. (2010) 'A recent decline in the global land evapotranspiration trend due to limited moisture supply', in *Nature*, vol 467, pp951–954, doi:10.1038/nature09396

Kabat, P., Schulze, R. E., Hellmuth, M. E., Veraart, J. A. and Lenton, R. (2003) 'Climate variability and change and freshwater water management', in *International Review for Environmental Strategies*, vol 3, pp294–302

Karl, T. R., Groisman, P. Y., Knight, R. W. and Heim, R. R. (1993). 'Recent variations of snow cover and snowfall in North America and their relation to precipitation and temperature variations', in *Journal of Climate*, vol 6, pp1327–1344

Khromova, T. E., Dyurgerov, M. B. and Barry, R. G. (2003). Late twentieth-century changes in glacier extent in the Ak-shirak Range, Central Asia, determined from historical data and ASTER imagery', in *Geophysical Research Letters*, vol 30, p1863, doi:10.1029/2003GL017233

Kim, D. W. and Byun, H. R. (2009) 'Future pattern of Asian drought under global warming scenario', in *Theoretical Applied Climatology*, vol 98, pp137–150

Kripalani, R. H., Kulkarni, A., Sabade, S. S. and Khandekar, M. L. (2003) 'Indian Monsoon variability in a global warming scenario', in *Natural Hazards*, vol 29, no 2, pp189–206, doi:10.1023/A:1023695326825

Krishna Kumar, K., Rajagopalan, B. and Cane, M. A. (1999) 'On the weakening relationship between the Indian Monsoon and ENSO', in *Science*, vol 284, no 5423, pp2156–2159, doi:10.1126/science.284.5423.2156

Kundzewicz, Z. W., Mata, L. J., Arnell, N., Doll. P., Kabat, P. et al (2007) 'Freshwater resources and their management', in M. Parry, O. F. Canziani, J. Palutikof, P. van der Linden and C. Hanson (eds) *Climate Change 2007: Impacts, Adaptation and Vulnerability. Contribution of Working Group II to the Fourth Assessment Report of the Intergovernmental Panel on Climate Change*, Cambridge University Press, Cambridge

Kushner, P. J., Held, I. M. and Delworth, T. L. (2001) 'Southern Hemisphere atmospheric circulation response to global warming', in *Journal of Climate*, vol 14, pp2238–2249

Labat, D., Godderis, Y., Probst, J. L. and Guyot, J. L. (2004) 'Evidence for global runoff increase related to climate warming', in *Advanced Water Resources*, no 27, pp631–642

Lau, K.-M., Shen, S., Kim, K.-M. and Wang, H. (2006) 'A Multi-model study of the 20th century simulations of Sahel drought from the 1970s to 1990s', in *Journal of Geophysical Research*, vol 111, D07111, p9, doi:10.1029/2005JD006281

Lehner, B., Doll, P., Alcamo, J., Henrichs, T. and Kaspar, F. (2006) 'Estimating the impact of global change on flood and drought risks in Europe: A continental, integrated analysis', in *Climatic Change*, vol 75, pp273–299

Li, H., Robock, A. and Wild, M. (2007) 'Evaluation of Intergovernmental Panel on Climate Change Fourth Assessment soil moisture simulations for the second half of the twentieth century', in *Journal of Geophysical Resources*, vol 112, D06106, p15, doi:10.1029/2006JD007455

Lin, J. L. (2007) 'The double-ITCZ problem in IPCC AR4 coupled GCMs: Ocean-atmosphere feedback analysis', in *Journal of Climate*, vol 20, pp4497–4525

Lin, J. L., Weickmann, K. M., Kiladis, G. N., Mapes, B. E., Sperber, K. R., Lin, W., Wheeler, M. C., Schubert, S. D., Del Genio, A., Donner, L. J., Emori, S., Gueremy, J.-F., Hourdin, F., Rasch, P. J., Roeckner, E. and Scinocca, J. F. (2006) 'Tropical intraseasonal variability in 14 IPCC AR4 climate models. Part I: Convective signals', in *Journal of Climate*, vol 19, pp2665–2690

Lu, J., Vecchi, G. A. and Reichler, T. (2007) 'Expansion of the Hadley cell under global warming', in *Geophysical Research Letters*, vol 34, L06805, p5, doi:10.1029/ 2006GL028443

Manabe, M., Spelman, J. and Stouffer, R. J. (1992) 'Transient responses of a coupled ocean–atmosphere model to gradual changes of atmospheric CO_2. Part II: Seasonal response', in *Journal of Climate*, vol 5, pp105–126

Maurer, E. P., Wood, A. W., Adam, J. C., Lettenmaier, D. P. and Nijssen, B. (2002) 'A long-term hydrologically-based data set of land surface fluxes and states for the conterminous United States', in *Journal of Climate*, vol 15, pp3237–3251

Maurer, E. P., Stewart, I. T., Bonfils, C., Duffy, P. B. and Cayan, D. (2007) 'Detection, attribution, and sensitivity of trends toward earlier streamflow in the Sierra Nevada', in *Journal of Geophysical Research*, vol 112, D11118, p12

Meehl, G. A., Stocker, T. F., Collins, W. D., Friedlingstein, P., Gaye, A. T., Gregory, J. M., Kitoh, A., Knutti, R., Murphy, J. M., Noda, A., Raper, S. C. B., Watterson, I. G., Weaver, A. J. and Zhao, Z.-C. (2007) 'Global Climate Projections', in S. Solomon, D. Qin, M. Manning, Z. Chen, M. Marquis, K. B. Averyt, M. Tignor and H. L. Miller (eds) *Climate Change 2007: The Physical Science Basis. Contribution of Working Group I to the Fourth Assessment Report of the Intergovernmental Panel on Climate Change*, Cambridge University Press, Cambridge and New York

Milly, P. C. D. and Dunne, K. A. (2001) 'Trends in evaporation and surface cooling in the Mississippi River basin', in *Geophysical Research Letters*, vol 28, pp1219–1222

Mote, P. W., Hamlet, A. F., Clark, M. P. and Lettenmaier, D. P. (2005) 'Declining mountain snowpack in western North America', in *Bulletin of the American Meteorological Society*, vol 86, pp39–49

Murphy, J. M., Sexton, D. M. H., Barnett, D. N., Jones, G. S., Webb, M. J. and Collins, M. (2004) 'Quantification of modelling uncertainties in a large ensemble of climate change simulations', in *Nature*, vol 430, pp768–772

Nakićenović, N., Davidson, O., Davis, G., Grübler, A., Kram, T., Rovere, E. L. L., Metz, B., Morita, T., Pepper, W., Pitcher, H., Sankovski, A., Shukla, P., Swart, R., Watson, R. and Dadi, Z. (2000) *Emissions Scenarios: Summary for Policymakers*, Intergovernmental Panel on Climate Change, Geneva, Switzerland

Neelin, J. D., Chou, C. and Su, H. (2003) 'Tropical drought regions in global warming and El Niño teleconnections', in *Geophysical Research Letters*, vol 30, no 24, p2275, doi:10.1029/2003GL018625

Neelin, J. D., Münnich, M., Su, H., Meyerson, J. E. and Holloway, C. E. (2006) 'Tropical drying trends in global warming models and observations', in *Proceedings of the National Academy of Sciences*, vol 103, no 16, pp6110–6115

NRC (2003) 'Understanding climate change feedbacks', in *Report of the Panel on Climate Change Feedbacks,* National Research Council, National Academies Press, Washington, DC

Palmer, T. N. and Räisänen, J. (2002) 'Quantifying the risk of extreme seasonal precipitation events in a changing climate', in *Nature*, vol 415, pp512–514

Peterson, A. T., Golubev, V. S. and Groisman, P. Y. (1995) 'Evaporation losing its strength', in *Nature*, vol 377, pp687–688

Peterson, B. J., Holmes, R. M., McClelland, J. W., Vorosmarty, C. J., Lammers, R. B., Shiklomanov, A. I., Shiklomanov, I. A. and Rahmstorf, S. (2002) 'Increasing river discharge to the Arctic Ocean', in *Science*, vol 298, pp2171–2173

Pinker, R. T., Zhang, B. and Dutton, E. G. (2005) 'Do satellites detect trends in surface solar radiation?', in *Science*, vol 308, pp850–854

Rind, D., Goldenberg, R., Hansen, J., Rosenzweig, C. and Ruedy, R. (1990) 'Potential evapotranspiration and the likelihood of future drought', in *Journal of Geophysical Research*, vol 95, pp9983–10,004

Robock, A., Vinnikov, K. Y., Srinivasan, G., Entin, J. K., Hollinger, S. E., Speranskaya, N. A., Liu, S. X. and Namkhai, A. (2000) 'The Global Soil Moisture Data Bank', in *Bulletin of the American Meteorological Society*, vol 81, pp1281–1299

Roderick, M. L., Farquhar, G. D. (2002) 'The cause of decreased pan evaporation over the past 50 years', in *Science*, vol 298, pp1410–1411

Ross, R. J. and Elliott, W. P. (2001) 'Radiosonde-based northern hemisphere tropospheric water vapor trends', in *Journal of Climate*, vol 14, pp1602–1611

Rowell, D. P. and Jones, R. G. (2006) 'Causes and uncertainty of future summer drying over Europe', in *Climate Dynamics*, vol 27, pp281–299

Ruiz-Barradas, A. and Nigam, S. (2005) 'Warm season rainfall variability over the US great plains in observations, NCEP and ERA-40 reanalyses, and NCAR and NASA atmospheric model simulations', in *Journal of Climate*, vol 18, no 11, pp1808–1830

Schubert, S. D., Suarez, M. J., Pegion, P. J., Koster, R. D. and Bacmeister, J. T. (2004) 'Causes of long-term drought in the US Great Plains', in *Journal of Climate*, vol 17, no 3, pp485–503

Seager, R., Ting, M., Held, I., Kushnir, Y., Lu, J., Vecchi, G., Huang, H.-P., Harnik, N., Leetmaa, A., Lau, N.-C., Li, C., Velez, J. and Naik, N. (2007) 'Model projections of an imminent transition to a more arid climate in the southwestern North America', in *Science*, vol 316, no 5828, pp1181–1184, doi:10.1126/science.1139601

Seidel, D. J. and Randel, R. J. (2007) 'Recent widening of the tropical belt: Evidence from tropopause observations', in *Journal of Geophysical Research*, vol 112, D23109, p13

Seidel, D. J., Fu, Q., Randel, W. J. and Reichler, T. J. (2007) 'Widening of the tropical belt in a changing climate', in *Nature Geoscience*, vol 1, pp21–24, doi:10.1038/ngeo.2007.38

Seneviratne, S. I., Koster, R. D., Guo, Z. C., Dirmeyer, P. A., Kowalczyk, E., Lawrence, D., Liu, P., Lu, C. H., Mocko, D., Oleson, K. W. and Verseghy, D. (2006a) 'Soil moisture memory in AGCM simulations: Analysis of global land-atmosphere coupling experiment (GLACE) data', in *Journal of Hydrometeorology*, vol 7, pp1090–1112

Seneviratne, S. I., Lüthi, D., Litschi, M. and Schär, C. (2006b) 'Land–atmosphere coupling and climate change in Europe', in *Nature*, vol 443, pp205–209, doi:10.1038/nature05095

Sheffield, J. and Wood, E. F. (2008a) 'Projected changes in drought occurrence under future global warming from multi-model, multi-scenario, IPCC AR4 simulations', in *Climate Dynamics*, vol 13, no 1, pp79–105, doi:10.1007/s00382-007-0340-z

Sheffield, J. and Wood, E. F. (2008b) 'Global trends and variability in soil moisture and drought characteristics, 1950–2000, from observation-driven simulations of the terrestrial hydrological cycle', in *Journal of Climate*, vol 21, pp432–458

Sheffield, J., Goteti, G. and Wood, E. F. (2006) 'Development of a 50-year high-resolution global dataset of meteorological forcings for land surface modelling', in *Journal of Climate*, vol 19, pp3088–3111

Simmons, A. J., Willett, K. M., Jones, P. D., Thorne, P. W. and Dee, D. (2010) 'Low-frequency variations in surface atmospheric humidity, temperature and precipitation: Inferences from reanalyses and monthly gridded observational datasets', in *Journal of Geophysical Research*, vol 115, D01110, p21

Sun, Y. and Ding, Y.-H. (2009) 'A projection of future changes in summer precipitation and monsoon in East Asia', *Science in China Series D: Earth Sciences*, vol 53, pp284–300

Trenberth, K. E. (1999) 'Atmospheric moisture recycling: Role of advection and local evaporation', in *Journal of Climate*, vol 12, no 5, pp1368–1381

Trenberth, K. E. and Shea, D. J. (2005) 'Relationships between precipitation and surface temperature', in *Geophysical Research Letters*, vol 32, no 14, p4

Trenberth, K. E., Fasullo, J. and Smith, L. (2005) 'Trends and variability in column integrated atmospheric water vapour', in *Climate Dynamics*, vol 24, pp741–58

Wang, G. L. (2005) 'Agricultural drought in a future climate: Results from 15 global climate models participating in the IPCC 4th assessment', in *Climate Dynamics*, vol 25, pp739–753

Wetherald, R. T. and Manabe, S. (1995) 'The mechanisms of summer dryness induced by greenhouse warming', in *Journal of Climate*, vol 8, pp3096–3108

Wetherald, R. T. and Manabe, S. (1999) 'Detectability of summer dryness caused by greenhouse warming', in *Climatic Change*, vol 43, pp495–511

Wetherald, R. T. and Manabe, S. (2002) 'Simulation of hydrological changes associated with global warming', in *Journal of Geophysical Research*, vol 107, 4379, p15

Wild, M. (2009) 'Global dimming and brightening: A review, *Journal of Geophysical Research*, vol 114, D00D16, p31, doi:10.1029/2008JD011470

Wild, M., Ohmura, A., Gilgen, H. and Rosenfeld, D. (2004) 'On the consistency of trends in radiation and temperature record and implications for the global hydrological cycle', in *Geophysical Research Letters*, vol 31, L11201, p4, doi:10.11029/2003L091188

Wild, M., Gilgen, H., Roesch, A., Ohmura, A., Long, C. N., Dutton, E. G., Forgan, B., Kallis, A., Russak, V. and Tsvetkov, A. (2005) 'From dimming to brightening: Decadal changes in solar radiation at earth's surface', in *Science*, vol 308, pp847–850

Willett, K. M., Jones, P. D., Gillett, N. P. and Thorne, P. W. (2008) 'Recent changes in surface humidity: Development of the HadCRUH dataset', in *Journal of Climate*, vol 21, pp5364–5383, doi:10.1175/2008JCLI2274.1

WWAP (World Water Assessment Programme) (2009) *The United Nations World Water Development Report 3: Water in a Changing World*, UNESCO, Paris, and Earthscan, London

Ye, H., Cho, H.-R. and Gustafson, P. E. (1998) 'The changes in Russian winter snow accumulation during 1936–83 and its spatial patterns', in *Journal of Climate*, vol 11, pp856–863

Yin, J. H. (2005) 'A consistent poleward shift of the storm tracks in simulations of 21st century climate', in *Geophysical Research Letters*, vol 32, L18701, p4, doi:10.1029/2005GL023684

Zhang, K., Kimball, J. S., Mu, Q., Jones, L. A., Goetz, S. J. and Running, S. W. (2009) 'Satellite-based analysis of northern ET trends and associated changes in the regional water balance from 1983 to 2005', in *Journal of Hydrology*, vol 379, pp92–110

Zhang, T. et al (2005) 'Spatial and temporal variability in active layer thickness over the Russian Arctic drainage basin', in *Journal of Geophysical Research*, vol 110, D16101, p14, doi:10.1029/2004JD005642

Zhang, X., Hogg, W. D. and Mekis, E. (2001) 'Spatial and temporal characteristics of heavy precipitation events over Canada', in *Journal of Climate*, vol 14, pp1923–1936

Zarriello, P. J. and Ries, K. G. (2000) 'A precipitation-runoff model for analysis of the effects of water withdrawals on streamflow, Ipswich River basin, Massachusetts, US', Geology Survey, Water Resources Division, no 00–4029

9
Summary and Recommendations

Natural disasters cause major disruption, economic losses, environmental damage and, sometimes, loss of life. Drought is one of many types of natural disaster, but is ubiquitous across the world, affecting every place where human habitation is possible. It can happen anywhere, at any time, and impact on anybody. By definition, droughts are impossible to prevent. Yet by studying their impacts, we can provide ample incentive to monitor their evolution, offer possible forewarning and, in the end, hopefully reduce future impacts. Although this may be an obvious course to take, it has not always been so. Up until recently drought was essentially an afterthought in the mélange of natural hazards, in some ways the ugly cousin of the more dramatic and tangible hazards, such as hurricanes and floods. There are many reasons for this. Drought is a slow-moving hazard that may take weeks to years to develop and as long to recede. Thus a drought may be in full swing before anyone has noticed, by which time it may be too late to reduce the impacts. Compare this to floods, hurricanes, landslides, earthquakes, winter storms and heat waves. These are fast acting (of the order of minutes to days) and cause immediate, tangible and quantifiable damage. Yet drought causes as much damage, if not more than hurricanes and floods, because droughts are generally long-lasting, spatially extensive and impact upon many sectors.

Despite growing acceptance that drought has tremendous impacts, actions are all too often taken as reactions, rather than pre-emptive acts. This is linked to both the creeping nature of drought and also the difficulty of predicting where and when drought might develop, and how it will impact. The challenge of predicting drought is an extremely difficult one. Drought development is very complex and its understanding requires interdisciplinary approaches that forge interactions between physical and social scientists, and between scientists, policy-makers and stakeholders. Although great strides have been made in quantifying the development of drought, there are still substantial gaps in our knowledge of the mechanisms that lead to its development, persistence and recovery. This is further complicated by the human element of vulnerability and response to drought, which is extremely difficult to understand. Nevertheless, it is at the axis of physical hazards and human progress that drought lies.

Central to understanding drought and reducing its impacts is the quantification of how it has occurred in the past and the reasons why it has arisen. As well as identifying the physical mechanisms that are associated with drought, it also allows us to put current and future projected droughts into context. This book has taken a journey through time to document and explore the occurrence of drought across the world and has revealed huge variability in drought occurrence and severity. If there is one message to be drawn from this study, it is that drought occurs repeatedly, but often unexpectedly, and will continue to do so in the future, regardless of anthropogenic influences. It would be impossible to summarize this book in such a short space. Nevertheless, it is useful to highlight some of the major points and to put this into the context of the broader issues of environmental climate change. The following is a list of key questions that underlie the objectives of this book and these provide a means to summarize the research presented. At the end of the chapter we give an outlook of future directions for drought research and a summary of the outstanding challenges left to the scientific community.

What is drought and what are its impacts?

Drought is a deficit in moisture that can manifest in different forms and impact in many different ways, depending on which part of the hydrological cycle it develops in. Generally, drought begins with a decrease in precipitation that may be exacerbated by increasing evaporation or overuse of water by human means. This will deplete moisture in the soil, leading to soil moisture or agricultural drought with various impacts on crops and livestock. At the same time, run-off of water over the land and through the soil into streams and rivers will decrease. Coupled with decreased soil moisture and subsurface flow, this will lead to hydrological drought, where river flow, lake and reservoir levels, and wetland area can decrease. At longer timescales, recharge to groundwater can fall, prompting lower well levels and reduced flow to rivers. Lower precipitation will reduce snowpack in cold regions, which can lead to drought in the spring when the snowpack melts, and even into the summer.

Each of these forms of drought will have various environmental and economic impacts, depending on their timing, duration and severity. Chapter 1 reviewed some of the most recent impacts across the world, and revealed the diversity and extent of damages experienced by local populations and the environment, and the ramifications of these impacts in a globally connected economy. An overview of drought impacts was provided in Chapter 2, which also summarized the consequences of some of the major drought events of the 20th century. Agriculture is usually the first sector to experience the impact of drought, followed by multiple others, some obvious such as water resources and hydropower, and some not so obvious, such as tourism and transportation. There are then knock-on impacts that filter through the economy. As such, the impacts are very difficult to quantify, both economically and environmentally. What is clear is that drought impacts differ greatly between developed and developing nations. For example, droughts in the US have been more severe in recent years, in terms of the physical moisture deficits as compared to, say, the Dustbowl drought of the 1930s. However, the population's increased coping capacity has reduced the impacts to purely economical terms, which can be subsumed within the regional and national economy, rather than causing land abandonment and migration. At the other end of the spectrum of impacts, droughts in developing regions can have serious consequences, hampering development, with impacts that are of the order of national economies, and even associated with famine and loss of life. For example, the Sahel droughts of the 1970 and 1980s led to hundreds of thousands of deaths, yet the 1988 US drought caused a staggering US$39 billion worth of mainly agricultural damage, but relatively few deaths.

What causes drought?

This is one of the key questions of the book, since it is the fundamental building block of drought prediction and thus reduction of impacts through forewarning. Chapter 3 discussed the hydrological cycle and how it interacts with the climate system. Essentially drought is an extreme state of the hydrological cycle. At a basic level, drought is caused by variations in climate that reduce precipitation and/or increase evaporation. This is related to the complex patterns of variability of climate in space and time that govern where and when drought occur. The main drivers of these patterns are the seasonal cycle and latitudinal variation of solar irradiance, the transport of energy from the equator towards the poles, and the arrangement of the continental land masses. This causes some places to be wetter (such as the tropical forests) or drier (such as the subtropical deserts) on average. At timescales that are longer and shorter than seasons, various processes act to bring wetter or drier weather to a region. At daily to weekly timescales, the random fluctuations of the atmosphere lead to what we experience as weather, which in terms of drought can be high-pressure systems that cause subsidence and blocking, thus preventing precipitation and generating dry and warm conditions. Feedbacks with the land surface can also influence drought conditions on subseasonal timescales, such as recycling of precipitation, which can be reduced if the soil is dry, thus intensifying the drought. The land surface also acts as a filter, delaying how drought conditions propagate through the hydrological cycle, from the land surface to the soil and so on, through the river and groundwater systems.

At seasonal timescales, the variation in solar irradiance is essentially known and this drives seasonal climate phenomena, such as monsoonal systems, winter storms, summertime convection and hurricanes. However, these processes are subject to variations in time and space because of coupling with other parts of the climate system that can lead to drought. For example, changes in snow cover extent across continental interiors can affect land–sea thermal gradients, which in turn can affect the timing and intensity of monsoons and thus cause drought. Weather processes that act on shorter timescales and coupled atmosphere–ocean processes that act on longer timescales shift these seasonal processes to increase or decrease drought occurrence. At interannual to decadal timescales, drought is associated with modes of internal variability in the coupled atmosphere–ocean system, some of which are characterized as the well-known climate oscillations. The most influential of these is the El Niño Southern Oscillation (ENSO), because of its widespread impacts from the tropics to mid-latitudes, which acts on interannual timescales. There are also various other climate modes of variability, such as the Northern and Southern Annular Modes, which are known by various other names, and the Decadal and Multidecadal Pacific and Atlantic Oscillations, which also compete to bring about drought (or wetness) to various regions. It is often a combination of these different modes that brings about the severest and longest droughts.

How do we measure and quantify drought?

Because of the wide-ranging impacts of drought that happen at different levels of water deficit, at different timescales and over different spatial scales, defining drought is difficult. The search for a universal drought index to cover all situations is probably fruitless, but much progress has been made in recent years in defining what a useful index might be and how it may be defined for particular situations. In general, an index should be transferable in time and space so comparisons can be made with other places and times. It should also have a robust probabilistic grounding so that risk may be quantified. Of great importance is the physicality of the index and the need for it to be based on underlying hydrological quantities, such as precipitation and soil moisture. Thus measurements of these quantities are needed at scales commensurate with the application, both in real time and historically. The latter forms the basis for putting current conditions into a context that helps interpretation of how severe conditions are. Chapter 4 discusses some of the most popular drought indices and describes the details of the soil-moisture-based percentile index that is used through the subsequent chapters to analyse the occurrence and severity of drought during the 20th and 21st centuries. Some of the most popular indices such as the Standardized Precipitation Index (SPI) are readily calculated from monthly precipitation data. Others such as the Palmer Drought Severity Index (PDSI) are more complex in what they represent, and as such suffer from simplifications in the physical processes that they simulate.

In the end, hydrological data are required to calculate these indices. Measuring hydrological quantities is wrought with problems for large-scale drought applications. In densely monitored regions such as Europe and the US, data are available in real time from meteorological stations, providing a relatively dense picture of the current state of drought, as well as long-term records. For less-developed regions, there can essentially be no hydrological measurements at all, yet the local need for high-resolution information on drought is just as vital as it is in developed regions. Chapter 4 highlighted some of the existing and emerging satellite remote-sensing tools that can provide high-resolution, large-scale coverage of hydrological quantities across the globe. Combining these data products, such as precipitation, evaporation, soil moisture, river flows, lake levels and total water storage, in a coherent drought-monitoring framework is an exciting way forward that has potential to alleviate problems of sparse data in regions of most need. If we want to look back further in time, beyond the instrumental period, then there is reliance on proxy palaeoclimate records that give indirect measurements of climate and hydrology. These records are shedding light on the large variations in climate over past millennia, and how 20th-century droughts compare to some of the very long and severe events that have occurred previously. Proxy data include tree rings, lake and marine sediments, ice cores and cave deposits, and also human documentary evidence in the form of written records, archaeological evidence and indirect records related to climate and hydrology, such as harvest records and religious ceremonies.

Additionally there have been tremendous increases in the complexity, accuracy and utility of modelling climate and hydrological processes at large scales, and this is now providing global-scale, century-long reconstructions of hydrology and drought occurrence. The advantage of these models is that they are able to ingest available measurements and proxy data, as well as provide spatially and temporally continuous and consistent hydrological data. Furthermore, they can be used to diagnose the mechanisms that lead to drought and its persistence, for example, in synthetic experiments, whereby parts of the physical system are fixed to see how this effects drought development.

How has drought varied over the Holocene?

It is clear from the palaeoclimate record that climate has varied greatly within the Holocene (the past 12,000 years or so). The picture that emerges from the plethora of clues that are hidden in lake and ocean sediments, tree-growth rings, speleothems and archaeological artefacts is that the Holocene was a time of great change. This epoch witnessed some of the greatest advances in human civilization, in part because of the conducive climate conditions, as embodied by the concept of a Holocene climate optimum around 4–6kya. However, periods of long-term and abrupt climate change were ever present and the Holocene was interspersed by severe drought events and long-term aridity. These events were driven by external forcings, such as changes in the Earth's orbit relative to the sun, and solar and volcanic activity, combined with internal variability in the coupled atmosphere–ocean system, such as changes in the frequency of the ENSO and the intensity of the Asian and African monsoons. Rapid glacial lake drainage and feedbacks with the land surface, such as major shifts in vegetation cover, have also played a role. Many of the drought events were far more severe and lengthy than anything we have experienced in the past century of the instrumental record, leaving their mark on the geophysical proxies and bringing associations with the decline and demise of cultures and great civilizations.

Given the current prominence of environmental sustainability and the threat of climate change, it is of no surprise that there is a growing interest in linking the fall of these great civilizations with climate change, and thus the potential for large impacts from future change. However, the concept of climate determinism (the direct attribution of cultural and economic change to climate variability) (Brázdil et al, 2005) is debatable within the cultural and social sciences. Recent work by a collaboration of the world's foremost experts on anthropology, archaeology and history of civilizations indicates that caution must be exercised when attributing civilization demise to environmental factors, including drought (McAnany and Yoffee, 2010). Climate change and drought impacts on human population are therefore one part of a more complex picture. Their message was to not underestimate the capacity of civilizations to adapt and make choices, such as abandonment or relocation, if this leads to a better life. It is therefore important to note the capacity of humans to adapt to changing conditions, despite the apparent enormity of the change (von Storch and Stehr, 2000). Furthermore, it should be noted that humankind made significant progress during the preceding ice age, despite more extreme and variable climate conditions that exhibited significant swings over short time periods. The capacity of humankind to adapt and cope with such disasters also changes over time and a 21st century megadrought would not have the impacts that it had 1000 years ago.

In the past 2000 years there has been much less climate variation, but there have still been major swings, manifesting in the Medieval Warm Period (MWP or Climate Anomaly) and the Little Ice Age, for example. The MWP, in particular, has been associated with dry conditions that caused problems to populations across the northern hemisphere and perhaps even further afield, although there is also increasing agreement that it was much more regionally and temporally complex than initially surmised. One of the oft-asked questions about the MWP is how it compares to the current climate, and if it is an analogy for present-day warming. The answer to this is probably that it is an imperfect analogy, being driven by increased solar irradiance and decreased volcanic activity rather than increased greenhouse gases. The MWP also manifested itself in summertime warming, whereas current warming trends have been greatest in the winter and spring.

How has drought changed over the 20th century?

Chapter 6 documented changes in drought during the 20th century, using a combination of measured hydrological data from the instrumental record and model simulations. The relative abundance of measurements in the 20th century has allowed for reasonable estimates to be made of the global hydrological cycle, including its variability and changes. Of course, the further back in time, the larger the uncertainties become as gauge measurements become sparser. The analysis reveals a large variation in time and space of drought, ranging from short-term but frequent relatively dry conditions in humid regions such as the tropical forests and in mid-latitudes, to long-term and widespread drought conditions, which last for multiple years in transitional regions between dry and humid environments.

The analysis shows that levels of soil moisture have probably declined since the mid-20th century, although this is highly dependent on the time period used to calculate the trend because of the large decadal variability in precipitation. The next few years will reveal whether this trend continues, or if there is a switch back to wetter conditions globally. Regionally, the long-term trends are more robust and there have been steady declines in soil moisture and increases in drought extent over West Africa, the Mediterranean and East Asia, whereas wetting is notable in Central Asia and the Tibetan Plateau. Recent droughts in the US West, Australia, southern South America, northern China and eastern Africa hint at the impact of climate change, but may just be part of the current decadal swing in global precipitation. The ENSO plays a leading role in shaping global climate variability at interannual timescales, and especially in the tropics and mid-latitudes, with El Niño events being associated with drier conditions overall and a preponderance of large-scale drought. On the other hand, La Niña is associated with dry conditions in certain parts of the world, including the southwest US and parts of Argentina. At decadal timescales, the Pacific Decadal Oscillation (PDO) and Atlantic Multidecadal Oscillation (AMO) are likely to exert an influence upon global drought occurrence.

What were the most severe drought events of the 20th century?

If we consider loss of life to be the most devastating of impacts, then the droughts in the Sahel in the 1970s and 1980s and early 1990s were probably the most severe droughts of the 20th century. Reported numbers of deaths that are linked to the droughts are in the hundreds of thousands. Nevertheless, there have been many other major events, especially in the earlier part of the 20th century, which are also associated with famines and the possible deaths of millions of people. However, these earlier droughts are inextricably linked to many other socio-economic and political factors that are likely to have played a much more important role. In economic terms, it is difficult to identify the most severe events because of escalating exposure of assets to drought with time and the various direct and indirect sectors that are impacted. The 1988 US drought was one of the most economically damaging, causing around US$39 billion dollars of damage. More recent droughts in Europe are also causing billions of euros in damages.

Overall, some of the longest events occurred in high latitudes of northern Eurasia, Alaska and northern Canada, yet their impacts were relatively small. From a physical standpoint, the severity–area–duration (SAD) analysis indicates that the 1975–1977 west Central Asian drought was the most severe for 12-month and 24-month durations. Other Central Asian droughts in the 1940s and early 1950s are ranked similarly. Severe events also include the 1954–1957 North American drought that affected much of the southern US, the 1963–1964 South America drought that centred mainly on the Amazon region, 1976–1977 North American wintertime event and the recent drought in southeast Australia. Interestingly, the African droughts of the 1980s, which had such devastating humanitarian impacts, do not appear on the global SAD curves that summarize the most severe droughts, highlighting the importance of local vulnerability to drought.

How does 20th-century drought compare to past millennia?

The occurrence of many severe and devastating droughts of the 20th century raises the question: how unusual are these types of events? This is particularly important in the context of what might be possible in the future from natural climate variation alone, irrespective of the impact of anthropogenic global warming. The short length of the instrumental record is insufficient to unearth the full range of natural variability of climate, but palaeoclimate proxy data extends the instrumental record and puts the 20th-century instrumental record into a long-term context. The palaeoclimate record indicates that there have been many severe and spatially extensive droughts across the world over the past several thousand years. In the Sahel, multiple droughts of similar duration and severity as those in the 1970s and 1980s have been identified in palaeoclimate records from lake sediments. Tree-ring reconstructions for the US West indicate large multiyear and multidecadal droughts that are outstanding when compared to the major events of the 20th century, including those in the 1930s, 1950s and 2000s. In Asia, similar data show multiple events that are widespread and severe, and associated with major impacts to human civilizations and activity. These types of reconstructions are repeated across the globe and many indicate more severe and lengthier drought events have taken place in the past, compared to anything experienced in the 20th century. This is important in the context of recent droughts that had large human, economic and water resources impacts because these droughts can be considered mild in comparison.

Has global warming already impacted drought?

It is tempting to assign recent droughts across the world to global warming, especially given the unique and rapid changes happening in the last decades in terms of CO_2 and rising temperatures. Many research studies have attributed drought in part to warming temperatures that are increasing evaporative demand, further decreasing soil moisture beyond any lack of rainfall. However, the evidence is sparse and sometimes spurious, and often statistically bound, rather than physically determined. Examination of the precipitation record, itself uncertain and not robust in some places, shows that the drying trends of the past ten years are comparable to previous changes on decadal timescales. Only time will tell whether the drying trends will continue or not, but if the past is any judge of the future, then we might expect a reversal at some stage. Despite this, temperature does have an effect that appears to have been exaggerated in the last few decades, especially in high northern latitudes, which have experienced warming that is twice the rate of the global mean. As well as direct impacts on snow and ice, warming also brings more rain than snow, which tends to bring drier conditions in the spring and summer. This is most pertinent within the context of potential continued temperature increases during the 21st century.

Will drought become more frequent and severe in the future?

Water is the principal medium through which climate change will affect economic, social and environmental conditions. These include sea-level rise, glacier melt, and more floods and droughts. Changes in water availability will have economy-wide impacts, none more so than in the developing world where population increase, especially in urban areas, is expected to heighten demand for fresh water. Current capacity to cope with natural water hazards such as floods and droughts is not adequate. A glance through the international section of any newspaper usually reveals a water-related disaster somewhere in the world, with serious direct impacts and long-lasting indirect effects. Therefore, it is likely that the impacts will get worse, given projections for future climate change. Our estimates of what might happen in the future are based on climate models and their projections of future climates, which explore plausible scenarios of future emissions of greenhouse gases. Confidence in these projections relies on the ability of coupled climate models to reproduce the statistics and mechanisms of historic drought occurrence. Globally, these models actually do reasonably well in representing regional differences in drought occurrence as a multimodel average, although the spread among models is large, and there is a tendency to over-predict the occurrence of long-term droughts. With the prospect of future global climate change, there is the expectation of increased drought occurrence and associated impacts, especially in semi-arid regions, such as the southwest US, the Mediterranean and southern Africa.

Efforts are now being mobilized to do something about climate change. The problem here is that climate change cannot be reversed, only slowed. Thus, adaptation must be at the forefront of the response to climate change. Lessons learnt from the past millennia indicate that humans have been severely impacted by drought but have the ability to adapt in some form, whether it is by economic means, as in the 20th century, or by abandonment and migration. The further back in time the drought and the more long-lasting its dry conditions, the more likely it is to have been a contributory factor towards the collapse and demise of the impacted civilization, but more cross-disciplinary research is needed in this area. It remains to be seen whether our current coping capacity and ability to adapt is sufficient for the potentially large changes projected for the future.

Looking forward

Drought is an ever-present hazard that will continue to test human aptitude. This book has shown that drought has affected every part of the habitable world, bringing serious and widespread impacts on the environment and society. This is likely to worsen through diminishing water resources and increasing sectoral conflicts, driven by population increase, land-use change and climate change. The lack of coping capacity in many regions underlies the prospect of increased hardship in the future. It is therefore a continuing goal to reduce these impacts through monitoring and timely prediction of droughts, and to prepare for adaptation in the face of longer-term changes.

One of the major tools in combating and preparing for the impacts of drought is the use of monitoring and forecasting tools. Recent advances in climate forecasting, hydrological modelling, remote-sensing and real-time telemetry and the availability of on-the-ground measurements have allowed governments and agencies to develop near real-time pictures of drought over large scales. In the US, recent droughts in the West have led to national priority initiatives, such as the federally mandated National Integrated Drought Information System (NIDIS; WGA, 2004), whose goals are to improve predictive and mitigation capabilities, and provide timely and integrated information on drought conditions at scales relevant to decision-making. Elsewhere, the European Drought Observatory (EDO; Niemeyer et al, 2009) has a similar mandate for Europe, and various other national meteorological services provide drought monitoring and seasonal forecasting for their local audience. Analyses of historic droughts, as described in this book, are central to quantitative drought monitoring, allowing current conditions to be put into perspective. Remote sensing, in particular, is becoming an essential tool in monitoring drought in developing regions, which have traditionally had the least investment in hydrological monitoring but have suffered the worst in humanitarian terms. Experimental global drought-monitoring systems are being developed by several institutions, mostly based on remote sensing, and there are emerging efforts to merge regional and national monitoring systems into spatially coherent tools that reflect the widespread nature of drought, across political boundaries.

Ideally, forewarning of droughts is required to prepare for and even mitigate their impacts. For example, response to drought-induced famine requires lead times of several months to put into place the chain of logistics to provide relief. The basis for forewarning, in addition to socio-economic information on food stockpiles and local economies, are seasonal climate forecasts and their translation into hydrological impacts. Climate forecasting at seasonal timescales has improved over the past few years, yet the skill of physically based models and statistical forecast tools remains modest at best. Understanding the mechanisms of how droughts develop and persist is central to improving this skill. Progress has been made in recent years, especially in terms of reconstructing past drought events and teasing out the physical mechanisms associated with them, and so the future looks promising. For example, the droughts of the 1930s in the US have been associated through model experiments to the state of the tropical oceans and better predictions of the ENSO will provide increased forecast skill in many parts of the world. Yet our current understanding is far from complete and new complexities of the coupled ocean–atmosphere–land system are being revealed all the time. Now work is needed to translate our emerging knowledge of drought mechanisms into predictive capability that can provide the necessary forewarning.

As we look to the long-term future, the potential impacts of climate change are expected to bring about more droughts of greater severity and duration. There are many uncertainties associated with this, the largest of which is the future greenhouse gas emissions pathway that we choose to take. Our current trajectory appears to be towards the higher end of the range of emission scenarios, which does not bode well if model projections are robust. However, there are also large differences in projections from different climate models, especially at regional scales, where the impacts will be felt, and reduction of these uncertainties is urgently needed for stakeholders to plan appropriately. Furthermore, detailed information is needed on the scale of the decision-making process, from seasonal to decadal, rather than just end-of-century projections that are beyond the planning horizon of most stakeholders.

Continued reconstruction of past drought via the instrumental and palaeoclimatic record is a central part of this endeavour, and this book has presented some of the latest research on large-scale drought occurrence. More work is required in this area to evaluate past droughts, their hydrological, ecological and human impacts, and to disentangle the various forcing mechanisms and feedbacks that generate such complex behaviour. A great deal of effort is required to bring together the diversity of palaeoclimate records and explanations of variability in a mechanistic sense. Currently, most work on attribution of climate variability and its impacts, such as drought, are done in a correlative statistical sense that may be a result of coincidence. Rather, the way forward is to use state-of-the-art coupled climate models with realistic land-surface processes of vegetation and hydrology to merge mechanistic climate understanding with palaeoclimate data to diagnose the complex spatial patterns, feedbacks and thresholds that force these climate events. This is beginning to happen with the merging of instrumental and palaeoclimate records and the use of climate models to assimilate these disparate but locally detailed pieces of palaeoclimate information into a regional to global picture of past climate variability and drought occurrence.

Conclusions

Our obsession with weather and climate is founded in the way it shapes our daily lives, no more so than when it brings extreme events such as drought, which causes not only inconvenience, but also inflicts ecological and economic hardship, and can ultimately bring about loss of life. The increasing recognition of drought as a major hazard and the potential for future changes in drought to go beyond current coping mechanisms in much of the developing world has provided the inspiration for this book. Its aim has been to synthesize the wealth of disparate data and information, and to provide a detailed picture of how drought has occurred in the past and how it might occur in the future. The numerous droughts that have impacted in living memory and back over the millennia are testament to its power to devastate crops, reduce water supply, curtail power generation and industry, to hamper economic development and to inflict famine. A concerted effort is required in the research community to understand why and how these events have occurred and to translate this knowledge into improved monitoring and prediction of drought in the future. At the same time, progress is needed to bridge the gaps between science and stakeholders, to provide useful information on drought risk and target best practices for alleviating the impacts. Drought will continue to cause problems, just as it has in the past, yet there is much hope that humans are adaptable to what may occur in the long-term, in much the same way as they have coped with many of the severe events of the past.

References

Brázdil, R., Pfister, C., Wanner, H., von Storch, H. and Luterbacher, J. (2005) 'Historical climatology in Europe – the state of the art', in *Climate Change*, no 70, pp363–430

McAnany, P. A. and Yoffee, N. (eds) (2010) *Questioning Collapse: Human Resilience, Ecological Vulnerability, and the Aftermath of Empire*, Cambridge University Press, Cambridge

Niemeyer, S., de Jager, A., Kurnik, B., Laguardia, G., Magni, D., Nitcheva, O., Rossi, S. and Weisstiener, C. (2009) 'Current state of development of the European Drought Observatory', in *Geophysical Research Abstracts*, vol 11, EGU2009–12802

von Storch, H. and Stehr, N. (2000) 'Climate change in perspective', in *Nature*, vol 405, p615, doi:10.1038/35015179

WGA (2004) *Creating a Drought Early Warning System for the 21st Century: The National Integrated Drought Information System*, Western Governors' Association, Denver, CO, available at www.westgov.org/wga/publicat/nidis.pdf

Index

For Product Safety Concerns and Information please contact our EU representative GPSR@taylorandfrancis.com
Taylor & Francis Verlag GmbH, Kaufingerstraße 24, 80331 München, Germany

www.ingramcontent.com/pod-product-compliance
Ingram Content Group UK Ltd.
Pitfield, Milton Keynes, MK11 3LW, UK
UKHW050755180926
475701UK00007B/1061

* 9 7 8 1 0 3 2 9 2 5 2 6 4 *